Comets

A Chronological History of Observation, Science, Myth, and Folklore

Donald K. Yeomans

Wiley Science Editions
John Wiley & Sons, Inc.
New York • Chichester • Brisbane • Toronto • Singapore

Library of Congress Cataloging-in-Publication Data

Yeomans, Donald K.
  Comets : a chronological history of observation, science, myth, and folklore / Donald K. Yeomans.
      p.   cm. — (Wiley science editions)
  Includes bibliographical references and index.
  ISBN 0-471-61011-9
  1. Comets.   I. Title.   II. Series.
QB721.Y46   1991
523.6—dc20   90-12657

Printed in the United States of America

91 92 10 9 8 7 6 5 4 3 2

Contents

Contents

Contents

Acknowledgments

Many authorities on cometary history and current cometary research have graciously contributed their time and wisdom in reviewing various chapters. Their suggestions and corrections have improved my initial efforts considerably. In alphabetical order, these authorities include

Michael F. A'Hearn, University of Maryland

Mark Bailey, The University of Manchester, England

Peter Broughton, Toronto, Canada

Ichiro Hasegawa, Nara, Japan

Hermann Hunger, Vienna, Austria

Jane L. Jervis, Dean, Bowdoin College, Maine

Gary W. Kronk, Troy, Illinois

Brian G. Marsden, Smithsonian Center for Astrophysics, Cambridge, Massachusetts

Malcolm B. Niedner, NASA/Goddard Space Flight Center, Greenbelt, Maryland

Ray L. Newburn, Jet Propulsion Laboratory, Pasadena, California

James A. Ruffner, Wayne State University, Detroit, Michigan

Peter F. Schloerb, University of Massachusetts

Hyron Spinrad, University of California at Berkeley

Acknowledgements

Craig B. Waff, Pasadena, California

Peter A. Wehinger, Arizona State University at Tempe

Paul R. Weissman, Jet Propulsion Laboratory, Pasadena, California

Susan Wyckoff, Arizona State University at Tempe

Professor Armand H. Delsemme, at the University of Toledo, Ohio, kindly reviewed the entire draft and offered several valuable comments and suggestions that have been incorporated into the final book. Ruth S. Freitag, one of the most remarkable sources of information at the U.S. Library of Congress, provided valuable assistance. Initially Ruth was to be a co-author of this work, but unfortunately her hectic schedule would not allow the time.

Librarians at three fine astronomical libraries were unfailing in their efforts to track down often obscure references on my behalf. They provided far more assistance than I had a right to expect. Thanks to

Brenda Corbin and Gregory Shelton,
U.S. Naval Observatory, Washington, DC

Joan Gantz, The Observatories, Pasadena, California

Angus R. Macdonald, Royal Observatory, Edinburgh

Finally, the assistance and encouragement of David Sobel at John Wiley & Sons, Inc., are gratefully acknowledged.

Preface

For millennia, observers have pondered the appearances of comets in the night sky. The concepts developed to explain these mysterious apparitions have a long and intriguing history. For years, the two-volume work by Alexandre Pingré remained the only general history of comets. While still an excellent reference, Pingré's history is now more than two centuries old. In this book, the development of cometary ideas is traced from antiquity until after a fleet of international spacecraft flew past comet Halley in 1986. Because the book's focus is on the cometary theories that were evident in each era, specific observations and individual observers do not receive detailed coverage.

Like the scientists they are concerned with, science historians are influenced by the data to which they have access. To some extent, this book reflects my personal view toward the development of cometary ideas. Because of a vast number of cometary treatises and cometary observations, I had to make decisions as to what ideas and observations were influential to the field. Although current knowledge of cometary phenomena was sometimes used to guide my historical studies, I have not ignored ideas from the past simply because they contributed nothing to the present. My overriding criterion for selecting topics was whether or not they were influential to subsequent inquiries.

The study of comets—tiny remnants from the solar system formation process—allows an examination of the conditions and mixtures from which the major planets formed four and one half billion years ago. Comets probably played a major role in delivering to the early Earth the veneer of carbon-based molecules and volatile gases that allowed life to form. Subsequent collisions of comets with Earth may have wiped out significant numbers of these early lifeforms, allowing only the most adaptable to develop further. The diminutive size of cometary bodies is in no way proportional to their scientific importance.

While researching the history of cometary ideas, I found that the wide diversity of views and the perception of comets as malefic signs or agents of destruction often made for interesting reading. I have included several vignettes that delve into colorful anecdotes and personalities.

The study of comets is historically important, scientifically compelling, and at the same time entertaining; I've tried to convey some of each in the following pages.

Donald K. Yeomans
Jet Propulsion Laboratory
California Institute of Technology
January 1991

1
The Origin of Cometary Thought

Aristotle presents cometary ideas of ancient Greeks and his own view of comets as terrestrial emanations. The comet and meteorite of 467 B.C. and Caesar's comet of 44 B.C. appear. Seneca outlines cometary ideas of ancient Greeks and Babylonians before expressing his opinion that comets are celestial phenomena like the planets. Pliny the Elder presents a scheme for classifying comets, and Ptolemy's work gives rise to the enduring notion of comets having astrological significance.

ONE TRUISM STANDS OUT in the history of comets—mankind has always been transfixed by their appearances. The Sun and Moon are far brighter, aurorae more impressive, and eclipses more startling. Yet it is comets, with their modest radiance and infrequent visits, that have commanded more concern. In the first century, the Roman sage Lucius Annaeus Seneca (4 B.C.–A.D. 65) explained this phenomenon in his *Natural Questions:*

> No man is so utterly dull and obtuse, with head so bent on Earth, as never to lift himself up and rise with all his soul to the contemplation of the starry heavens, especially when some fresh wonder shows a beacon-light in the sky. As long as the ordinary course of heaven runs on, custom robs it of its real size. Such is our constitution that objects of daily occurrence pass us unnoticed even when most worthy of our admiration. On the other hand, the sight even of trifling things is attractive if their appearance is unusual. So this concourse of stars, which paints with beauty the spacious firmament on high, gathers no concourse of the nation. But when there is any change in the wonted order, then all eyes are turned to the sky. . . . So natural is it to admire what is strange rather than what is great.

The same thing holds in regard to comets. If one of these infrequent fires of unusual shape have made its appearance, everybody is eager to know what it is. Blind to all the other celestial bodies, each asks about the newcomer; one is not quite sure whether to admire or to fear it. Persons there are who seek to inspire terror by forecasting its grave import. And so people keep asking and wishing to know whether it is a portent or a star.

Right up to the seventeenth century comets were considered to be omens. We still refer to the appearance of a comet as an apparition—a ghost or phantom. The scientific study of comets got off to a late start because the ancients believed them to be divine warnings, usually malefic, of things to come. Scholarly inquiry was limited to such topics as whether comets actually caused the events that followed or served merely as precursors. One did not study ghosts in a scientific fashion.

Aristotle's Meteorologica

Aristotle (384–322 B.C.) was born in Thrace and grew up at the Macedonian court, where his father was the king's physician. In 367 B.C., at the age of 18, he went to the Academy at Athens to study with Plato and remained there until the latter's death 20 years later. Plato referred to him as the intellect of the school. In 343 B.C. Aristotle returned to Macedonia as tutor of Alexander the Great, then a lad of 13. When Alexander died at the age of 33, his conquests had spread Greek culture and learning throughout most of the known world. Thus Aristotle's association with the young prince, which lasted four years, was to have far-reaching effects on history in general and his own life in particular.

Aristotle returned to Athens in 335 B.C. and founded a school of philosophy and rhetoric near the temple dedicated to the god of shepherds, Apollo Lyceus. Hence his school was called the *Lyceum,* and his thought became known as the peripatetic philosophy because of the covered walk, *peripatos,* through which he liked to stroll with his students. He was identified with the Macedonian rulers of Athens and their supporters, and upon Alexander's death in 323 B.C. he fled the city to escape the outbreak of anti-Macedonian sentiment. He died a few months later, in 322 B.C., leaving a body of writings whose importance in the history of Western thought cannot be overestimated. However, he would have been the first to deplore the stagnation of the physical and natural sciences resulting in part from the veneration in which his views were held for centuries.

The compilation of Aristotle's *Meteorologica,* in which his cometary ideas were expressed, was probably completed during his years at the Lyceum. As well as outlining his own views, Aristotle presented and dismissed those of Pythagoras (ca. 560–480 B.C.), Hippocrates of Chios (fl. 440 B.C.), Anaxagoras of Clazomenae (ca. 500–428 B.C.), and Democritus of Abdera (fl. 420 B.C.).

Although Pythagoras was born in Greek Samos and spent 30 years studying with priests and sages in Asia Minor, he established a school in southern Italy when he was about 50 years old. The Pythagorean ideas of beauty required that the heavenly bodies move along simple curves, and they were responsible for establishing the notion that all celestial bodies move in circles, the circle being the perfect curve. Until the work of Johannes Kepler (1571–1630) in the early seventeenth century, belief in the circular motion of heavenly bodies remained largely unquestioned. According to Aristotle, the Pythagoreans believed there was only one comet and that it was a planet. It appeared infrequently and, like Mercury, it rose only a little above the horizon.

As related by Aristotle, Hippocrates of Chios and his pupil Aeschylus believed that a comet's tail was formed when the comet drew up moisture from the Earth below. According to the peculiar optical theories then current, the tail was seen not when sunlight was reflected from the moisture to the eye, but rather when one's sight was reflected to the Sun from the moisture. In the north the comet would assume a visible tail, but in the tropics, where the Sun dries most available moisture, tails could not form. Although the comet had the necessary moisture south of the tropics, it was generally below the horizon and hence invisible. Hippocrates stated that the comet appeared at greater intervals than other stars (planets) because it was slowest to move clear of the Sun. It seems evident from Aristotle's writing that Hippocrates shared the Pythagorean belief in a single comet.

According to Aristotle, Democritus and Anaxagoras declared that each comet was a conjunction of planets approaching one another and so appeared to touch. Although Aristotle easily dismissed this idea, noting that comets were observed in regions of the sky where planets do not travel, he did not subdue it altogether. The notion that comets are caused by planetary conjunctions was popular well into the seventeenth century. Democritus wrote several books on geometry and derived formulae for the areas of cones and pyramids. He is best remembered for his atomist philosophy that all matter is composed of atoms differing in size and weight, with like atoms combining to produce the planets and stars. Anaxagoras, a friend and teacher of Pericles, is chiefly remembered in astronomical history for correctly explaining eclipses and for his cosmology, which posits that the universe was formed from chaotic, diverse seeds pervaded by a world mind that gave them order and set them spinning into a vortex-like motion. This rotary motion of the ether tore away stones from the Earth and kindled them into stars.[1] Charges

3

The Comet and Meteorite of 467 B.C.

Both Aristotle and Pliny the Elder (A.D. 23–79) mentioned that in a year corresponding to 467 B.C., a meteorite fell at Aegospotami, in Thrace, on the European side of the Dardanelles, and that a comet was seen the same year. In his *Natural History* (Book II, Chapter 59), Pliny noted that the meteorite fell during the daytime: it was brown in color and the size of a wagon load. Pliny also mentioned that Anaxagoras prophesied the event by predicting that, in a certain number of days, a rock would fall from the Sun. In his life of the Spartan general Lysander, the Greek biographer Plutarch wrote of the same prediction and added that before the stone fell, a vast fiery body was seen in the heavens for 75 days continually "but when it afterwards came down to the ground . . . there was no fire to be seen . . . only a stone lying big indeed. . . ."

Anaxagoras considered the sky to be made up of whirling stones. Since Anaxagoras believed that both the Sun and comets were made up of burning stones, the cometary apparition of 467 B.C. may have indicated to him that the furniture of the heavens was slipping and that a large stone could arrive on Earth several days after the comet appeared. It is equally likely that the account of the prophecy is apocryphal.

The early Greeks and Romans had no trouble believing that stones fell from the sky, but Aristotle's cosmology demanded they be first

of impiety were brought against Anaxagoras in Athens for his refusal to admit that the Sun was anything but a mass of stone on fire—not a god.

Aristotle's own views are of the utmost importance in the history of comets. For nearly two millennia his teachings dominated almost every aspect of Western thought. As Galileo witnessed in the early seventeenth century, to doubt the Church-endorsed Aristotelian ideas was heresy. Aristotle's universe was finite, spherical, and geocentric. The first four elements (earth, water, air, and fire) moved naturally along straight, finite—rectilinear—lines confined to the imperfect sublunar world. Later, a fifth element or essence, *quintessence*, became associated with Aristotle's cosmology; it permeated supralunar space where the planets, stars, and Moon moved timelessly in perfect circles. The four concentric spheres were ordered by their nature and density. The first sphere, earth, was the most dense, and the watery sphere lay just above it. These two regions contained the heaviest and coldest elements. Above the watery sphere were located first the airy, then the fiery spheres.

4

lifted off the Earth's surface by strong winds and then thrown back to the ground. Nevertheless, these thunderstones from the sky were held in awe and in some cases were venerated as thunderbolts or weapons hurled by angry gods. The Roman emperor Elagabalus, when he began his four-year reign in A.D. 218, entered Rome in triumph and brought with him a black stone, probably a meteorite, from Emesa, Syria. It was thought to represent the Sun god, and Elagabalus insisted that it be worshiped publicly.

Silver coin, denarius, struck during the reign of the Roman emperor Elagabalus (A.D. 218–222). The obverse shows the emperor's portrait and the reverse, a four-horse triumphal carriage carrying the stone of Emesa. *(Courtesy of the Natural History Museum, Vienna, Austria.)*

Aristotle made it clear that the fiery sphere was not a region of actual fire, but rather of potential fire. It was a warm, inflammable region, containing fuel-like material so that the slightest agitation, or friction, would set it blazing at its most inflammable point. This combustible material was the fuel of cometlike phenomena. Aristotle described the heavenly bodies and supralunar space in his *De Caelo* (On the Heavens). Comets, however, were treated in his *Meteorologica*, which dealt with the sublunar, or terrestrial world.

Aristotle believed comets would form when the Sun, or planets, warmed the Earth causing the evaporation of dry, warm exhalations (like those in the fiery sphere) from the earth itself. At the same time, the cooler moisture contained in and on the Earth was also evaporated. This cool, moist vapor remained in the lower region of the airy sphere while the warm, windy exhalations rose up through it. At the border of the fourth or fiery sphere, the friction of their motion ignited them and the resultant comet, along with

5

neighboring dry exhalations, was carried about the Earth by the circular motion of the heavens in the fifth sphere. The comet's form and duration depended on the amount and form of the exhalation. If the exhalation, or fuel, happened to take on a diffuse appearance, the resulting comet would be a *fringed star*. Similarly, a *bearded star* resulted from an exhalation stretched in one direction. A comet with little fuel was soon extinguished and if the exhalation was excessively inflammable, it would burn quickly and form a meteor. Meteors, or shooting stars, also formed when the air was condensed by cold, squeezing out the hot, combustible exhalations; in this case, the meteor's motion was rapid, like a slippery fruit stone squeezed between the fingers. A comet most often formed independently, but if the exhalation was ignited by the motion of a fixed star, it could form as a halo. In this case, the halo or fringe only appeared to accompany the star, much as a lunar halo appears to follow the Moon in its motion.

Comets, when frequent, foreshadowed winds and drought because they formed under these conditions. Aristotle considered this to be an observed fact and that supported the fiery nature of a comet's constitution. For an example, he noted in his *Meteorologica* that the meteorite of Aegospotami had to be carried aloft by a wind before it fell back to Earth and "then too a comet happened to have appeared in the west." Like Hippocrates, Aristotle believed that comets are seen more often outside the tropic circles than within, because the Sun and stars moving within the tropic zone not only caused the warm exhalations to be secreted from the earth, but also dissolved them when they were gathering. Since the Milky Way extended far outside the tropic zone, the exhalations tended to gather there undissipated by the motion of the Sun, Moon, and planets. In order to consider the Milky Way as a collecting area for exhalations, Aristotle had to reject the ideas of Anaxagoras and Democritus that the Milky Way was made up of the light from certain stars, visible only in the Earth's shadow.

Aristotle's theories were physical, not metaphysical, in nature. Superstition and astrological predictions were noticeably absent. Although his cosmology of concentric spheres and circular, geocentric, heavenly motions was quickly superseded by Ptolemy's epicyclic motions, Aristotle's views on the nature of comets went mostly unchallenged for the next two thousand years.

Seneca's Natural Questions

Like Aristotle, Lucius Annaeus Seneca was deeply affected by the rulers he served. While Aristotle had the good fortune to serve as tutor to Alexander the Great, Seneca was sentenced to death by two Roman emperors and exiled by another. Born in Cordoba, Spain, about 4 B.C., he was educated in Rome and became one of its leading orators, writers, and teachers. Although a millionaire by today's standards, Seneca's writings and personal habits reflected the *Stoic* philosophy. He was abstemious in food and drink and kept a stoic calm even though his life seemed constantly threatened. Perhaps jealous of Seneca's influence in the Roman Senate, the third emperor, Caligula, condemned him to death. The sentence was not carried out because he was thought to be so sickly that he would soon die anyway.

Messalina, the third wife of the next emperor, Claudius, arranged for her rival Julia Livilla, Caligula's sister, and Seneca to be exiled to Corsica in A.D. 41 on adultery charges that had more to do with the political climate than any moral lapse on Seneca's part. Eight years later Agrippina, who had replaced Messalina as the wife of Claudius, dispatched to Seneca some good news and some bad news. He was to return to Rome to tutor her 12-year-old son—the good news; her son was soon to become the infamous emperor Nero—the bad news. Seneca taught Nero literature and morals, the latter with a notable lack of success. With Seneca's help, Nero ruled competently for a few years, but his mind became more and more unbalanced thereafter. By A.D. 62 Seneca's criticism of Nero's behavior, and worse, his poetry, led to his removal from court.

Seneca's *Natural Questions*, a work devoted primarily to meteorology and astronomy, was written about A.D. 63 during an especially trying period. In an attempt to mollify Nero—who by this time was living a life of total depravity and had arranged for the murders of his stepbrother, his mother, and his wife—Seneca wrote about two contemporary comets:

> . . .the recent one which appeared during the reign of Nero Caesar—which has redeemed comets from their bad character . . .

and,

> . . . the recent one which we saw during this joyous reign of Nero . . .

Viewed with hindsight, his attempt to vindicate the bad reputation of comets to flatter Nero seems outrageous, though admittedly the passages may have been ironic. Nonetheless, in A.D. 65 Seneca was accused of involvement in a conspiracy against Nero and ordered to commit suicide. This time the sentence was carried out.

The earliest ideas about comets come to us primarily from two treatises, the *Meteorologica* of Aristotle and the *Natural Questions* of Seneca. Aristotle presented the teachings of his Greek predecessors from the sixth to the fourth century B.C. Although Seneca included the newer cometary ideas of

7

the Greeks from the fourth to the first century B.C., he was also the source of what we know about the earliest views on comets—those of the Chaldeans or Babylonians. He wrote that Epigenes and Apollonius of Myndus (both fl. fourth century B.C.) had studied among the Babylonians. Unfortunately, they offered conflicting views. The Babylonians, according to Apollonius, classified comets among the wandering stars—the planets—and had determined their orbits. However, Epigenes reported that they had no understanding of comets, considering them to be fires produced by a kind of eddy of violently rotating air.

Both Aristotle and Seneca related their own views on comets only after they had discussed and rejected those of their predecessors, thus taking the remarkably modern stance of mentioning their sources only when they disagreed with them. However, had they not troubled to refute the ideas of their predecessors, we would now be largely ignorant of them. From Seneca we learn the cometary ideas of four Greek scholars: Ephorus of Cyme (fl. 340 B.C.), Epigenes, Apollonius of Myndus, and Posidonius (135–51 B.C.).

Seneca was particularly unkind to the historian Ephorus, calling him a mere chronicler who was often duped and tried to dupe others. For example, he ridiculed Ephorus' observations of a comet that split into two separate pieces, an event that cometary historian Alexandre Guy Pingré (1711–1796) discussed with regard to the comet of 372 B.C. Although Ephorus was the only source to mention the phenomenon, Seneca might have treated him with more respect had he known that several comets have since been observed to split (see Chapter 11).

In Seneca's outline, Epigenes' ideas were but those of Aristotle slightly modified. According to Epigenes, there were two types of comets: one was stationary and the other was in motion among the stars. The stationary type shed light symmetrically and formed when dry and moist exhalations were driven out through narrow apertures in the Earth, forming a whirlwind and setting fire to the surrounding atmosphere. Because of the heavier, moist exhalations, this type of comet was located in the lower regions. The second class of comets, Epigenes suggested, could result from an abundance of dry exhalations that sought the higher regions and were driven by the north wind. Epigenes' cometary fires, like those of Aristotle, remained until the combustible exhalations were consumed.

According to Seneca, Apollonius of Myndus considered comets as heavenly bodies; diaphanous, unequal in size, unlike in color, and waxing and waning like other planets. They were not illusions nor the result of planetary conjunctions. Their orbits, though not visible, intersected the upper part of the universe, and each comet was seen only when it reached the lowest part of its course. Although indicating that comets followed specified orbits, Apollonius maintained there was no reason to believe that the same comet reappeared; they were as numerous as they were varied. Seneca countered Apollonius' ideas by noting that comets cannot be planets because they were not always seen in the zodiac and because stars could often be observed through them. He also pointed out that if its orbit was other than circular, a comet's brightness would increase as it came closer to the Earth. He then observed that some comets attained their maximum brightness on the first day of their appearance, then began to fade. Seneca quoted Apollonius as stating that some comets are "blood-stained and threatening, bringing prognostication of bloodshed to follow in their train." Apart from this observation, however, Apollonius' ideas were quite modern. During the seventeenth century they were cited to support the view that comets were distinctive heavenly bodies traveling in highly eccentric orbits, and as a counter argument to the Aristotelian views that had been dominant until that time.

Seneca mentioned that Posidonius once observed a comet during a solar eclipse and concluded that many comets may be hidden by the Sun's rays.[2] Posidonius, a tutor to Cicero and a friend of Pompey the Great, had considerable influence on contemporary Roman thought. The comets of Posidonius, although transitory, lasted longer than other luminous objects because their motions were higher in the warm region of the ether. Since comets required the combustion of dry exhalations, they appeared during times of drought and, upon their disappearance, heavy rains could be expected. His theories were not particularly original, but helped to spread the Aristotelian views. Much of Seneca's own *Natural Questions* may be attributed to Aristotle, via Posidonius.

Although the Romans had captured the Greek city-states in the second century B.C., the Greeks had captured Roman minds. Typical of Roman scientific writing, Seneca's *Natural Questions* was a popularizing work primarily derived from Greek sources. In general, Seneca's views on meteorology were based on speculation and analogy. However, his views about comets were quite rational, original, and modern. He first asserted that his views differed from those of the Stoic sage Zeno and other Stoic brethren. Comets were not sudden fires but were among nature's permanent creations, and while their orbits generally differed, the two comets seen during his age had circular orbits much like the planets. Seneca suggested that comets moved in closed orbits, traveling in a uniform manner and disappearing only when they passed beyond the planets. To the argument that comets cannot be celestial bodies because their appearance and courses differed so greatly from the planets, he replied in his *Natural Questions* that

> Nature does not turn out her work according to a single pattern; she prides herself upon her power of variation. . . . She does not often display comets; she has assigned them a different place, different periods from the other stars, and

9

motions unlike theirs. She wished to enhance the greatness of her work by these strange visitants whose form is too beautiful to be thought accidental.

In an often quoted passage, Seneca addressed posterity and noted some questions that remain unanswered:

> The day will yet come when posterity will be amazed that we remained ignorant of things that will to them seem so plain. . . . Men will some day be able to demonstrate in what regions comets have their paths, why their course is so far removed from the other stars, what is their size and constitution. Let us be satisfied with what we have discovered, and leave a little truth for our descendants to find out.

Going even further, he suggested the scientific method for future cometary studies:

> . . . it is essential that we have a record of all the appearances of comets in former times. For, on account of their infrequency, their orbit cannot as yet be discovered or examined in detail, to see whether they observe periodic laws, and whether some fixed order causes their reappearance at the appointed day.

Beginning with the work of Edmond Halley (1656–1742) in the late seventeenth century, Seneca's advice was heeded and the periodic nature of cometary motion established (see Chapter 6). Most of Seneca's discussions were summed up with a main point or moral. His book on comets pointed to the neglect of learning in contemporary Rome, and as an example he noted that existing cometary ideas were formed with little attention to observational evidence; they were out of touch with reality. While Seneca's views did little, in a quantitative way, to advance understanding of the phenomenon, his rejection of the prevailing Aristotelian theory inspired eventual rethinking of the nature of comets. Unfortunately this rethinking had to wait some 15 centuries.

As a Stoic, Seneca believed in astrology and divination, and while his treatment of comets was rational—even modern—in its approach, he did make one quick bow to superstition by warning that a cometary apparition threatened the whole year with wind and rain. The seeds of superstition had been planted by Aristotle and Seneca and the subject was to flourish under the guidance of Pliny and Ptolemy.

The Natural History of Pliny the Elder

Caius Plinius Secundus, better known as Pliny the Elder, was a lawyer, traveler, administrator, head of the western Roman fleet under the emperor

Vespasian, and one of the most prolific writers of antiquity. He wrote treatises on oratory, grammar, the javelin, and Roman history, but his sole surviving work is *Natural History* in 37 books. His intent was "to give a general description of everything that is known to exist throughout the Earth." He covered 20,000 topics and apologized for omitting others. According to his nephew, the elder Pliny was able to carry out his official duties in addition to his voluminous writings because he had extraordinary zeal, an incredible devotion to study, and needed little sleep. Considering the nature of his writings, perhaps he should have slept longer, written less, and spent more time being critical or, at least, less gullible.

Pliny gathered superstitions, portents, love charms, and magic cures into his work indiscriminately, and his comments on comets are important only because the *Natural History* was so well known and respected during the Middle Ages. Without citing specific sources, he noted that some comets move while others do not, and that they were usually seen toward the north, chiefly in the region of the Milky Way. He did not acknowledge the contemporary views of his fellow Roman, Seneca, and relied almost exclusively on the earlier views of Aristotle. According to Pliny, comets could appear in any direction, but those in the south had no tails. He noted that the shortest and longest periods of a comet's visibility on record were seven and eighty days respectively. He gave more credence to comets as portents than did either Aristotle or Seneca and bolstered his view of comets as terrifying apparitions by noting the disasters that followed a few cometary returns. Pliny went so far as to mention how a particular comet's location, tail direction, and appearance could be used to predict imminent disasters and regions of the world that might be affected. For example, a comet resembling a flute with its tail rays pointing toward the east would indicate a malefic influence on music in the eastern territories.

In addition to Pliny's rules for predicting the nature of disasters following cometary apparitions, another questionable legacy was his system for classifying various types of comets—which would be repeated and often elaborated throughout the Middle Ages and well into the seventeenth century. Seneca had already mentioned a few types of comets, so the classifications did not originate with Pliny. But his writings, by their great popularity, were responsible for the transmission and long survival of the scheme. The 10 types of comets, according to Pliny, were

1. Pogonias: comet with a beard or mane hanging down from the lower part
2. Acontias: vibrating like a javelin with very quick motion
3. Xiphias: short and pointed like a dagger

11

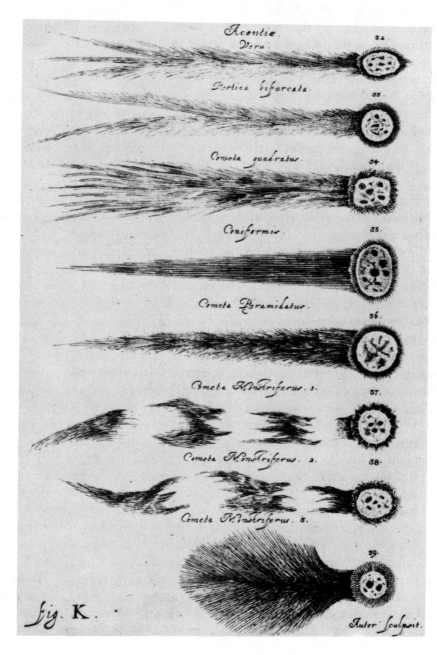

Even in the late seventeenth century, some of the cometary forms of Pliny the Elder are recognizable in serious scientific works. These eight types of comets were illustrated in Johannes Hevelius' *Cometographia* (Danzig, 1668).

The Comet of 44 B.C.

In his *Natural History* (Book II, Chapter 23), Pliny tells of a temple dedicated by the emperor Augustus (63 B.C.–A.D. 14) to a comet that appeared during athletic games he sponsored in 44 B.C. just after the assassination of Julius Caesar, his father by adoption. Pliny cited Augustus as saying that the comet was seen everywhere as a bright star in the north for seven days, rising an hour before sunset. The common people assumed that it was the soul of Caesar on its way to the region of the immortal gods. The emblem of a star was added to a bust of Caesar that was dedicated in the forum.

Augustus used an emblem of a comet on some of the coins struck during his rule, perhaps to remind his subjects that the reins of power had passed to him from the hands of his now deified father.

This Roman silver coin, denarius, was struck during the reign of the first emperor Augustus (27 B.C.–A.D. 14). The head of Augustus appears on the obverse with the partly obliterated inscription "Caesar Augustus." The reverse shows a stylized comet and the inscription "DIVVS IVLIVS," Divine Julius.

4. Disceus: like a quoit or discus, amber in color
5. Pitheus: figure of a cask, and emitting a smoky light
6. Ceratias: appearance of a horn
7. Lampadias: appearing as a burning torch

8. Hippeus: like a horse's mane in rapid motion
9. Argenteus: silver in color, so bright that it is difficult to look at
10. Hircus: goat comets ringed with a cloud resembling tufts of hair

Pliny's work was an extraordinary catalog of truths, half-truths, myths, and outright nonsense. He was more an encyclopedist than a scientist; his knowledge was almost exclusively derived from the writings of others rather than from personal observations. One notable exception occurred when he was in command of a Roman fleet in the Bay of Naples in A.D. 79. When Mount Vesuvius suffered a volcanic eruption on August 23 and 24, overwhelming the towns of Pompeii and Herculaneum, Pliny sailed across the bay to investigate and was killed by the poisonous fumes.

The Tetrabiblos *of Ptolemy and the* Centiloquy

Claudius Ptolemaeus (ca. A.D. 100–175), last of the great astronomers of antiquity, lived and worked in Alexandria, but virtually nothing more is known about him. He is celebrated for his masterwork, the *Almagest,* a comprehensive theory of celestial motions based on the observations of other astronomers as well as his own. His theory rested on a system in which the Sun, Moon, planets, and fixed stars moved daily around a spherical, stationary Earth. In his theory of *epicycles,* Ptolemy assumed that a planet moves upon a circle, an epicycle, the center of which describes a larger circle, a deferent, about a central point. This central point was slightly offset from the immobile Earth. Although the epicycle's center moved nonuniformly upon the *deferent,* the apparent motion of this center appeared to be uniform as seen from the *equant* point.

Ptolemy's system was complex, with enough adjustable variables that planetary motions could be predicted surprisingly well. In the seventeenth century, the accuracy with which one could predict them with the incorrect Ptolemaic system was not much inferior to that available using the correct, heliocentric system of Kepler. Ptolemaic models, using epicycles and equants, can produce planetary positions in longitude differing by only 10 arc minutes from a Keplerian ellipse with the same eccentricity. Though incorrect, the Ptolemaic system was founded on strictly scientific principles and enabled the user to accurately predict the motions of the heavenly bodies. The *Almagest* remained the most influential and widely read work on theoretical astronomy until the seventeenth century.

In a companion volume, the *Tetrabiblos,* Ptolemy explained the astrological influence of the heavenly bodies on earthly matters. Ptolemy believed in, and defended, astrology in his *Tetrabiblos,* in which he treated comets. As

14

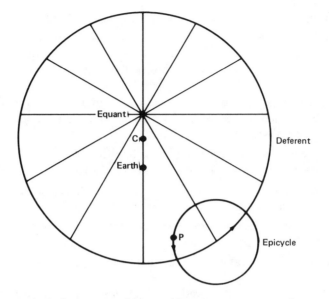

In the Ptolemaic cosmology, a planet, P, moves on a circle or epicyle, the center of which describes a larger circle or deferent about a central point, C. This central point has on one side the Earth and at an equal distance on the other side, an equant point. The center of the epicycle moves nonuniformly around the point C, but as seen from the equant point, the motion of this center appears to be uniform. The eccentricity is defined as the ratio of the distance from C to the Earth with respect to the radius of the deferent circle.

a result, Ptolemy's influence placed comets in the realm of astrology and ushered in more than 16 centuries of cometary superstition.

If Ptolemy had considered comets as heavenly bodies, they would have received scientific treatment in the *Almagest*. However, because they were unusual phenomena, he regarded them as mysterious signs or portents that provoke discord among men and give rise to wars and other evils. Unfortunately, he was even more specific and expanded somewhat on the astrological implications that Pliny had outlined. The part of the sky in which a comet first appeared and the direction the tail pointed indicated the geographic area that was threatened. The shape of the comet suggested the nature of the malefic event and the persons to be affected. Its position relative to the Sun foretold when disaster would strike. Evil was nigh if the comet first appeared near the rising Sun but might be delayed if it was first seen in the west. The length of time during which the comet remained visible was directly proportional to the duration of its ill effects.

Because of his great reputation, many astrological works were incorrectly attributed to Ptolemy in the Middle Ages. One such anonymous tract was the *Karpos*, or *Centiloquy*, which often accompanied the *Tetrabiblos* in medieval Latin manuscripts. It was long accepted as one of Ptolemy's writings and was thus influential. The last of its hundred aphorisms laid down specific rules for predicting comet-related disasters.

- An appearance of a comet at a cardinal point 11 astronomical signs from the Sun implied that the king of a particular kingdom, or one of the princes or chief men, would die.
- An appearance in a succeeding house implied prosperity for the kingdom's treasury, but the governor or ruler would change.
- An appearance in a cadent house (one that had passed the meridian) would be followed by diseases and sudden deaths.
- If the comet's direction moved west to east, a foreign foe would invade the country.
- If the comet was stationary, the foe would be provincial or domestic.

In the centuries that followed, Ptolemy's guidelines, and those in the *Centiloquy*, were used repeatedly to correlate cometary apparitions and terrestrial disasters. During the Middle Ages and in Renaissance times, the bulk of cometary literature was superstitious in nature. Writers seemed unwilling to deviate from Aristotle's ideas on cometary origins or Ptolemy's insistence on the relationship between comets and adversity.

Summary

Among the ancients, Aristotle and Ptolemy were the two dominant scientific authorities and neither one regarded comets as celestial phenomena. Aristotle did not include comets in his work on heavenly bodies *(De Caelo)*, but rather in his treatise on terrestrial phenomena *(Meteorologica)*. Ptolemy's *Almagest* treated the motions of the heavenly bodies but did not mention comets at all; they were left to his companion work on astrology *(Tetrabiblos)*. Under the dominant influence of Aristotle and Ptolemy, the more rational views of Seneca were quickly submerged. Thus the study of comets in the Middle Ages was handicapped by prevalent views that were incorrect, yet very authoritative.

NOTES

1. The notion of an invisible, intangible substance in space, *ether*, which takes a causal part in the motions of planets and transmission of light, was postulated by Aristotle, Anaxagoras, and other ancient philosophers. The idea of a single,

all-pervasive ether received its greatest support in Descartes' cosmology of vortices, a theory that was widely believed prior to the Newtonian era.

2. If Posidonius personally observed a comet during a solar eclipse, it might have been during the total eclipse of either 115 or 103 B.C. or perhaps during the partial eclipse of 94 B.C. Each of these events was observable in the Mediterranean area.

2
Medieval Views

A diffusion of Greek and Roman ideas into western Europe occurs during the Middle Ages. The Christian church embraces the cometary views of Aristotle. Accurate cometary positions begin in the west with Paolo Toscanelli's fifteenth century measurements of several comets. First attempts are made to determine the distance of comets above the Earth's surface. The comet of 1577 is shown to be well above the Moon by Tycho Brahe and his colleagues. Early Chinese astronomers make contributions to the study of comets.

The Period to About 1200:
Theology and Superstition

With the disintegration of the Roman empire, general knowledge of the Greek language disappeared and with it went the widespread knowledge of early Greek science. Latin translations of early Greek science were not widely circulated prior to the twelfth century; even so, contemporary views on comets, handed down from church scholars, were clearly derived from Aristotle and Ptolemy. The medieval period was extremely church oriented. It was an age of faith and medieval authorities did not look forward, they looked back. When physics was discussed, it was Aristotelian physics, and whenever computations were required to predict the positions of heavenly bodies, it was Ptolemy's *Almagest* that provided the necessary techniques.

Perhaps the first medieval astronomer was the English Benedictine theologian Bede the Venerable of Yarrow (ca. 673–735). Bede, one of the most enlightened figures of his time, stated that comets portend changes of rule, pestilence, wars, winds, or heat. They never appeared in the western sky; they nearly all were in the north, usually in the Milky Way. Some comets moved with planetlike motions, while others remained stationary. The longest interval a comet remained visible was 80 days, the shortest 7 days. Bede's views were clearly derived from those of Ptolemy and Aristotle.

Around A.D. 1000, a number of previously uncirculated astronomical treatises were translated into Latin from the original Greek. They were generally done by wandering scholars who visited Spanish monasteries and used existing Arabic translations of the early Greek writings. The diffusion of early Greek science into Europe was aided by the crusades of the twelfth century, which created an intercourse between Western Europe and the culturally more advanced Middle East. Aristotle's *Meteorologica* was first translated into Latin in 1156 by Henricus Aristippus of Sicily, while the translation of Ptolemy's *Almagest* did not occur until four years later. By 1200, the works of Ptolemy and Aristotle were in circulation in Latin and hence available to a larger readership in Western Europe. Contemporary churchmen embraced these ideas because of their strict logic, wide knowledge, and scriptural compatibility. The Church had unquestioned authority and the astronomy of Aristotle and Ptolemy was not only approved for teaching, it was virtually sanctified and rendered unassailable.

The theologian Albert Magnus (1193–1280) wrote that a comet was a coarse, terrestrial vapor that gradually rose from the lowest part of the airy region to the upper part where it touched the concave surface of the fiery sphere. The comet then ignited and remained visible until the fuel was exhausted. Albert considered comets as signs, not causes, of malefic events. They cannot cause the death of magnates, since vapor no more rises in a land where a rich man lives than where a pauper resides. Albert and his pupil, Thomas Aquinas (ca. 1225–1274) labored to work out a synthesis of Aristotelian science and Christian theology. The mathematics of Ptolemy was combined with the conceptually simpler and more qualitative astronomy of Aristotle. Aquinas added to the prevalent fear of comets by writing that they were among the 15 signs preceding the Lord's coming to judgment.

Another pupil of Albert Magnus was the English scientist Roger Bacon (1214–1294), who advanced the scientific method by insisting on objective observations and experiments as guides to knowledge. He observed a comet in July 1264, described it in some detail, but drew conclusions that had more to do with contemporary superstition than with the scientific method. Bacon noted that the comet of 1264 appeared in Cancer and moved toward Mars, which—due to the warlike nature of Mars—presaged discord and wars. The comet was described as dreadful and followed by vast disturbances and wars in England, Spain, Italy, and other lands in which many Christians were slaughtered. Martin Luther (1483–1546) went even further, referring to comets as harlot stars and works of the devil. He is quoted as saying "the heathen write that the comet may arise from natural causes, but God creates not one that does not foretoken a sure calamity" (see White, 1910, 1:182).

"The Opening of the Fifth and Sixth Seals." This 1511 woodcut by Albrecht Dürer illustrates John's description of the Apocalypse in Revelation 6:9–17: "And I beheld when he had opened the sixth seal, and lo, there was a great earthquake, and the Sun became black as sack cloth, the full moon became like blood, and the stars of the sky fell to the Earth . . . ." Dürer used the analogy of a meteor shower, and possibly eclipses by the Sun and Moon, to indicate the disorder and terror associated with the apocalypse. (*Courtesy of the New York Metropolitan Museum of Art.*)

A typical example of medieval cometary thought, written in 1578 by the Lutheran bishop Andreas Celichius, was entitled *Theologische Erinnerung von dem Neuen Cometen. . . .* (The Theological Reminder of the New Comet). In an opinion that represented the majority view of comets during the late medieval period, Celichius wrote that comets are

> . . . the thick smoke of human sins, rising every day, every hour, every moment full of stench and horror, before the face of God, and becoming gradually so thick as to form a comet, with curled and plaited tresses, which at last is kindled by the hot and fiery anger of the Supreme Heavenly Judge.

In a rational reply, the Hungarian scholar, Andreas Dudith (1533–1589), expressed the minority opinion by noting that if comets were caused by the sins of mortals, they would never be absent from the sky. Dudith's common sense views were published in 1579. Another blow to cometary superstition was struck a year later by Blaise de Vigenère. His treatise provided rational counter arguments to many foolish beliefs, including examples of the deaths of many great monarchs with no comet to herald the event.[1] However, the works by Dudith and Vigenère were exceptions, not the rule, and the ripples of rationalism they caused were largely ignored in the wave of contemporary superstition.

During the medieval period to approximately A.D. 1200, there was virtually no original work on comets in the Western world. When comets were ob-

Woodcut showing destructive influence of a fourth century comet from Stanislaus Lubienietski's *Theatrum Cometicum. . . .* (Amsterdam, 1668).

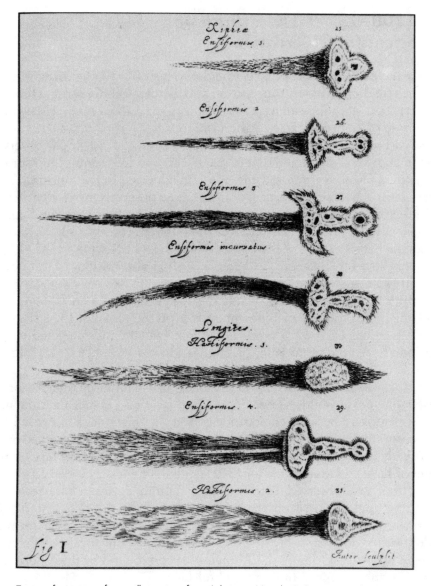

Types of cometary forms, illustrations from Johannes Hevelius' *Cometographia* (Danzig, 1668).

served, care was taken only to note in which constellation they first appeared so that appropriate astrological predictions could be made. Comets were considered terrestrial phenomena, balls of fire thrown at a sinful Earth from the right hand of an avenging God. Cometary superstition reigned supreme and no significant advances were made in understanding the phenomena.

23

1200–1577: The Rebirth of Scientific Observations

In the period from 1200 to before the comet of 1577, the views on comets were still dominated by superstition and astrological nonsense. However, a few hesitant steps forward were taken and an occasional observer began to make note of comets in a scientific fashion.

The French Dominican Aegidius of Lessines (ca. 1230–1304) observed the comet of 1264 and noted that it was first seen in the evening after sunset; then after a few days it crossed the Sun and appeared in the morning. This observation is noteworthy because, at the time, observations of a particular comet before and after passing the point on its orbit closest to the Sun, its perihelion, were generally attributed to two different comets. Indeed, prior to his comprehensive study of the comet of 1680, Isaac Newton (1642–1727) himself believed that the pre- and postperihelion observations of this object were due to two separate comets.

Peter of Limoges (fl. 1300) was the author of a tract on the comet of 1299 and quite probably an anonymous tract on the comet of 1301 (Halley) as well. The comet of 1299 appeared in late January and remained visible into March. According to Peter of Limoges, it was of moderate size, had a long tail, and was dark blue. As recorded in the anonymous tract, the comet of 1301 appeared September 1 and remained visible for more than a month. The similar ideas from these two tracts suggested that a comet's natural motion was eastward because comets, located in the upper air, slip behind the sky's westward, daily motion. That these two comets were not observed moving directly eastward was due to the influence of Mercury and Mars for the comet of 1299 and Mars alone for 1301. For both comets, celestial longitude and latitude were observed directly using a sighting device, a *torquetum*, for measuring angles. These observations were perhaps the first instance in which astronomical instruments were used in the West. Both treatises concluded with a discussion of the comet's malefic significance, but Peter of Limoges was careful to note the Almighty was not bound by these portents; they were only warnings from a watchful God.

The Florentine physician and astronomer Paolo Toscanelli (1397–1482) made descriptive and positional observations of 6 fifteenth century comets: 1433, 1449–1450, 1456 (Halley), 1457 I, 1457 II, and 1472.[2] These observations were apparently unknown and remained in manuscript form until discovered in 1864 at the National Library in Florence. Giovanni Celoria, director of the Milan Astronomical Observatory, successfully used Toscanelli's observations to compute their orbits. While the surviving details

Paolo Toscanelli

of Toscanelli's life are few, a contemporary biographer described him as modest, religious, a vegetarian, and learned in Greek, Latin, and geometry.

Toscanelli is known primarily for the role he may have played encouraging Christopher Columbus prior to his discovery of the New World in 1492. Interested in navigation and geography, Toscanelli constructed a nautical map of the Atlantic Ocean. Because of extremely poor longitude data, China and Japan were located 100 degrees east of their correct positions. In 1474, Toscanelli is believed to have forwarded the map to the Portuguese canon Ferdinando Martini, who in turn transmitted it to Columbus. In constructing his own navigation charts, Columbus probably used Toscanelli's incorrect map to verify his own, which was even less accurate. Columbus figured he had 2400 miles of ocean between the Canary Islands and Japan. Toscanelli's map suggested 3000 miles. The actual distance was more like 10,000 miles. If the maps of Toscanelli and Columbus had been correct, the voyage might not have been attempted.

During the nearly 40 years that Toscanelli made cometary observations, he was concerned with finding accurate cometary positional data, perhaps to facilitate his use of astrology for his medical profession. After studying Toscanelli's manuscript charts, the science historian Jane Jervis concluded that for the comet of 1433, the positions were determined using crude stellar alignments on his hand-drawn star charts. For the comet of 1449–1450, the positions were determined more accurately with a straight edge on his star charts and for the comet of 1456, his measures were made directly with a torquetum. His series of manuscript pages may be the first time celestial charts were used as an integral part of celestial measurements and not just as pictorial representations of stellar positions.

The Viennese astronomer Georg von Peurbach (1423–1461) was the first to attempt a cometary *parallax determination* using his observations of the 1456 apparition of comet Halley. Parallax determination is a technique for determining the distance of a celestial object by observing it simultaneously from different locations and noting the resulting angular change of the object's apparent position referred to the distant stars. The greater the distance between the observer and the object, the smaller this parallax angle becomes. A parallax determination can also be observed at one location if the measurements are made several hours apart so that the Earth's rotation changes the observer's position between the first and second set of observations. In this latter case, a medieval scientist would attribute the position changes to the rotation of the celestial sphere, rather than the Earth's rotation. The parallax of the Moon can be defined as the largest angle subtended by the Earth's radius, as seen from the distance of the Moon. It is 57 arc minutes, just less than one degree. An object twice the distance of the Moon would have a parallax angle just less than one-half degree.

In the mid-fifteenth century, Peurbach taught astronomy at the University of Vienna and wrote an often reprinted theory of the planets and a theory of eclipses. He composed a number of tables of trigonometric sines and chords, constructed sundials, calculated tables of planetary positions, *ephemerides,* and made positional observations of stars and planets. He also wrote a number of lyrical love poems in Latin, all of them addressed to men. Peurbach wrote a brief German treatise when the comet of 1456 was still visible, and followed it with a more formal Latin treatise when it departed. Neither treatise was formally published but a number of manuscript copies exist. His ideas on the physical nature of comets were Aristotelian; comets were hot, dry, terrestrial exhalations rising to the upper regions of air or region of fire where they were ignited under the influence of the stars and planets. Using an unspecified instrument, Peurbach first observed the comet of 1456 on June 10 and followed its motion for most of that month. His parallax determination was not a precise measurement. In fact, he apparently assumed the comet was more than an Earth radius from the Earth's surface and used the comet's slight parallax to verify his assumption. Jane Jervis studied Peurbach's Latin treatise at the Vienna State Library and gave the following outline of Peurbach's reasoning:

- From Ptolemy, the radius of the sublunar sphere is 33 Earth radii.
- One degree on the Earth's surface subtends 16 German miles.
- The Earth's circumference is then $16 \times 360 = 5760$ German miles.
- The Earth's radius is $5760 \div 2\pi = 1000$ German miles (approximately).
- Hence the distance from the sublunar sphere to the Earth's surface is $33,000 - 1000 = 30,000$ German miles (very approximately).
- If comets are in the upper region, below the sphere of fire, and the sphere of fire is less than say 27,000 German miles measured from the lunar sphere, then the comet must be $30,000 - 27,000$ or more than 1000 German miles above the Earth's surface.
- The comet's parallax angle was found to be slight and this was taken as proof that the comet was at least 1000 German miles above the Earth's surface.

There is a real question as to whether Peurbach's cometary parallax determination was an attempt to determine the comet's distance or the height of the sphere of fire. Peurbach also used the comet's distance and apparent tail size to compute the linear length of the tail. He computed that the comet's 10-degree tail would make it 80 German miles long. Once again, Peurbach's estimates were crude. At a distance of 1000 German miles, a 10-degree tail would be larger than 175 German miles. Peurbach's treatise also included a brief discussion on the quantity of matter necessary to sustain a fire of this size and a rather lengthy discussion on the comet's portents.

In approximately 1335, the mathematician Levi ben Gerson (1288–1344) suggested that Ptolemy's parallax method for the Moon be applied to determine the distance to a comet. However, it was Peurbach's student Regiomontanus, or Johannes Müller (1436–1476) of Königsberg, who is generally given credit for making known the techniques required for cometary parallax determination. The German city of Königsberg means "King's Mountain." In Latin, Regiomonte translates as "from the royal mountain." Hence, Regiomontanus took his name from his home town. He received his bachelor's degree at age 15 from the University of Vienna and was appointed to the faculty at age 21. Upon Peurbach's death in 1461, Regiomontanus took over his teacher's condensation and explication of Gerard of Cremona's twelfth century translation of Ptolemy's *Almagest*. Although this work was completed sometime before 1463, it was not published until 1496. In 1471, Regiomontanus left for Nürnberg, where the patronage of Bernard

Walther (1430–1504) allowed him to set up an observatory. In 1475, Pope Sixtus IV summoned him to Rome to work on calendar reform, but Regiomontanus died within the year.

Contemporary accounts attribute his death variously to poisoning, the plague, or a comet. If the events surrounding the death of Regiomontanus seemed strange, the events surrounding the fate of his effects were even more peculiar. After his death, Regiomontanus' books, papers, and instruments were acquired by Bernard Walther. Walther used the instruments to continue the observations that he and Regiomontanus had started, but the books and papers were locked up and everyone denied access to them. After Walther's death in 1504, the executors of Regiomontanus' estate apparently started selling his books, all the while denying that they were doing so. Several law suits were initiated over the remains of the estate. Finally, most of the remaining books and papers were acquired by Johannes Schöner, who published a few of them in the 1530s and 1540s. Hence it was an unusually long time before the works were published.

Regiomontanus' major work on comets was first published by Schöner at Nürnberg in 1531 to take advantage of the interest generated by the comet (Halley), which appeared in that year. This work in entitled *Ioannis de Monteregio Germani, viri undecunque doctissimi, de comete magnitudine, longitudineque ac de loco eius vero, problemata XVI*. An English translation of the Latin title would read "Sixteen problems on the magnitude, longitude and true location of comets by the German Regiomontanus, the most learned among men." This work provided several techniques for determining a comet's parallax, as well as determining a comet's position and size. A guide book, it was general in scope with no specific examples of actual cometary measurements. Another work entitled *De cometis. . . .* was first published under Regiomontanus' name by Jakob Ziegler in 1548 as part of Ziegler's history of the universe. It was often reprinted in the sixteenth and seventeenth centuries, including a reprinting in 1668 in the *Cometographia* of Polish astronomer Johannes Hevelius (1611–1687). Although this work was often attributed to Regiomontanus, Jane Jervis made a convincing case for its authorship by someone of Peurbach's school. The *De cometis. . . .* was concerned with the position, size, and distance of the comet of 1472. The qualitative path of the comet through the constellations was described with notes on its apparent speed and tail direction. Based on an estimate of its apparent motion with respect to the star Spica, a parallax of not more than six degrees was determined. The comet was thought to be at least nine Earth radii from the surface of the Earth, putting it in the region of air, but not in the region of fire.

Girolamo Fracastoro (ca. 1478–1553), an instructor of logic and anatomy at Padua, Italy also described the comet of 1472 and observed those

Johannes Müller, or Regiomontanus

seen in 1531 (Halley), 1532, and 1533. In a work that was designed to re-place the Ptolemaic epicycles with a system of 76 homocentric spheres, Fracastoro was the first European to report, in print, the antisolar nature of comet tails. His work of 1538 was entitled *Homocentricorum, sive de stellis.* However, the credit for the first European to draw attention to this phenomenon goes to Peter Apian (1495–1552), professor of mathematics at Ingolstadt. Apian's work on the 1531 apparition of comet Halley *Practica auff dz. 1532 Jar . . . .* had a diagram on the title page showing the antisolar nature of comet tails. In this work, as in his later work entitled *Astronomicum Caesareum,* Apian drew diagrams showing the extended radius vector from the Sun to the comet passing through its tail. Perhaps too much has been made of these diagrams since the dust tails, most easily seen with the naked eye, are nearly always strongly curved and rarely appear directly along the extended Sun-comet radius.

When investigating Apian's observations of the comet of 1531, over the interval August 13–23, Edmond Halley (1656–1742) himself mentioned their relative inaccuracy. All eight of Apian's longitudinal observations were discrepant by at least one degree and two were in error by nearly five degrees. While he lacked observational precision, Apian excelled in cre-

29

ating magnificent books. The *Astronomicum Caesareum* is the most spectacular contribution of the bookmaker's art to sixteenth century science. Its pages are large, brilliantly hand colored, and filled with ingeniously contrived mechanisms, up to six layers of paper disks arranged to give planetary positions in addition to calendrical and astrological data. Printed on Apian's private printing press, the book was designed for Charles V and his brother Ferdinand. Flattered by the book, Charles V handsomely rewarded Apian by raising him to the rank of hereditary nobility, crossing his palm with 3000 gold coins, granting him the right to appoint notaries, to award the titles of doctor and poet laureate, and to legitimize children born out of wedlock.

Peter Apian's August 1531 observations of a comet (Halley) in the constellation Leo were used to demonstrate the antisolar nature of cometary tails. Woodcut illustration from Apian's *Practica auff dz. 1532 Jar. . . . (Landshut).* (Courtesy of the Crawford Library, Edinburgh, Scotland.)

Fewer than one hundred copies of this book remain extant, making it more rare than a first edition of Copernicus' *De Revolutionibus.*

Nine years after Apian's beautiful book was published, the first cometary catalog was issued by Antoine Mizauld (1520–1578). Mizauld's book was neither accurate nor beautiful. Entitled *Cometographia crinitarum stellarum . . . .* this work included a catalog of comets and their attendant disasters until 1539. Mizauld was considered gullible and referred to as *Mizzaldus ineptus* by one of his contemporaries. Although Mizauld's own work was somewhat inaccurate, it does contain a reasonably accurate catalog of 46 comets. Alexander Pingré attributed this latter catalog to Paul Eber, Protestant minister and professor of belles-lettres at Wittenberg.

Girolamo Cardano (1501–1576), a teacher at Padua and Rome, acknowledged the importance of parallax measurements for the study of comets but decided upon a supralunar position for a comet seen in 1532, not by parallax measurements, but rather by noting that its apparent speed was less than that of the Moon. Cardano, known chiefly for his mathematical works on probability, was imprisoned by the Inquisition in 1570 for having cast Christ's horoscope and asserting that the events in his life were governed by the stars. He prudently recanted.

Cardano's cometary views were presented in his 1550 work *De Subtilitate* and in his *De Rerum Varietate*, published seven years later. These two works were later republished in his collected works. Cardano thought of comets as globes, or spherical lenses, illuminated by the Sun with the tail formed from sunlight shining through the comet and brought to a focus behind it. He believed the antisolar nature of the tail was common to all comets. When the air became dry, comets appeared, and while they may not have been the cause, they certainly were signs of dryness, corruption, famine, and death. Unlike Fracastoro and Apian, Cardano believed in the supralunar position of at least one comet. Unfortunately, his views were still the minority opinion.

Johannes Vögelin, professor of mathematics at Vienna, attempted a parallax measurement for the comet of 1532. Vögelin's work was outlined in his *Significatio cometae qui anno 1532 apparuit*, first published in 1533 at Vienna and later in a 1574 work by Prague astronomer Thaddeus Hagecius (ca. 1525–1600). Following Regiomontanus' parallax technique, Vögelin took two altitude and azimuth observations 42 minutes apart on October 6 for the comet of 1532. He determined extraordinarily large parallax measurements of 35 degrees, 31 minutes, and 1 second (35° 31′ 1″) and 34 degrees, 58 minutes, and 32 seconds (34° 58′ 32″). On October 6, 1532 the comet was approximately 0.76 *astronomical units,* AU, from the Earth so the correct parallax value for that day should have been approximately 11.5 seconds.[3]

Note that Vögelin, in a custom that is still frequently employed, wrote his result to the nearest arc second when his error was far larger than the result itself. His method appears to be sound and his computations correct, but the observations were not precise enough to give meaningful results. The computed value of the parallax was extremely sensitive to the uncertain values of azimuth measured.

Nicholas Copernicus (1473–1543) observed the comet of 1533 and wrote a brief treatise about it published by M. Curtze in 1878. Copernicus mentioned comets only once in his 1543 masterwork *De revolutionibus orbium coelestium*. Ironically, he assumed comets were terrestrial objects and used their motion in the upper air to demonstrate that these regions take part in the daily celestial rotation:

> It is said. . . . that the highest region of the air follows the celestial motion. This is demonstrated by those stars that suddenly appear—I mean those stars that the Greeks called cometae or poganiae. The highest region is considered their place of generation, and just like other stars they also rise and set. We can

"An awesome, wondrous sign (portending) two earthquakes, seen at Rosanna and Constantinople in the year 1556." Broadside by Herman Gall, printed by Valentin Neuber. (*Courtesy, Department of Prints and Drawings of the Zentralbibliothek, Zurich.*)

say that this part of the air is deprived of the terrestrial motion because of its great distance from the Earth.

Part of book I of Copernicus' *De revolutionibus* was translated into English in 1576 by Thomas Digges (ca. 1546–1595), the sixteenth century leader of the English Copernicans. Digges was nearly alone in being a Copernican with a pre-1577 belief in the celestial nature of comets. This statement is supported by Digges' description of the new star, or *supernova*, seen in 1572. While arguing for its celestial nature, Digges consistently identifies it with the cometary region. However, by 1576, Digges spoke of comets as terrestrial phenomena in his work *A Perfit Description of the Caelestiall Orbes*. This Copernican treatise was appended to his father Leonard's Ptolemaic treatise *A Prognostication Everlasting*. Thomas Digges' switch to treating comets as terrestrial objects in 1576 was particularly untimely because the next year a great comet arrived and subsequent observations would prove that it was definitely a celestial object beyond the Moon.

The Comet of 1577

The great comet of 1577 was an extraordinary apparition. Contemporary descriptions note that it was seen through the clouds like the Moon, and that it rivaled Venus in brightness. It was first recorded on November 1, 1577 in Peru and last recorded January 26, 1578 by Danish astronomer Tycho Brahe (1546–1601). The comet reached perihelion on October 27, 1577 when it was 0.18 AU from the Sun, well inside the orbit of Mercury. The comet's nearly parabolic motion around the Sun was opposite to that of the Earth and planets, or *retrograde* motion, and it approached closest to the Earth, at 0.63 AU, on November 10, 1577. The works published on this comet form a turning point in the history of astronomy because precise observations were used to demonstrate that the Aristotelian views on comets, which had remained dominant for nearly two millennia, were incorrect. The comet of 1577 was shown to be well above the Moon.

First to publish his observations was the German astronomer Michael Mästlin (1550–1631). Educated at Tübingen, as a pupil of Apian, his observations were simplicity itself. Using a thread, he aligned the comet with two neighboring stars, so that they were all on the same great celestial circle. He repeated the process with two different stars. Then using the star catalog in Copernicus' *De revolutionibus*, the stellar positions were noted and the comet's position determined from the intersection of the two great circles connecting the stellar pairs. Some hours later, the observations were repeated and apart from the comet's actual motion on the sky, which he estimated,

Comet of 1577 as seen in a woodcut broadside by Jiri Daschitzsky and published by Peter Codicillus. An artist is seen drawing the comet and is aided by men holding his sketchbook and lantern. The heading reads, "Concerning the fearful and wonderful comet that appeared in the sky on the Tuesday after Martinmass (November 12) of this year 1577." (*Courtesy of the Department of Prints and Drawings of the Zentralbibliothek, Zurich.*)

Mästlin noticed no difference in its apparent position with respect to the neighboring stars. His null parallax determination was based on measurements a few hours apart for three nights in early December 1577.

Using observations from November 12, 1577, until January 8, 1578, Mästlin devised a theory to account for the comet's apparent path. From the observed maximum angle between it and the Sun, the comet was assumed to move on a circular path slightly outside the orbit of Venus—concentric with the orbit but not quite heliocentric. The observed deviations from uniform circular motion were explained by assuming the comet moved on a small epicycle perpendicular to its orbit plane so that it librated back and forth about a mean position. Mästlin believed that the origin of comets was a mystery known only to God, but once created they were celestial phenomena and could be treated as such. A confirmed Copernican by 1572, his heliocentric orbit for the comet of 1577 may have been conjecture. While the true orbit

Michael Mästlin

for this comet was parabolic, he assumed it to be circular. However, his observations covered only 60 degrees of its path, an interval short enough that his theory could roughly explain its observed motion. Mästlin's observations showed the comet to be a celestial object. Nevertheless, he declared it a new and horrible prodigy and devoted a chapter of his book to its portents, concluding that it was a type that betokened peace, but peace purchased by a bloody victory.

The work of astronomer Tycho Brahe is most closely associated with the comet of 1577. A member of Danish nobility, Tycho was sent to Leipzig in 1562 to study law. However, his interest in astronomy was unsuppressed and he left Leipzig in 1565 and began studying at Rostock the following year. He continued at the University of Basel three years later. Tycho's observations of the new star in 1572 did much to spread his fame, and in 1575 King Frederick II of Denmark offered Tycho the island of Hveen in the Danish sound and provided him with enough money to erect a residence and observatory. In fact, two observatories, Uraniborg and Stjerneborg, were erected and the island soon had a windmill, a paper mill, herbaries, flower gardens, and several fishing ponds. Tycho's residence had running water, a library, a chemical laboratory and room for eight assistants. He had at his disposal something like one percent of the annual income of the Danish government, which supported astronomical research to a percentage level never reached since. In return for his princely benefits, Tycho had only to act as a consultant to the Danish royal court on astrological and astronomical matters. He compiled a horoscope at the birth of each royal son and supplied the royal family with annual astrological predictions. Tycho's belief in astrology was genuine, but he took issue with contemporary astrologers whose predictions were unwarranted or too specific.

During the evening of November 13, 1577, Tycho was out by his ponds catching fish for the evening meal when he noticed a comet in the western sky. For the next two and a half months, he observed it whenever the weather was clear. Early in 1578, he wrote a brief manuscript in German setting down his opinions and conclusions on comets in general and the comet of 1577 in particular. This important treatise existed only in the form of two manuscript copies until it was published by Johann Louis Emil Dreyer in 1922 and translated into English in 1979 by J.R. Christianson.

In his German treatise, Tycho quickly dismissed the Aristotelian theory of the cosmos by citing the new star of 1572 as evidence against the immutability of the heavens. He also objected to the Aristotelian notion of comets and concluded, from his own measurements, that the comet's parallax at the horizon could not have been greater than 15 arc minutes. Tycho thought the comet to be at least 230 Earth radii above the Earth's surface and certainly above the Moon's sphere, which he took to be 52 Earth radii. He also recorded his physical observations of the comet, noting that, on November 13, 1577, the head was whitish or Saturnlike in color and 8 arc minutes in diameter. On the same date, the tail had an apparent length of 21 degrees 40 minutes and appeared a reddish dark color similar to a flame seen through smoke. Tycho observed that the tail was directed in an antisolar direction due to the Sun's rays passing through the comet's rarefied and porous head. The head itself was visible because it did not pass all the Sun's light; it was translucent.

Fully one half of Tycho's German treatise was concerned with the astrological implications of the comet. Because it was first seen in the west, the most affected regions would be Western Europe. However, the eastern regions were not exempt from the comet's influence because, as Tycho stated,

> Although this comet appeared in the west and will realize its greatest significance in those lands that lie toward the west, yet it will also spew its venom over those lands that lie eastward in the north, for its tail swept thence.

When Tycho first observed the comet on November 13, it was Saturnlike in color and its apparent position on the sky was not far from Saturn itself. The comet first appeared in the eighth house, which astrology ascribes to death. From its initial position and color, Tycho concluded that the comet augured

Tycho's Noble Nose

This portrait of Tycho Brahe clearly shows something peculiar about his nose. In fact, a portion of it was artificial. While studying astronomy at Rostock, on the North Sea, the 20-year-old Tycho became engaged in a heated quarrel at a Christmas party with another nobleman named Manderup Parsbjerg. According to a contemporary source, they argued over who was the better mathematician. The two hotheads met on December 29, 1566 at seven o'clock in the evening to settle the dispute. Although the most appropriate choice of weapons would have been paper and pens, they chose swords instead. Tycho came out second best

Tycho Brahe

when Parsbjerg's sword sliced off a piece of his nose. Thereafter he wore a metal replacement, said to have been made of gold and silver. Apparently the noble nose had a high copper content as well, since an examination of his skeleton in June 1901 revealed a green stain surrounding the upper end of his skull's nasal opening.

Although Tycho and Parsbjerg became friends after their fateful encounter, Parsbjerg always claimed the duel was a fair one—history does not record whether he boasted of winning the contest by a nose.

an exceptionally great mortality among mankind. Since it was intentionally qualitative and popular in nature, Tycho's German manuscript was probably intended as a report to the Danish royal court. The detailed data and scholarly analysis would be published in a larger, Latin treatise entitled *De mundi*.

Tycho's masterwork on the comet of 1577, published in 1588, is most often given as *De mundi*; the full title of the work is *Tychonis Brahe Dani de mundi: aetherei recentioribus phaenomenis liber secundus qui est de illustri stella caudata ab elapso fere triente Nouembris anni 1577, usg; in finem Ianuarij sequentis conspecta*. Sixteenth century authors seemed to vie for the longest book title, which often served as the book's abstract. Tycho's title would loosely translate to "Two Books Concerning the Quite Recent Phenomena of the Aethereal Region, seen from November 13, 1577, until the following January, by Tycho Brahe, the Dane."

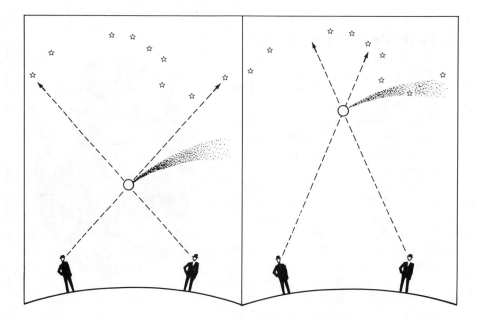

Schematic illustration showing parallax determination. Tycho Brahe and his colleagues observed the comet of 1577 simultaneously from two different locations on the Earth's surface. If the comet was below the Moon and close to the Earth, each observer would see it appearing against an entirely different stellar background. Since each observer noted the comet appearing against nearly the same background stars, the comet must have been quite distant and beyond the Moon.

Tycho's most important conclusions were presented in the first chapter, where he began by deducing new positions for the 12 reference stars used in determining the comet's positions. They were deduced from the measured angular distances, or *offsets*, from neighboring stars whose positions were known. Differing observations, as well as different parallax techniques, were used to determine the comet's supralunar location. On the evening of November 23, its position relative to the star ε Pegasi was determined twice, the second observation was three hours after the first. Assuming the comet was at the same distance as the Moon, its parallax and actual motion on the sky would have had the effect of making the second angular measurement from the star equal to the first. However, the observed difference between the measurements was actually 12 minutes, and Tycho concluded that the comet must have been at least six times more distant than the Moon. Comparing the observations by Hagecius at Prague with his own at Hveen, he found a difference of 1 to 2 minutes whereas it should have been 6 to 7 minutes if the comet were as near as the Moon. Although the observations of Cornelius Gemma (1535–1579), at Louvain, Belgium, were less accurate, a similar comparison with Tycho's own observations also implied a supralunar position.

Tycho examined his physical observations of the comet's tail and concluded that its axis never passed through the Sun but seemed to pass much closer to Venus. He felt the anti-Venus direction of the tail to be an illusion because it would more naturally turn away from the Sun. As did Mästlin before him, Tycho believed that the comet moved in a circular path just outside the orbit of Venus. The observed discrepancies from uniform circular motion were noted and Tycho was the first to suggest that a comet's orbit may not be circular, but *ovoid*. Alternatively, its circular motion might not be uniform. Also included in Tycho's *De mundi* was his Tychonic world system, whereby the Sun revolved around the stationary Earth and the remaining planets revolved about the moving Sun. Stimulated by his conclusions on the comet of 1577, Tycho was unable to accept the original Copernican concept and sought a compromise between the geocentric and Copernican systems. The last portion of this work was devoted to a detailed discussion of the observaions and writings of contemporary astronomers, including the Landgrave

Allegorical woodcut of an eagle with a comet in its beak. A poem by the King's poet, Jean Dorat, identifies the eagle as a minister of God with fire at the head and feet, the former to warn humans, the latter to punish them if they do not mend their ways. From Blaise de Vigenère's *Traicté des comètes, ou Estoilles chevelues apparoissantes extraordinairement au ciel.* (Paris, 1578).

of Hesse Cassel, Hagecius, Cornelius Gemma, and Helisaeus Roeslin (1544–1616).

The astronomer William IV, Landgrave of Hesse Cassel (1532–1592), also believed in the supralunar position of the comet of 1577. His observations, which appeared in Tycho's *De mundi*, included the length and width of the comet's tail and position measurements. He made redundant observations, with his sextant reversed, to account for instrumental errors and noted, but did not correct for, errors due to the refraction of light in the Earth's atmosphere. While he made no parallax determinations himself, the Landgrave believed the comet to be supralunar. Tycho's comparison of the Landgrave's observations with his own showed they were consistent with a location of the comet above the Moon.

The Prague astronomer Thaddaeus Hagecius presented his own work in a tract published in 1578. Hagecius determined a parallax of 5–6 degrees for the comet of 1577, which would have placed it well below the Moon. However, Tycho showed that Hagecius had misinterpreted his own observations, and in a work published in 1580 Hagecius corrected the error and concluded that the comet of 1577 was located above the Moon.

Cornelius Gemma, a doctor and astronomer like his father Gemma Frisius, believed some comets were below the Moon and some above. In his publication of 1578, he gave day-to-day observations of the comet of 1577, noting its position and appearance. From his parallax determination of not more than 40 minutes, he placed the comet above the moon on an orbit near that of Mercury.

Helisaeus Roeslin, a physician and astrologer, also considered the comet of 1577 to be above the Moon in his book *Theoria nova coelestium*. However, Roeslin's inference was based more on conjecture than observations.

In an unpublished work, the young English astronomer Jeremiah Horrocks (1618–1641) suggested that the comet of 1577 originated in the Sun and issued forth on a rectilinear path. The solar rotation deformed the path into a nearly circular orbit and eventually the comet returned to the Sun. It would be unfair to criticize Horrocks' ideas because he left few notes on the subject and he may not have entertained his ideas seriously.

In the few centuries prior to the comet of 1577, leading astronomers like Peter Apian, Georg von Peurbach, and Johannes Vögelin expressed the prevailing belief that comets were sublunar. The latter two used inaccurate parallax measurements to bolster their viewpoints. While Girolamo Cardano did offer a dissenting opinion, his argument for the comet of 1532 being located above the Moon was not based on parallax measurements but rather on its apparent motion in the sky, which he observed to be slower than that of the Moon.

Smashing the Crystalline Spheres

The placement of the comet of 1577 above the Moon by Tycho Brahe and his colleagues was but one step in the eventual discarding of the Aristotelian cosmology. Yet the strong grip of Aristotle's ideas on Tycho was evident in the world system that he developed for the solar system and the comet of 1577. Tycho was convinced that the Earth occupied the center of the Universe and was not, as Copernicus would have it, whirling about the Sun. In Tycho's world system, the Earth remained stationary at position A while the Sun, C, revolved about it. The orbits of the interior planets Mercury, KLNM, and Venus, PORQ, encircled the moving Sun, as did the orbit for the comet of 1577, TSXV.

In 1577, the planets were thought to revolve in solid crystalline spheres and Tycho carefully placed the comet's sphere so that it did not smash into the neighboring planetary spheres. Tycho balked at representing the sphere of Mars in this system, since it would necessarily encircle that for the comet of 1577 and intersect, or smash, the Moon's crystalline sphere about the stationary Earth. However, by 1584 Tycho accepted as possible the intersection of the planetary and lunar orbits, thus casting aside the notion of solid crystalline spheres for the planetary orbs.

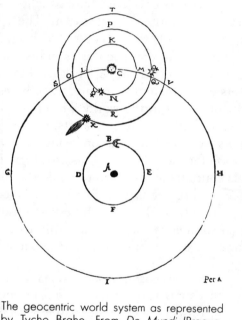

The geocentric world system as represented by Tycho Brahe. From *De Mundi* (Prague, 1603).

Parallax measurements by Michael Mästlin, Tycho Brahe, Thaddaeus Hagecius, and Cornelius Gemma clearly placed the comet of 1577 above the Moon. However, only Mästlin, Brahe, and Horrocks attempted to explain the comet's observed motion. Each considered its orbit to be closed upon itself, but none suggested the motion was periodic. Mästlin and Brahe thought the comet to be a temporary phenomenon that faded from sight as it expired. According to Horrocks, the comet returned to the Sun and disappeared.

Accurate observations of the comet of 1577 enabled Tycho Brahe and his colleagues to finally determine the distance, or parallax, of a comet. The determination of this comet's supralunar position, coming as it did on the heels of the Copernican revolution, helped turn scholars away from the entrenched cometary views of Aristotle and Ptolemy. At the end of the sixteenth century, the intelligentsia generally believed that comets were celestial phenomena. However, their paths and physical nature were still very much in question.

Early Chinese Contributions to the History of Comets

Histories of science have usually been rather chauvinistic toward European contributions to astronomy. The astronomical progress of Eastern cultures, especially the Chinese, has been ignored or dismissed perfunctorily. Fortunately, recent discoveries, translations, and analyses have begun to reveal their extraordinary scientific achievements. Prior to the Arabic astronomers of the eleventh century, the Chinese were the most accurate and prolific observers in the world. Through the tenth century, their astronomical observations of comets, planets, novae, meteors, aurorae, eclipses, and sunspots were virtually the only quantitative measurements being made. The reason for their diligence was astrology. The Chinese astronomer/astrologer was continually called on to guide the ruling sovereign with astrological advice. Unlike the Greek, Roman and medieval European astronomers, who were largely academics, Chinese astronomers were intimately connected to their sovereigns and resided within the walls of the imperial palace. The Viennese historian Franz Kühnert said it well in 1888 (see Needham, 1959):

> Probably another reason why many Europeans consider the Chinese such barbarians is on account of the support they give to their astronomers—people regarded by our own cultivated western mortals as completely useless. Yet there they rank with Heads of Departments and Secretaries of State. What frightful barbarism!

There are fundamental differences between early Chinese astronomy and that passed down from the early Greeks. The Hellenistic tradition was more theoretical than observational, emphasizing geometrical formulations for the motions of the celestial bodies. Early Chinese astronomy was empirical, and while this observational approach prevented them from making significant advances in theoretical astronomy, it did allow them to escape the Greek notions of perfect circular motion, immutability of the heavens, and concentric spheres for the planetary regions.

While Hellenistic astronomy was based on a coordinate system defined by the Sun's apparent motion on the sky, an *ecliptic* system, and Arabic astronomy was based on the altitude and azimuth, or *horizon* system, Chinese astronomers employed a coordinate system based on the apparent rotation of the stars about an axis, one end of which was the north celestial pole. It is this equatorial coordinate system that is used by modern astronomers, and the two celestial coordinates (analogous to Earth's longitude and latitude) are right ascension and declination. The right ascension of a celestial object is its Earth-centered angle measured eastward from the vernal equinox, and the declination of the object is its Earth-centered angle measured north or south from the celestial equator, which is simply the extension of the Earth's equator onto the celestial sphere. In this system, the *north polar* distance of an object is defined as an angle equal to 90 degrees minus the value of the object's declination.

The Chinese celestial sphere was divided into *asterisms*, each one consisting of a few neighboring stars, much like smaller versions of our familiar constellations. By the second half of the third century A.D., there were 283 asterisms composed of 1464 stars. Twenty-eight lunar mansions were spread around the celestial sphere in unequal strips of right ascension and the boundary, or wall, of each was established by a reference, or determinative star. The celestial position of an object was denoted as being a certain distance within a particular lunar mansion and occasionally a north polar distance, in Chinese degrees, would also be given. There were 365.25 Chinese degrees in a circle so that one ancient Chinese degree equals 0.9856 western degrees. This ancient Chinese system of expressing celestial coordinates is entirely equivalent to the modern equatorial coordinate system, except the Chinese had no single origin for their right ascension, *lunar mansion*, system. Although they had a perfectly good angular measure in the Chinese degree, it was rarely used to express distances between celestial objects, the size of a comet's tail, a meteor trail, or the zodiacal light. A linear unit such as a *chhih*, a foot, was used. From a seventh century account of the positions of the ecliptic with respect to the 28 lunar mansions, Tao Kiang of Dunsink Observatory determined that 1 *chhih* equals 1.5 western degrees approximately.

In 1973, a book of silk pages was discovered in an ancient Han tomb at Mawangdui, China, dating from about 168 B.C. The silk book has approximately 250 drawings representing such things as clouds, halos, rainbows, lunar occultation, and star groups. Some 29 of the drawings represent types, or forms, of comets; they are the earliest surviving illustrations of comets. Two of the forms are unrecognizable. An example of the original illustrated silk is accompanied here by 27 recognizable types redrawn for clarity. The drawings do not represent specific comets but rather general cometary types, and each is briefly described with the associated portents. Some of the

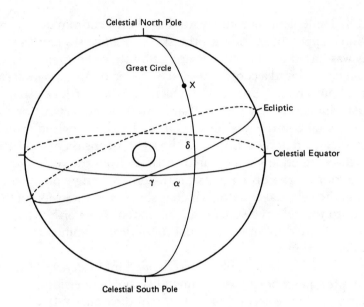

Schematic illustration showing equatorial coordinate system. The celestial equator is formed by extending the Earth's equator onto the imaginary celestial sphere. The celestial ecliptic is the apparent path of the Sun on the celestial sphere. The point where the Sun appears to cross the celestial equator going north is called the *vernal equinox*, γ. It occurs every year about March 21 and marks the beginning of spring in the northern hemisphere. For a celestial object located at point X, its right ascension, α, is measured eastward along the celestial equator from the vernal equinox to the great circle passing through the object and the north and south celestial poles. The object's declination, δ, is then measured along the object's great circle, north or south from the equator to the object.

names are used for more than one comet illustration. The comet type and associated portents follow:

1. Chi-Guan: War comes, general dies

2. Bai-Guan: Five-day rebellion in the state

3. Tian-shuo: The small man cries (*shuo* is a ceremonial dance pole)

4. Chan: State perishes

5. Hui-Xing: Army that gains the direction will win

6. Bai-Guan: It appears for 5 days and goes, death in the state

7. Chi-Guan: Death among generals

8. Pu-Hui: Disease in the world (*pu* means water weeds)

9. Pu-Hui: Calamity in the state, many deaths

10. Gan-Hui: War for years (*gan* means the stem of a cereal)

11. Gan-Hui: Same portents as for number 10

12. Zhou-Hui: Bumper harvest but internal war

13. Li-Hui: Small war, corn plentiful

14. Zhu-Hui: There will be death of kings

15. Zhu-Hui: Same as number 14

16. Hao-Hui: Armies arise, war and famine

17. Hao-Hui: Revolt in the army, otherwise same as number 16

18. Shan-Hui: Arms are raised in the world, army abroad will return

19. Shan-Hui: Same as number 18

20. Shan-Fa-Hui: War, famine; same as numbers 18 and 19

21. Shen-Xing: It leads to war, many calamities, fear of defeat and worry about result of conflict (*shen* is the fruit of the mulberry)

22. Qiang-Xing: Three small battles, seven large battles (a *qiang* is a screen placed next to coffins)

23. Nei-Xing: War, large battle

24. Gan-Hui: Means war

25. Shan-Hui-Xing: Raising of arms, famine during the year

26. Chi-You-Qi: Army abroad will return

27. Di-Xing: Appearing in spring means good harvest, in summer means drought, in autumn means flood, in winter means small battles (*di* is a long-tailed pheasant)

From internal evidence, it appears that the contents of the silk book were composed in the fourth century B.C. There is little evidence to suggest intercourse between Greek and Chinese cultures at that time, yet both civilizations considered comets to be malefic signs with similar disastrous consequences.

Concerning cometary astronomy, the ancient Chinese astronomers have two "first observations" to their credit. They were the first to note the antisolar nature of comet tails and they may be the first to have noted the unusual *antitail* phenomenon, one that appears to be directed *toward* the Sun.

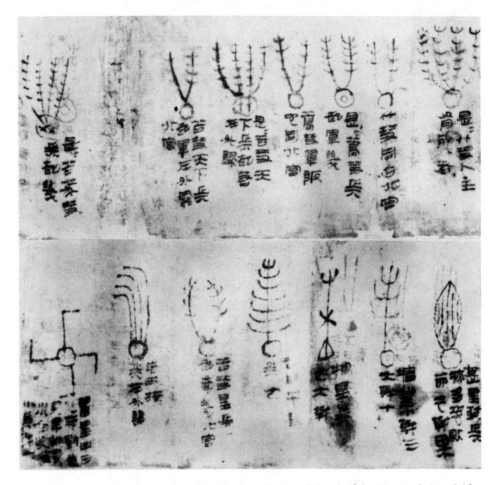

Some examples of the cometary types displayed in the silk book of the Han tomb (ca. 168 B.C.). The original images have degraded with time. (*Photograph courtesy of F. Richard Stephenson.*)

In about A.D. 635, the Chinese history of the Chin Dynasty (A.D. 265–419) was completed and Li Chung-feng, who was responsible for the astronomical portion of this history, recorded the following paragraphs (see Needham, Beer, and Ho, 1957):

Among ominous stars the first are the hui-xing, commonly known as broom stars. The body is a sort of star while the tail resembles a broom. Small comets measure several inches in length, but the larger ones may extend across the entire heavens. The appearance of a comet predicts military activities and great

46

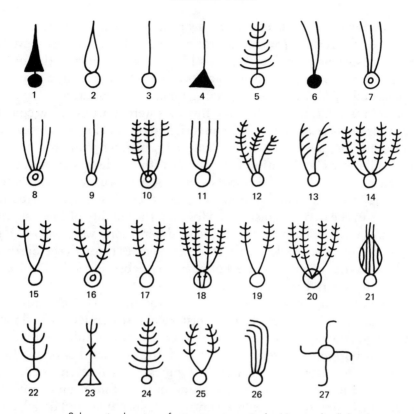

Schematic drawing of cometary types in the Han tomb silk book.

floods. Brooms govern the sweeping away of old things and the assimilation of the new. A comet can appear in any one of the five colors, depending on the essence of that one of the five elements which has given birth to it.

According to the official astronomers, the body of the comet itself is non-luminous but derives its light from the Sun, so that when it appears in the evening, it points toward the east while in the morning, it points toward the west. If it is south or north of the Sun, its tail always points following the same direction as the light of the Sun—then suddenly it fades. The length of the rays is a measure of the calamity foretold by the comet.

These paragraphs suggest that by approximately A.D. 635, when this history was written, Chinese ideas on the nature of comets had a Hellenistic influence. The notion of five elements giving birth to comets came by some circuitous route, but it originated with Aristotle. The antisolar nature of cometary tails, mentioned in the second paragraph, was the first statement of this phenomenon. This effect was again mentioned in the Chinese descriptions of the A.D. 837 apparition of comet Halley, so it must have been well known by then. Thus Peter Apian's 1532 diagram, showing the antisolar nature of a comet's tail, as well as Fracastoro's 1538 statement to that effect, were preceded by Chinese records some nine centuries earlier.

As additional Chinese dynastic histories were written, a few cometary types, or forms, were added to those mentioned in the Han Tomb silk book. One of these forms was termed a *Chang-keng*, or long path, and may have been the Chinese classification for a comet with two horns or tails, one on either side of the nucleus. The comets seen in A.D. February 467 and November 886 may have exhibited antitails because they were both referred to as *Chang-keng* in the dynastic histories. However, the validity of these sightings remains in question because the Chinese descriptions were vague. A true cometary antitail is seen when an active comet passes through the Earth's orbital plane so that an edge-on view of cometary debris gives the illusion of a sunward spike, or tail (see Chapter 9).

In-depth study of ancient Chinese astronomy has been undertaken only relatively recently. There is much historical research left to be done before the significance of ancient Chinese astronomy is fully realized. The major contribution to date is the work of the science historian, Joseph Needham. To mention only a few of the Chinese contributions is enough to indicate that these astronomers were ahead of their European counterparts up to and including the fourteenth century. Quantitative and detailed Chinese star maps are extant from A.D. 940 through the fourteenth century, a period when Europe had nothing comparable. During the thirteenth century, after a continuous development of complex astronomical instruments, the Chinese invented the equatorial mounting. Even earlier, in 1090, an armillary sphere designed by Su Sung was equipped with a sighting tube and the apparatus was driven with a water-powered clock drive. However, the most important ancient Chinese contributions are dynastic records containing nearly continuous observations of celestial phenomena such as eclipses, novae, sunspots, and comets. The ancient Chinese observations of comets are outlined in the Appendix.

Summary

With the decline of Greek and Roman cultures, scientific empiricism gave way to Aristotelian dogma. The medieval period through 1200 saw the cometary ideas of Aristotle and Ptolemy first translated into Latin, then widely circulated and finally embraced by the European Christian Church.

In the thirteenth century, Albert Magnus (Albert the Great) and Thomas Aquinas brought the science of Aristotle into the theological fold. By so doing, they not only supported these incorrect viewpoints, they granted them sanctuary.

The medieval Church found the Aristotelian views to be safe because there were no serious contradictions with biblical texts. This compatibility is not surprising since many of these texts were written in lands conquered by Alexander the Great and at a time when Aristotle's views were well known and widely believed. With the encouragement of the Church, comets were considered atmospheric phenomena, or signs, sent by an angry God to warn a sinful Earth to repent. Medieval views considered comets analogous to dreadfully close thunder clouds, all the more frightening because they were direct messages from an angry, almighty God. When comets were discussed, it was invariably to question their meaning rather than their nature.

There is a tendency to unfairly condemn the medieval period for the lack of scientific advancement and to dismiss the views during this time as ignorant and superstitious. Science historian Jane Jervis makes the point that medieval scholars asked different questions than their counterparts today; what they saw was affected by what they believed. We still don't know what comets mean, we simply don't ask that question. Modern scientists have learned to ask only questions that can be answered by the scientific method. The early medieval period knew little of empiricism and the scientific method.

The period from 1200 to 1577 saw less than complete devotion to comets as portents, and there were isolated examples of scientific observations in Europe. Beginning with the fifteenth century, the scientific study of comets was reborn and useful observations were made by Paolo Toscanelli, Georg von Peurbach, Regiomontanus, and others. Astronomers began to test the terrestrial nature of comets by attempting to determine their distance from the Earth using parallax measurements.

Although Peurbach had tried a cometary parallax determination as early as 1456, it was not until 121 years later that techniques and instruments were accurate enough to decide whether comets were terrestrial or celestial phenomena. For the comet of 1577, Michael Mästlin, Tycho Brahe, and Cornelius Gemma all made parallax measures and each correctly inferred the comet's position as being beyond the Moon.

The period between the fall of the Roman empire and prior to the comet of 1577 witnessed very little progress in the way of quantitative observations of comets in the West. Fortunately, Chinese astronomers were recording accurate cometary observations for this entire interval.

NOTES

1. In this regard, the French astronomer and historian Alexandre Pingré wrote that the comet preceding the death of Charlemagne, in January 814, may well have been invented by contemporary authors. In the ninth century, great men simply did not die without a great comet in the sky.

2. When comets are first discovered, or recovered, they are given a provisional letter designation indicating the order of discovery within a particular year. Later, when their orbits have been computed, they are given a Roman numeral designation to indicate the order of their *perihelion passage* (the order in which they pass closest to the Sun). For example, comet Halley was the eighth one in 1982 to be discovered or recovered; its provisional designation was 1982h. However, its permanent designation is now 1986 III, since it was the third comet in 1986 that was known to have passed perihelion. To indicate an orbital period less than approximately 200 years, periodic comets are often designated with a P/ prior to their names (i.e., P/Halley).

3. One astronomical unit, AU, is defined as the mean distance between the Earth and Sun. It is approximately 150 million kilometers or 93 million miles.

3

Johannes Kepler, Galileo, and the Comets of 1607 and 1618

Johannes Kepler believes comets are ephemeral. Although his laws explain the motions of the planets around the Sun, Kepler considers comets to move on straight line paths. Galileo and the Roman Jesuits argue over the comets of 1618. Galileo criticizes the Jesuit views and those of Tycho Brahe but does little to advance the understanding of cometary phenomena. The first telescopic observations of comets are made in 1618.

THE CHINESE WERE THE first to observe the comet of 1607 on September 21 as a morning object in the northeast. Toward the end of September, the comet became a northwest evening object attaining a maximum brightness of approximately first or second magnitude.[1] The comet was last seen by Johannes Kepler on October 26, one day before it reached perihelion as it moved into the glare of the Sun. Obvious to the naked eye, its appearance prompted several treatises, the most important of which was Kepler's.

Three comets appeared in 1618, the last of which was the most impressive.[2] Often called *the* comet of 1618, it was first seen in mid-November exiting from the solar glare as a tail projected above the horizon. The tail reached impressive proportions in mid-December and the comet was last seen on January 22, 1619. The appearance of the comet of 1618 also prompted several contemporary treatises and touched off a heated controversy between the Roman Jesuits and Galileo Galilei (1564–1642).

The Ephemeral Comets of Johannes Kepler

In a letter dated July 4, 1603 to his friend David Fabricus, Johannes Kepler stated that when he was a boy of six, his mother led him to a high place to view the comet of 1577. Kepler's future teacher, Michael Mästlin, observed this same comet and determined it to be supralunar. Mästlin used circular orbits within a Copernican system to represent the motion of comets, and it is reasonable to assume that Kepler initially shared his teacher's views. If so, he later abandoned this nearly correct concept for his own erroneous theory. By 1602 Kepler believed that comets were ephemeral, rocket-like bodies having rectilinear motions above or below the Moon. From 1602 on he never deviated from his assumption of rectilinear paths for comets.

Kepler's first work about comets was published in 1604 and entitled *Astronomiae pars optica*. In a short appendix to this work, Kepler noted that comets, like rockets, travel most slowly in the beginning and end of their trajectories. Kepler's logic was Aristotelian but straightforward. Unlike permanent celestial objects, comets were observed to be ephemeral—that is, they moved on straight line trajectories. To admit circular motion, Kepler would have had to concede their divine or permanent nature and comets were simply not observed to be periodic. Within the main text of his book, Kepler outlined an experiment that could be used to create a phenomenon akin to a comet's tail. If a narrow beam of light were directed toward a solid, or water-filled, globe in a darkened room, a taillike image would appear on the opposite wall. He later rejected this conjecture because it was necessary to suppose that some mechanism existed to deflect the light passing through the head toward an observer's eye. In his experiment, the wall played this role.

Kepler first observed the comet of 1607 in the northwest evening sky on September 26 as he stood on a bridge over the Moldau River in Prague. He had just finished watching a fireworks display. He observed the comet for the month that it was visible, and the following year he published a short treatise written in German that was intended for a popular audience. In this work, Kepler set down his ideas on the origin of comets and their tails. According to Kepler, comets were spontaneously created from impurities, or fatty globules, in the ether.[3] This process was likened to an ocean's ability to spontaneously generate whales and sea monsters. Space was as full of comets as fish in the sea but they only became visible when near the Earth. When a comet was created, a special spirit or intelligence formed to guide it. The comet and the attendant spirit were created together, and they dissipated together.

Engraved portrait of Johannes Kepler. Frontispiece from *Johannes Kepler* by Edmund Reitlinger, Stuttgart, 1868. Engraving based on original oil painting by an unknown artist.

In this same work, Kepler stated that his previously published experiment for creating a cometlike tail phenomenon was premature and offered only as a conjecture. He went on to present his revised ideas on cometary tail formation. The Sun's rays passed through the comet, penetrated its substance, and drew a portion of the head in the antisolar direction. Much like clouds dissipated by the Sun, cometary particles drawn away from the head dissipated in the tail region. The comet lasted only as long as it took the Sun to do this. Although not exactly what Kepler had in mind, this suggestion has been referred to as the first mention that solar radiation has pressure. Kepler believed that comets became visible as sunlight was reflected by the head and tail. Tail curvature was due to the refraction of the sunlight in the somewhat transparent head and tail material. His treatise on the comet of 1607 also presented his ideas on its astrological significance. It contained a few general astrological predictions and the statement that contact with the tail, although extremely unlikely, would render the Earth's atmosphere im-

The Three Laws and World View of Johannes Kepler

Although Kepler's work on comets added little to the understanding of them, his three laws of planetary motion would prove important in Isaac Newton's subsequent solution to their paths. By simply listing Kepler's three laws, we implicitly couch them in our own modern cosmology—wherein Earth is a smallish planet orbiting one modest Sun amongst billions of others in a dark, infinite void. With the aid of the diagram, Kepler's laws are introduced below in the cosmology he embraced.

While Kepler was one of the very first scientists to support the Sun-centered cosmology of Copernicus, it would be a mistake to assume he disavowed a geocentric viewpoint. Like Copernicus, he thought of the Sun as the center of a finite universe with the stars being lights (not suns) on a very large spherical shell. Kepler assumed that the ratio of the starry sphere's distance to that of the outermost planet Saturn was the same as the ratio of Saturn's distance to the radius of the Sun. He had no evidence for this assumption, but the analogy pleased him. In Kepler's mystical cosmology, special celestial guiding spirits were spontaneously generated and the planets created music as they orbited the Sun. These divine sounds were not physically audible but were heard by the spiritual ear just as the spiritual eye notes the harmony of the sizes of the planetary spheres. Kepler's adoption of the Copernican system in no way invalidated his belief that the Creator's handiwork was done for mankind alone. Far from being humiliated by the Sun's central location, the inhabitants of Earth retain a favored position among the planets, enabling them to view a lovely variety of apparent motions provided by the Earth's five sisters, two of which were closer to the Sun and three farther away. The Creator moved the Earth from the center of the physical system to provide for more interesting journeys around the immobile Sun.

pure and cause widespread mortality. Some took this notion quite seriously when this same comet (Halley) returned three centuries later in 1910.

Although Kepler's German book on the comet of 1607 was first scheduled for publication at Leipzig, he was forced to publish at Halle instead. Lutheran theologians at Leipzig objected to Kepler's creation of special cometary guiding spirits. As was the custom, Kepler intended to publish a

In Kepler's world view, solar magnetic emanations arising from a rotating Sun drive the planets about the Sun, S, in elliptical orbits with the Sun at one focus. This was Kepler's first law. The Sun's motive force on a planet decreases linearly with the planet's distance from the Sun, not as the square of the distance, so that distant planets move most slowly. A particular planet moves slowest in its own orbit when farthest from the Sun, at H, and fastest when closest, at A. Kepler's second law states that this motion can be described by imagining a line drawn from the Sun to the planet sweeping equal areas of the ellipse in equal amounts of time—area ASB = area BSC. Kepler's third law states that the squares of the planetary orbital periods are proportional to the cubes of their orbital semimajor axes (i.e., the semimajor axis in the figure would be one-half the distance AH). Newton would later show that the validity of Kepler's third law depends on a force between the Sun and a planet that varies as the inverse square of the distance between them.

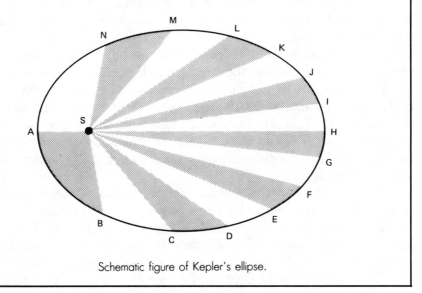

Schematic figure of Kepler's ellipse.

more comprehensive and scientific exposition in Latin shortly after the 1608 German edition. However, the Latin edition was never published separately. There were still theological objections to Kepler's cometary spirits, and noblemen to whom various parts of the book were dedicated kept dying and substitutes had to be found. The printer caused further delays, then maintained that people had lost interest in comets. The problem was finally solved in 1618. Three comets appeared, and Kepler—taking advantage of renewed public interest—published a new essay on comets and worked into it his Latin treatise.

Kepler's 1619 Latin essay, *De cometis libelli tres,* was his cometary magnum opus. Similar to his German tract of 1608, this work was divided into three sections: astronomical, physical, and astrological. In the first section, Kepler adhered to his view that a comet moved along a straight path while the Earth traveled in a circular orbit around the Sun. He outlined 30 propositions on cometary parallax measurements that could be used to determine the difference between a comet's true motion, curved or straight, and its apparent, observed motion. His propositions were applied to the comet of 1607 and the three comets of 1618. After setting down and rejecting rectilinear or circular cometary motion in conjunction with a fixed Earth, he attempted to represent the available cometary observations by assuming the Earth circles the Sun while the comets move along straight paths.

Kepler's inability to satisfy the observations by assuming a fixed Earth strengthened his Copernican views. Since his observations were crude, and the observation interval short, he managed to fit them into a straight line orbit. However, discrepancies between observation and theory reached nearly one degree in celestial longitude and three degrees in latitude. While realizing that his theory did not fit the observations, he believed the agreement could have been improved with more complicated calculations. Referring to the comet of 1607, Kepler noted that additional computations were not warranted because comets do not reappear. As if to underscore the blunder, the comet of 1607 was later identified as comet Halley, the first comet shown to reappear periodically.

Kepler's ideas on the physical nature of comets were essentially the same as those expressed in his German tract 11 years earlier. He surmised that the second and third comets of 1618 were originally part of the same object, but he didn't base this conclusion on observational evidence. According to Kepler, the influence of comets on the terrestrial world was threefold. The first influence was the defiling of the air when a comet's tail touched the Earth. The second was their ominous significance. Employing a classic astrological subterfuge, Kepler backed away from specific predictions based on the comets of 1618, noting that signs or meanings were apparent to those for whom they were intended. The third cometary influence was due to a disruption in the sympathy of nature. According to Kepler, there exists a sympathy of the heavens with that living force that resides within the Earth and regulates its inner works.

When something unusual arises in the heavens, whether from strong constellations or from new hairy stars, then the whole of nature, and all living forces of all natural things feel it and are horror stricken. This sympathy with the heaven particularly belongs to that living force which resides in the Earth and regulates its inner works. If it is alarmed at one place it will, in accordance with its quality, drive up and perspire forth many damp vapors. From there arise long lasting rains and floods, and therewith (because we live by air) universal epidemics, headaches, dizziness, catarrh (as in the year 1582), and even pestilence (as in year 1596).

While Kepler insured his place in the history of science by establishing the elliptical nature of planetary orbits in a heliocentric system, his ideas on comets consisted of a basic Ptolemaic framework interwoven with his own brand of mysticism. If Kepler was a mathematical mystic on the subject of comets, then his equally famous contemporary, Galileo, was a mathematical skeptic.

The Devil's Advocate: Galileo Galilei

As had been the case for bright comets in the previous century, the three comets of 1618 witnessed the publication of numerous tracts, most of which were astrological in nature. One of the most reasonable of these treatises was written anonymously in 1619 by the Jesuit Horatio Grassi, professor of mathematics at the Collegio Romano. In this work, entitled *De tribus cometis annus MDCXVIII*, Grassi dismissed the common fear of comets as groundless and applied Tycho's ideas to the comets seen in 1618. He argued for their celestial nature by citing parallax measurements made in Rome and Antwerp that "scarcely ever exceeded one degree." According to Grassi, comets moved on great circles with constant motion, and their brightness was due to reflected or refracted solar light. As one possible mechanism, he noted Kepler's 1604 experiment with a taillike phenomenon being formed on a wall by sunlight reflected off a crystalline sphere. The comets of 1618 should be placed between the Sun and Moon because their speed was midway between the two.

The newly invented telescope, not yet widely understood, was also mentioned in an erroneous argument for the comet's supralunar position. Grassi believed that a telescope would magnify objects in direct proportion to their distance—the closer the object, the more the telescope would magnify it. Hence, comets should be placed beyond the Moon because the telescope did not perceptibly magnify them as it did the Moon.

Galileo, confined to his bed with severe arthritis and a double hernia, may not have even seen the comets of 1618. Yet, through his friend and student, Mario Guiducci, he replied to Grassi's work in the form of two published lectures. Containing the thinly disguised views of Galileo, Guiducci's work was entitled *Discorso delle comete*. Galileo did not present a coherent

Portrait of Galileo Galilei from his work on sunspots entitled
Istoria e dimostrazioni intorno alle macchie solari. . . . *(Rome,
1613).*

cometary theory of his own. Instead, he jabbed away at the minor flaws in
Grassi's arguments while occasionally adding a rapier thrust at his major
blunders. Galileo relished his role as devil's advocate. In scientific polemics,
he was unrivaled. In reading Galileo's reply, one gets the impression he rather
enjoyed verbally jousting with Grassi, who was a poor match for the master.

Beginning his role of devil's advocate, Galileo pointed out that comets
were not periodic. The only other comet in recent history that could match
the splendor of the 1618 comet was in 1577. Assuming that these two comets were the same object traveling in a circular orbit, it would not have traveled even one degree in the observed interval. In fact, the comet of 1618 was
observed to move over 90 degrees in its orbit. If the comets of 1577 and 1618
were not the same object, the latter's 90-degree motion in a few months
would require that it return in less than a year if its orbit were circular, as
Grassi suggested.

Continuing to attack Grassi's ideas, Galileo stated that one must first prove that comets are real objects, rather than reflections, before using the slight parallax observations to infer a supralunar position. For example, rainbows exhibit no parallax because they move as the observer moves. Galileo correctly pointed out that an object's enlargement in a telescope is independent of its distance. Using one contradiction after another, he attacked Grassi's arguments that comets were more distant than the Moon because the telescope showed them to be enlarged less than the Moon. For example, if an annular eclipse is visible with the naked eye and the Sun is farther away than the Moon, the eclipse should appear total when viewed through a telescope!

Galileo also suggested explanations for cometary phenomena. Comets might move vertically upward from the Earth's surface at a uniform rate and in a rectilinear fashion. An observer at O observes the comet slowing down

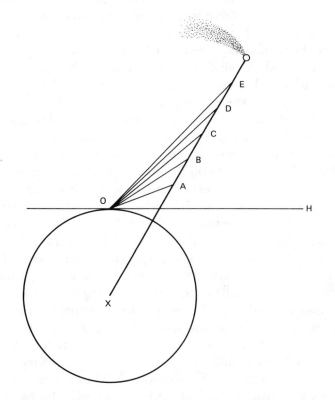

Galileo's scheme for representing observations of the comet of 1618 without assuming its extraterrestrial nature. The Earth's center is at X, OH is the observer's horizon, and AE is the comet's path.

and decreasing in size as it moves from A to E. H is the eastern horizon, which points to the rising Sun. Galileo felt this explained why comets first appear very large, then grow little or not at all. A comet could shine by sun-light reflecting off a cloud of vapors, and the tail curvature could be due to the refractive effect of the Earth's atmosphere. Galileo also criticized Tycho Brahe directly by pointing out that his explanation for tail curvature was er-roneous. Tycho suggested that it was an effect of perspective because one end of the tail was farther from the eye than the other. Galileo pointed out that a straight tail could appear foreshortened but never curved as a consequence of perspective. He also criticized Tycho's suggestion that some comet tails appeared to be directed away from Venus. How could Venus, with its feeble il-lumination, provide the light source for bright comets? he asked. This, of course, was unfair since Tycho had stated that the Sun was the most likely source of a comet's light. Although Galileo's criticism of Tycho would even-tually bring Kepler into the fray, the Roman Jesuits were heard from next.

Offended by Galileo's rebuttal, Grassi and his fellow Jesuits published a direct response to Galileo with a treatise entitled *Libra astronomica*. The Jes-uits used the pseudonym Lothario Sarsi for the author. While Galileo did not explicitly state that comets were terrestrial exhalations, Grassi assumed that was what Galileo had in mind. He then outlined the ancient arguments against terrestrial exhalations; for example, they would be impossible to maintain in the midst of the raging north winds. Grassi also argued that if comets were phenomena akin to rainbows, they should—but don't—reflect the Sun's motion. Responding to Galileo's objection to his suggested circu-lar orbit for the comet, Grassi stated that comets could move in ellipses about the Sun. While touching on the truth, Grassi's suggestion was made within a framework that was Tychonic rather than heliocentric. According to Grassi, the Copernican system was "in no way permitted to us Catholics."

The controversy over the comets of 1618 continued with Guiducci him-self publishing a long letter to the Reverend Father Tarquinio Galluzzi. Galileo, using his own name this time, responded in 1623 with his *Il saggia-tore*, "the assayer." It was written in response to Grassi's *Libra astronomica*, or "astronomical balance." Galileo began by defending himself against Simon Marius, who had claimed prior discovery of four satellites of Jupiter. In fact, Galileo had first seen them; ironically, the names finally adopted for the four so-called Galilean satellites of Jupiter—Io, Europa, Ganymede, Callisto— were suggested by Marius. Galileo returned to the controversy at hand and admonished Sarsi for following Tycho's foolish fabrications in every detail. He referred to a recently published work attacking Tycho by the "distin-guished" Scipio Chiaramonti, which he considered a clear rebuttal of Tycho's ideas. Retreating from his previous attack, Galileo pointed out that he never affirmed the location of the comet nor denied that it might be above the Moon; he said only that other authors' theories were not immune from objections. A comet moving rectilinearly away from the Earth was considered because it would explain cometary phenomena more simply and in better agreement with the observations. In his *Il saggiatore* Galileo dismissed Tycho's parallax measurements by stating:

> Tycho himself, among so many disparities, chose those observations which best served his predetermined decision to assign the comet a place between the Sun and Venus, as if these were the most reliable.

Galileo was not one to mince words. In response to Sarsi's assertion that he borrowed the explanation of tail curvature from Kepler, Galileo stated that his ideas required refracted light in the Earth's atmosphere, whereas Kepler's required refraction in the tail itself. He seemed very careful not to step on Kepler's toes after trampling Tycho's.

Although Galileo wrote nothing derogatory about Kepler, he was very critical of Kepler's teacher, Tycho. When Kepler received a copy of Galileo's *Il saggiatore,* he had just finished a defense of Tycho against the peripatetic Italian astronomer Scipio Chiaramonti. The defense against Chiaramonti's *Anti Tycho* was entitled *Tychonis Brahei Dani hyperaspistes* (Shieldbearer of Tycho Brahe, the Dane). It was published at Frankfurt in 1625. Defending Tycho without responding to Galileo's charges would have been unsuitable, so Kepler added an appendix to his work. While avoiding the controversy between Grassi and Galileo, Kepler's appendix answered, point by point, issues related to Tycho or himself. He also straightened out some of Grassi's misconceptions, pointing out that the experiment outlined in his 1604 work *Astronomiae pars optica* was not meant to apply to real comets. It was pure conjecture, an academic exercise to demonstrate how one could simulate a comet's tail by allowing a beam of light to pass through a glass globe. Kepler then expressed his revised views on cometary tail formation:

> The head is like a conglobulate nebula and somewhat transparent; the train or beard is an effluvium from the head, expelled through the rays of the Sun into the opposed zone and in its continued effusion the head is finally exhausted and consumed so that the tail represents the death of the head.

While Galileo suggested that the curvature of cometary tails was due to refraction of light in the Earth's atmosphere, Kepler countered that if this were the case, the curvature would be slight and always upward toward the zenith. In addition, the curvature would exist only when the comet was near the horizon. Kepler stated that his idea of nonuniform motion was not a result of direct cometary observations. Rather, it arose in analogy to observations of rockets and meteors. He went on to explain that after considering the observations of the comets of 1618, he could find no strong reason for establishing that rectilinear motion was slower at each end of the trajectories, so he left the nonuniformity of a comet's motion in doubt.

Returning to his defense of Tycho, Kepler emphasized the extreme care with which Tycho made his observations and implied that Galileo resented Tycho's well-earned authority. Grassi and the Jesuits were also subjected to Kepler's sharp rebuke. To Sarsi's rejection of the Copernican system because "these things are in no way permitted to us Catholics," Kepler responded, "perverse and querulous at best, servile at worst."

Again using the pseudonym Lothario Sarsi, Grassi answered Galileo's *Il saggiatore* with a weak, heavy-handed response in 1626. Typical of Grassi's clumsy wit was the attempt to suggest that Galileo had been imbibing too freely of wine, pretending that *saggiatore* really meant "winetaster," *assaggiatore*. In the controversy over the comets of 1618, Grassi and his fellow Jesuits were polemical lightweights compared to Galileo. One gets the impression that Galileo could have argued effectively for a flat Earth against Grassi. One also wonders what the outcome might have been had Galileo's battle been, not with Grassi, but Kepler.

Early Seventeenth Century Views on Comets

Although eclipsed by the authoritative views of Galileo and Kepler, contemporary cometary ideas were also expressed by the notable scientists Willebrord Snel (1580–1626), Pierre Gassendi (1592–1655), Seth Ward (1617–1689), and René du Perron Descartes (1596–1650).

Willebrord Snel began his astronomical career under Tycho Brahe's tutelage at Prague. Although best known for his law of refraction, which he formulated about 1621, Snel also published descriptions for the comets of 1585 and 1618 while he was a professor of mathematics at Leiden. He eliminated terrestrial vapors as a possible origin of comets because his own parallax measurements placed the comet of 1618 above the Moon. He believed the Sun constantly threw off exhalations from which sunspots and comets formed. In contrast to cometary material, sunspots were made of a less pure, and hence more opaque, substance. As ejected solar material, comets glowed by their own light. However, Snel did not break with contemporary superstition and professed a belief in comets as portents.

The French scientist Pierre Gassendi left a short essay on comets embedded in his posthumously published, six-volume work *Syntagma philosophicum*. Like Kepler, Gassendi believed that comets moved uniformly along rectilinear paths. Although he was uncertain about their nature, Gassendi was a central figure in the fight against cometary superstition. His common

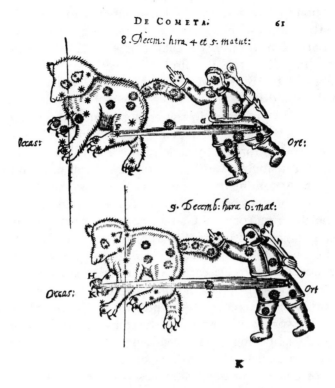

Comet of 1618 in constellation of Boötes, the herdsman, on December 8, 1618. The long tail is directed toward Ursa Major (the great bear). From Johann Baptist Cysat's *Mathemata astronomica de loco, motu, magnitudine et causis cometae* ... (Ingolstadt, 1619).

sense arguments formed the beginning of a French movement toward a rational view of comets that would be carried forward by Pierre Petit in 1665 (see Chapter 4).

In his *De cometis*, published in 1653, the Englishman Seth Ward expressed the opinion that comets were eternal and returned periodically on closed orbits, which were circles or ellipses that could either include or exclude the Earth. So great were the extent of these orbits, they could only be seen when nearest the Earth, at perigee. Ward's views were not based on observational evidence but rather on his belief in their eternal nature. The opposite argument was used by Kepler to infer their rectilinear motion.

The possible elliptic nature of cometary orbits was mentioned first by Tycho in connection with the comet of 1577, and was brought up occasionally in the early seventeenth century. In a letter dated February 6, 1610, addressed to Thomas Harriot, the Welsh amateur astronomer Sir William Lower wrote that ellipses "shews a way to the solving of the unknown walkes of comets." Baron Franz Xaver von Zach (1754–1832) discovered a portion of this letter in 1784; the remainder was subsequently found among Harriot's papers in the British Museum. The complete letter was published by Stephen P. Rigaud in 1833.

Of the theories presented in the early seventeenth century, the most original was that of the French scientist-philosopher René Descartes. Before the acceptance of Newtonian cosmology, the Cartesian vortices ruled the day, and—not surprisingly—Descartes' cometary ideas were an integral part of his cosmology. According to Descartes, each star was surrounded by a vortex, supported by the light pressure of the central star and the pressure from particles of neighboring vortices. Caught up in the surrounding ether, planets revolved around the central star like a chip of wood in a whirlpool. Planets and comets were both formed from dead stars, the only distinction being that comets were more dense. In Descartes' imaginative system, both planets and comets were formed when sunspots linked together and completely covered the central star. Once covered, the light pressure supporting the surrounding vortex ceased, the vortex collapsed, and the star—now dead—was caught up in the strongest neighboring vortex. Dead stars of lesser density acquired momentum equal to the surrounding vortex particles and became planets in stable periodic orbits. If the dead star was sufficiently dense, the vortex particles, through continued *agitation*, gave it an escape velocity, allowing it

Comet of 1618 moving northward through the constellations of Libra and Boötes. From John Bainbridge's *An Astronomicall description of the late comet.* . . . (London, 1619).

to move tangentially out of the vortex and become a comet. The comet then wandered from one stellar vortex to another, temporarily taking the velocity and direction of the vortex through which it passed. Its orbit was slightly curved in each vortex. For our own solar system, comets were found at the outer edge—at the distance of Saturn.

While Descartes considered comets more dense than planets, presumably he was comparing the density of the planets with that of the cometary head, or nucleus. The tail and coma phenomena were explained using an optical refraction hypothesis. The nucleus of a comet was seen when solar rays were reflected directly to an observer, while an illusion of a tail was formed when solar rays were refracted in the ether of the vortex before reaching the observer's eye. The number of refracted solar rays was small compared with the number reflected directly to the observer's eye, so the nuclear region appeared brighter than the coma or tail. Descartes' theory, contrived as it was, provided one of the few contemporary explanations of how sunlight could optically pass from the comet's head, form a tail, and then be directed toward an observer's eye. However, among other shortcomings, his theory did not require the tail to point in an antisolar direction. According to Descartes, planets should also exhibit a tail, which was in fact visible in countries where the air was clean and pure. In the pre-Newtonian era, acceptance of Descartes' cosmology was widespread. However, his cometary views were not as well received.

The comets of 1618 prompted influential cometary theories by both Kepler and Galileo and were the first to be observed telescopically. Although Kepler observed the first comet of 1618 on September 6 with the aid of a telescope, it was the last comet seen that year that was extensively studied with the recently invented device. While Galileo was the first to view the heavens with the telescope, there is no record that he ever used it to observe the comet of 1618; that honor goes to the Swiss Jesuit Johann Baptist Cysat (ca. 1586–1657) and the English astronomer John Bainbridge (1582–1643).

After assisting Christoph Scheiner in his observations of sunspots, Cysat became professor of mathematics at Ingolstadt, where he observed the comet of 1618. Cysat measured its position with respect to two neighboring stars and employed a wooden sextant of six-foot radius for most of his measurements from December 1, 1618, through January 22, 1619. He attempted to determine the comet's true motion on the sky by noting its position on consecutive days at the same altitude. The apparent motion was observed over a few hours then compared with the true daily motion. Although his technique was questionable, Cysat believed his observations revealed no sensible parallax. On this basis, he proposed two theories, both of which assumed a fixed Earth. The first theory postulated a Tychonic circular orbit around the Sun located between Venus and Mars. The second theory fit a

Title page of woodcut engraving showing comet in celestial, not terrestrial, space. From John Bainbridge's *An Astronomicall description of the late comet. . . .* (London, 1619).

straight line trajectory to his observations of December 1, 20, and 29. Using a telescope, Cysat also observed the comet's physical appearance. The head and tail dimensions were measured, and it was noted that the head was composed of several condensations after December 8. Cysat likened the comet's appearance to nebulae seen in Cancer, Sagittarius, and the sword of Orion. He believed that it received its light from the Sun and recorded that its tail undulated as if blown by a wind.[4]

John Bainbridge published his observations of the comet of 1618 in a small tract entitled *An Astronomicall Description of the Late Comet.* His observation interval was November 28 to December 26, 1618. Using a telescope in the second week of December, Bainbridge observed the comet with respect to two neighboring stars. By comparing their relative positions near the horizon and the zenith, he determined that its distance from the Earth was more than 10 times the Earth-Moon distance.[5] Its apparent size at this enormous distance precluded a make-up of terrestrial vapors. Although he confessed his ignorance to the Almighty, Bainbridge denied that the comet was formed from the Milky Way or by planetary conjunctions. He ended his tract by arguing against comets as portents. This work was remarkable, not only for the telescopic parallax determination, but for its attack on the Aristotelian system and his clear preference for a heliocentric cosmology. Bainbridge's treatise on the comet of 1618 did not go unrewarded. It was at least partially responsible for his appointment to the first Savilian professorship at Oxford University in 1619.

Summary

The early seventeenth century witnessed the first use of the telescope for cometary observations. This significant step forward in cometary science must be measured against the two steps backward made by the period's two most influential astronomers, Kepler and Galileo. From 1602 on Kepler never deviated from his belief in the straight line motion of comets. However, a few of Kepler's views on the *physical* nature of comets were surprisingly modern: they shine by reflected sunlight and solar rays draw out a portion of the head material in an antisolar direction, so that the tail represents the death of the head.

The controversy over the comets of 1618 began when the Jesuit Horatio Grassi published a reasonable treatise on comets that relied mostly on the ideas of Tycho Brahe. Playing the role of devil's advocate, Galileo and his student, Mario Guiducci, responded with a work that did not really present a coherent cometary theory. Rather, Galileo chose to rebut and ridicule Grassi's arguments and his reliance on Tycho. Galileo, an argumentative, sarcastic, and mocking adversary, was clearly the polemical victor. However, Grassi's position is physically more defensible.

The contemporary views of Willebrord Snel, Pierre Gassendi, Seth Ward, and René Descartes were indicative of the more enlightened viewpoints in the early seventeenth century, and by the end of this period a transition from terrestrial exhalations to celestial objects had been made. However it was still questionable whether comets were eternal or ephemeral, or whether they moved on closed orbits or straight lines. If they moved on closed orbits, no preferred central body was identified. Unfortunately, in the latter half of the seventeenth century, the confusion over the nature and paths of comets was to become worse before getting better.

NOTES

1. The north star, Polaris, is a second-magnitude star. A celestial object whose magnitude is one would be two and one-half times brighter than Polaris while an object two and one-half times fainter would have a magnitude of three. An object whose magnitude is equal to six would be 100 times fainter than a first-magnitude object and just at the limit of naked-eye visibility.

2. See Appendix for observation summaries of the three comets of 1618.

3. The notion of an invisible, intangible substance in space, the ether, which takes a causal part in the motions of planets and transmission of light, had been postulated by the ancient philosophers. The idea of a single, all-pervasive ether received its greatest support in René Descartes' cosmology of vortices, a theory that was widely believed prior to the Newtonian era.

4. The observed wave phenomena in cometary ion tails are believed due to traveling interplanetary magnetic fields that originate with the Sun.

5. At the time of Bainbridge's parallax determination, the comet was actually as far away as 145 times the Earth-Moon distance.

4

The Comet of 1664

Confusion Reigns

Jean Dominique Cassini suggests that the comet of 1664 orbits the star Sirius. Pierre Petit dispels cometary superstition and offers an incorrect prediction for a comet's return. Robert Hooke's comets dissolve in interplanetary ether. Johannes Hevelius introduces the concept of cometary frisbees, and Christiaan Huygens develops a theory of rocketlike comets. Confusion reigns in the contemporary ideas used to explain the comets of 1664 and 1665.

THE COMET OF 1664 was discovered in Spain on November 17, 18 days prior to perihelion. It reached its greatest apparent brightness on December 29, when it passed within 0.17 astronomical units, AU, of the Earth. It was followed telescopically in Spain until March 20, 1665. Christiaan Huygens (1629–1695) first observed it on December 2 from Leiden, and Johannes Hevelius made his first observation 12 days later in Danzig, now Gdansk, Poland. The comet was also observed by Jean-Dominique Cassini (1625–1712) and Giovanni Alfonso Borelli (1608–1679) in Italy, by Adrien Auzout (1622–1691) and Pierre Petit (ca. 1594–1677) in France, Robert Hooke (1635–1702) in England, Samuel Danforth (1626–1674) in North America, several scientists in Spain, and by just about anyone who glanced skyward on a clear night in late December 1664. Due to its proximity to Earth at that time, the comet reached an impressive apparent magnitude of −1, with its tail reaching nearly 40 degrees in length. To the few who could remember the great comet of 1618 and the attendant popular concern, the appearance of this comet was even more impressive. There were many contemporary treatises written about this comet but the majority were astrological or religious in nature.

69

Only one week after the comet of 1664 was last sighted retreating from the Sun, the less impressive comet of 1665 was discovered at Aix in southern France. It quickly moved into the glare of the Sun and was last seen on April 20, 1665, four days before reaching perihelion. Hevelius observed it from April 6 to 20, and observations were made by Auzout, Petit, Borelli, Hooke, and others. The appearance of the comets of 1664 and 1665 prompted treatises from many important astronomers, and their widely varying conclusions and theories underscored contemporary confusion concerning the nature of comets.

Closed Orbits: The Comet of 1664 as a Permanent Object

As the first of four in the great family of Italian and French astronomers, Jean-Dominique Cassini was an exacting observer of comets. However, he was not a theoretician, and his resistance to new ideas led French science historian Jean-Baptiste Delambre to accuse him of having found his best ideas in the writings of his predecessors, and directing French astronomy backward. Even the Copernican ideas of a century earlier were rejected by the conservative Cassini. Born in Italy, Cassini's early work was carried out in Panzano, Italy, at the private observatory of a rich amateur astronomer, the Marquis Cornelio Malvasia. In 1669 he continued his work in Paris at the newly formed Academy of Sciences.

In his account of a comet seen in 1652 and 1653, Cassini began by stating that comets were situated beyond Saturn and formed from terrestrial and planetary emanations. However, after comparing his observations with those of others, he considered comets analogous to the planets, except that their motions were highly eccentric with respect to Earth.

After observing the comet of 1664, Cassini published his ideas on cometary motions in his *Hypothesis motus cometae novissimi*. He correctly noted that the comet of 1664 passed closest to the Earth on December 29, 1664. Using the Earth-Moon and Jupiter-Galilean satellite systems as analogies, Cassini sought this comet's central body. From the position of the comet near *perigee*, he selected the star Sirius. According to Cassini, the center of the comet's circular epicycle was Sirius and the star-comet system itself turned upon a deferent around the Earth. The comet was invisible during most of its orbit due to its very large distance from Earth, approaching Earth only in the lowest part. Although Cassini previously located comets beyond Saturn, there is some evidence that the comet of 1664 passed within Earth's solar orbit. While he theorized a circular orbit about Sirius, his computations for the comet's motion assumed a rectilinear trajectory. To make theory

Jean-Dominique Cassini. (*Courtesy of the Paris Observatory.*)

agree with observations, Cassini employed a computational device and introduced a daily advancement of 6 arc minutes in the comet's perigee and the points, or *nodes,* where its orbit intersected the *ecliptic.* The ecliptic is the plane defined by the Earth's motion around the Sun (or as Cassini would have it—the plane formed by the Sun's motion around the Earth).

Cassini also tried to explain the motion of the great comet of 1680. Pre- and postperihelion observations of this comet were considered observations of two separate comets. Together, these observations could not fit with a great circle, whereas separately they could. For the preperihelion observations before December 18, 1680, Cassini postulated a circular orbit around Earth that intersected the Sun's orbit. He also suggested that cometary orbits were restricted to a certain band on the celestial sphere—a cometary zodiac.

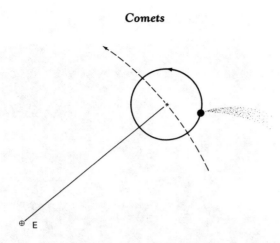

Cassini's theory for the comet of 1664 had the star Sirius as the center of the comet's circular orbit, the epicycle. The star then revolved around the immobile Earth, E, in its orbit, or deferent.

Like Cassini, Adrien Auzout approached astronomy using instruments rather than mathematics. Although he invented neither, the Frenchman Auzout is best remembered for making significant improvements to the final development of the micrometer and telescopic sights.

After observing the comet of 1664 four or five times between December 22 and December 31, 1664, Auzout predicted cometary positions on the sky, an *ephemeris*, for dates from the previous November through February 1665. Finished on January 2, 1665, his *search ephemeris* was the first such prediction to be published. The predicted path of the comet was generated by marking the observations on a celestial globe and then orienting the globe until the observation points fell on its horizon circle. The path along the horizon circle then represented the motion of the comet on the sky, which could be extended forward in time. The position of the nodes and inclination angle that the comet's orbit made with the ecliptic were then easily read directly from the globe. Both Cassini and Auzout used this device to generate a crude search ephemeris, but while Cassini took time to make additional computations based on spherical trigonometry, Auzout rushed his ephemeris into print.

Commenting on Cassini's notion that the comet of 1664 moved in a circular orbit about Sirius, Auzout mentioned that he, too, thought of this idea and noted that the comet of 1651 had about the same speed as Sirius and came to perigee opposite the bright star. Auzout reasoned that if the comets of 1664 and 1651 were one, its reappearance could be expected around 1676. However, as he could find no recollection of a comet that appeared every 12 years and there was no evidence that well-observed comets arrived at perigee in conjunction with a notable star, Auzout could not endorse Cassini's theory. Although Auzout could find no evidence of reappearance, he did not abandon the idea that comets might be permanent celestial bodies subject to return.

Fearing that cometary superstitions were a hindrance to his kingdom, young Louis XIV of France requested a rational book on comets from Pierre Petit. It was a fortunate choice because Petit was an empiricist among the doctrinaire Cartesians. Rising to the royal edict, Petit wrote a work entitled *Dissertation sur la Nature des Comètes*. Much of the book attempted to dispel the common fear of comets. In a geocentric, infinite universe, Petit thought of comets as universal garbage collectors, created to pass into the solar neighborhood and collect waste gases and fumes exhaled by planetary atmospheres. They were permanent celestial bodies that returned at periodic intervals. In the distant reaches of their orbits beyond Saturn, they were invisible, and remained unrecognizable when they had returned in the past.

Turning to the comet of 1664, Petit attempted—but could not detect—a parallax. Using Grassi's invalid argument against Galileo, Petit placed part of its orbit beyond those of the planets because he found that a telescope magnified the planets more than the comet. Noting that celestial bodies have revolutions lasting from 1 day to 30 years, he saw no reason why comets could not have periods of 100 or 1000 years or more. While Petit favored elliptic motion for the comet path, he was not sure whether the ellipse should surround the Sun, the Earth, or neither. For the comet of 1664, he did not place the ellipse around either one.

The following interpretation of Petit's theory for the comet of 1664 was given by James A. Ruffner in 1966. According to Petit, the comet of 1664 should have appeared and disappeared at the same distance from the Earth. He argued that its motion could not be circular since then the observed arc A′P before perigee should have equaled the observed arc PB after perigee. The perigee point P was well determined, and the arc observed with the naked eye before perigee appeared to be longer than the one after perigee. To represent the observations, Petit introduced an ellipse whose major axis was skewed with respect to the line from the Earth to the comet's center of motion. The comet appeared at A and disappeared at B, and its final apparent motion, as seen from the Earth, was nearly stationary near B. Hence, the observed arcs before and after perigee, AP and PB, did not appear to be equal. Petit assumed the comet appeared and disappeared at the same geocentric distance, EA = EB, so that the skewed ellipse satisfied both the observations and his assumptions.

From their physical appearance and similar apparent positions on the sky, Petit claimed that the comets seen in 1664 and 1618 were the same object. He then proceeded back in time at 46-year intervals and identified pos-

An Oversight by the Astrologer John Gadbury

The importance of the comets of 1664 and 1665 was not overlooked by astrologers who attempted to presage their influence by noting the appearance and initial locations of the two visitants. The London astrologer John Gadbury had a particularly fertile imagination. In 1665, he published a booklet entitled *De cometis* in which he outlined the possible unpleasantries that would arise if a comet first appeared in any of the 12 zodiacal constellations. Mr. Gadbury's list follows:

Comet first appears in	Expected events
Aries	Diseases affecting the head and eyes, detriment unto rich men, sorrows and troubles to the vulgar
Taurus	Sickness and great earthquakes, death of a great man, detriment to cattle, rotting of fruit
Gemini	Grievous diseases for children, men given to commit fornications, many abortions, prodigious winds
Cancer	Famine, pestilence, wars, abundance of locust or caterpillars and such worms that destroy the fruit
Leo	Vermin, rats, detriment to great ladies, dogs run mad in multitudes, corn destroyed by worms
Virgo	Detriment to merchants, noble women subject to scandals, infamies, and disgrace
Libra	Portends thieves, housebreakers, and highwaymen, extremes of heat and cold, death of some king
Scorpio	Wars and controversies among men, rebellion, scarcity of grain and fruit
Sagittarius	Denotes depression in noblemen
Capricorn	Fornication and adulteries to be rife and common among men, persecution of religious men
Aquarius	Plague sweeps away a multitude, terrible and durable wars, death of an eminent prince or great female
Pisces	Air replete with prodigies, destruction of fish

These rules for predicting events by noting a comet's celestial position would have failed miserably for the comets of 1664 and 1665. The former was seen first in southern Virgo on November 17, 1664, and last on March 20, 1665. As if to continue the cometary portent, the latter was first recorded above Capricorn on March 27, 1665.

Gadbury's rules would have predicted scandals, persecution, and fornication, but London was about to be beset with a different catastrophe. In May 1665, the black plague broke out. Among those who fled the city was Isaac Newton, who would soon help diminish the popular fear of comets. Deaths in May 1665 numbered a few dozen, in June a few hundred, in July over 6000, in August 17,000, and in September over 31,000 persons died. By the end of 1665 nearly 90,000 people had died in London—nearly one-fifth the total population. As a lasting testament to astrological prediction, the comets of 1664 and 1665 first appeared in two of the few zodiacal constellations that did *not* predict some sort of disease or pestilence.

A woodcut by Diebold Schilling, done in 1508–1513, showing the disastrous effects upon the populace after the appearance of a comet (Halley) in 1456. Note the classic symptoms of the black plague, which included splitting headaches, pain to the back and limbs, and fever. (*Courtesy of the Swiss National Museum, Zurich.*)

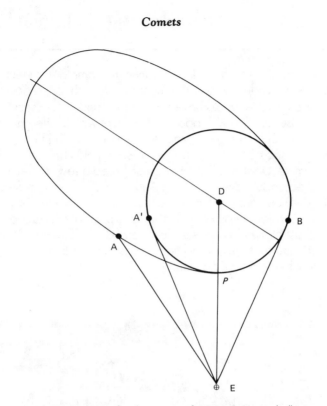

Pierre Petit's theory for the comet of 1664 suggested elliptic motion for the comet. The comet's ellipse, which enclosed neither the Sun nor the Earth, was skewed with respect to the line ED drawn from the Earth to the comet's center of motion. The total observed path of the comet on the sky was APB.

sible apparitions of the same comet during preceding centuries. He predicted it would be seen approximately every 46 years, the next return would be 1710. Petit gloried in being the first to announce the return of a comet, a prediction based on previous observations and his theory of cometary motions. Just as Columbus fulfilled Seneca's prophesy that in time someone would discover a new world by sea, Petit immodestly felt that he had fulfilled Seneca's prophesy that the way of comets would one day be known. Alas, the world would have to wait another 40 years for a correct prediction: Petit's comet of 1618 and 1664 were entirely different. Writing a century later, Alexandre Pingré put Pierre Petit's work into perspective with an outrageous pun. Keeping in mind that *pierre* is French for stone, Pingré recorded that Petit's system was only "rough chiseled," it failed to decide the nature and position of the cometary orbit and that was the "stumbling stone of Petit," *la pierre d'achoppement de Petit.*

Writing at the same time as Petit, another Frenchman, Claude Comiers (d. 1693), wrote a rational treatise on comets. In addition to a review of existing ideas and a long, balanced discussion of comets as portents, Comiers developed Kepler's theory of tail formation. In what was a rather modern stance, Comiers stated that comets reflect the Sun's light from their atmosphere, their tails form when solar heat rarefies and dissipates the coma material, and the solar rays push this material behind the head in an antisolar direction.

The versatile Italian scientist Giovanni Borelli observed the comet of 1664 from December through the beginning of February 1665. Toward the end of this period, Borelli made parallax measurements and concluded that the comet was beyond the Moon. From his attempts to compute its path, Borelli was forced to conclude that its geocentric distance varied with time. For the sake of computation, he assumed the heliocentric theory to be correct and concluded that the comet probably traveled in an elliptical orbit or another curved path. The Church had forbidden the Copernican system so Borelli referred to it obliquely, giving it a classical name. When his results were published as a letter to Stefano degli Angeli entitled *Del Movimento della Cometa Apparasa il mese di Dicembre 1664*, Borelli further covered his tracks by using the pseudonym Pier Maria Mutoli.

Borelli's attention was also directed to the comet of 1665 and on May 4, 1665, he wrote a letter to Duke Leopold de Medici noting that its motion could not be represented by a straight line. Borelli's observations were best represented with a curved line resembling a parabola. On a trip to Florence Borelli hoped to demonstrate his hypothesis to Leopold by mathematical calculations and a mechanical device. Unfortunately the particulars of the device, if they ever existed, have been lost. A portion of the letter to Leopold was published in 1817 by Baron Franz von Zach.

The comet of 1664, extensively observed and reported in Europe and China, also witnessed the first original astronomical contribution from North America. Samuel Danforth, a puritan clergyman in Roxbury, Massachusetts, published his observations and analysis in *An Astronomical Description of the Late Comet or Blazing Star*. Observing the comet from December 15 to February 14, Danforth noted its varying apparent motion and suggested its circular orbit was offset from the Earth's center. He considered the comet's supralunar position was evident from its slow apparent motion, visibility to all countries, and from his own null parallax determination. Noting that the comet's motion appeared most rapid on December 28, 1664, Danforth stated that it passed closest to the Earth on that day. Attempting to explain its physical appearance, Danforth suggested that "sunbeames" are refracted in the transparent body, congregate together, and become visible as a blazing stream.

The Poetic Use of Comets

After the fact, the two comets of 1664 and 1665 were commonly thought to have presaged the London plague of 1665 and a subsequent fire in late summer 1666. The fire completely destroyed 436 acres of the city within a few days. One year later, the English poet John Milton published *Paradise Lost* and his allusion to a comet becomes all the more meaningful in view of these disasters. Describing the hostile meeting between Satan and Death before the gates of hell, Milton writes:

> On the other side,
> Incensed with indignation, Satan stood
> Unterrified, and like a comet burned,
> That fires the length of Ophiuchus huge
> In the arctic sky, and from his horrid hair
> Shakes pestilence and war

Milton's poetic use of comets is reminiscent of a quote from the Greek poet Homer in the late eighth century B.C. In the nineteenth book of the *Iliad*, speaking of the helmet of Achilles, we read that it shone

> Like the red star, that from his flaming hair
> Shakes down diseases, pestilence and war

Rectilinear Orbits: The Comet of 1664 as a Transitory Object

It was unfortunate that Robert Hooke was a contemporary of Isaac Newton and Edmond Halley. In any other era, he could have stepped out of their shadows and dominated contemporary science. A true Renaissance scientist, Hooke's influence was felt in nearly every branch of the exact sciences.

A woodcut by Diebold Schilling in 1508–1513 illustrates the dreadful effects of the comet of 1472. (*Courtesy of the Swiss National Museum, Zurich.*)

If it were not for continued badgering by the Royal Society of London, Hooke's views on comets might not have been recorded. He observed the comets of 1664 and 1665 and had studied both the observations and other people's theories. He also studied published results on the comet of 1577 and developed a theory that he hoped to test against the observations of a new comet. Hence, Hooke waited until the appearance of the comet of 1677 before he finally published his ideas in 1678 as a Cutler lecture entitled *Cometa or Remarks about Comets*. Hooke first saw the comet of 1677 on May

1 and observed it through May 4 using a six-foot telescope. With characteristic ingenuity, he measured the apparent diameter of the coma by noting its dimensions with respect to the support pole of a neighboring weather cock. Based upon his own and other observations of historic and contemporary comets, Hooke put together a coherent theory to explain their physical appearance and motion.

Hooke suggested that cometary nuclei were solid bodies made up of the same magnetic material residing within the Earth's interior. A nucleus might dissolve by chemical action of the surrounding ether and internal agitation, and the dissolution process would reduce its gravitation. As an analogy, Hooke mentioned an eruption of Mt. Aetna in Sicily, where the internal parts of the Earth were so agitated that matter changed state so as to "confound the gravitating principle." Hooke also used lodestone as an example of a substance that can lose its attracting qualities if it changes state. The contrary effect, repulsion, was also possible. Hence the nucleus dissolved equally on all sides, and the coma and tail material flew away from the center in all directions. Material shot toward the Sun obtained a repulsive virtue and was deflected back in parabolic curves. Single particles continued in their motion until they burned out or dissolved into the ether. In an aside, Hooke noted that the Earth's atmosphere was a result of the ether dissolving parts of its crust.

Since the cometary nucleus does not exhibit a shadow, Hooke concluded that not all of a comet's light was due to the Sun. Some light might be analogous to that produced by the Sun or stars or to the light of such things as decaying filth, rotten wood, or glow worms. He then cautioned the reader against seeking common analogies for a comet's light source since it might differ from other known sources. Any deflection of the tail from the antisolar direction was due to the resistance of the ether. With regard to age, Hooke asked, How could so vast a body be generated? How could it supply a constant stream of ascending parts? How could a newly generated body receive so great a degree of motion? Answering his own questions, he responded that comets were likely as old as the world, retained their original motions, and were still slowly dissolving in the ether. The apparent disappearance of a comet was not due to its consumption but rather to its removal to a great distance from the observer.

Hooke's ideas on cometary motion were intertwined with their physical behavior. By internal agitation and external dissolution, a comet's gravitational principle was disturbed over time. Then the comet no longer circled the central body but tended toward a straight line as it lost its attractive, magnetic virtue. The remaining nucleus would retain some of this attractive virtue so the resultant motion was slightly curved, concave toward the central body. Earth or planetary attractions distorted the comet's path, but as long as it moved faster past any point in the system than a planet would move there, its path would be less sharply curved than the planet's. Although vague as to whether a comet's orbit was once closed, Hooke emphasized the central role of gravitation in its orbital motion:

> I cannot imagine how their various motions can with any satisfaction be imagined . . . without supposing a kind of gravitation throughout the whole Vortice or Coelum of the Sun, by which the Planets are attracted, or have a tendency toward the Sun, as terrestrial bodies have toward the center of the Earth.

In a letter dated January 6, 1680, Hooke wrote to Isaac Newton stating his conviction that gravity decreased in a power proportional to the square of the distance between the two bodies. Thereafter, Hooke was convinced that Newton had stolen the inverse square relation from him. For his part Newton cited Hooke's *Cometa* as proof that Hooke did not know of this relationship in 1678. Newton believed Hooke could not even claim second place because Christopher Wren knew of it in 1677. In fact, Ismael Boulliau suggested it even earlier. In his 1645 work *Astronomia Philolaica*, Ismael Boulliau asserted that if a planetary moving force did exist—which he denied—it should vary inversely as the square of the distance.

Concerning the distance to the comet of 1664, Hooke could only conclude that it had no demonstrable parallax. He expressed dismay at the inaccuracy and inconsistency of contemporary position observations. Different observers recorded the comet at different positions at the same time; the more observations he collected, the more inconsistent they became. Hooke could not obtain a meaningful parallax determination and concluded only that the comet's distance was very great. Although he was a firm believer in the Copernican system, contemporary observations were so inaccurate they supported varied hypotheses concerning Earth and cometary motions. In fact, the observed motion of the comet of 1664 could be represented best with an immobile Earth and the comet moving on a circular path. The observations could also be satisfied by assuming the Earth and comet were moving on circular paths, and these were not the only two hypotheses that could represent the observations. He pointed out that elliptical orbits and diverse other hypotheses would also explain the observed phenomena. At the end of his discussion on the comet of 1664, Hooke concluded that the comet's path was a little curved by the Sun.

In the history of science, comets in particular, the number of ideas anticipated by Hooke is surprising. They include universal gravitation, the dynamical effects of planetary perturbations and nongravitational forces (i.e.,

Portrait of Johannes Hevelius from his book *Machinae coelestis pars posterior* (Danzig, 1679).

drag effects of the ether), cometary self-luminescence, and a slowly dissolving cometary nucleus. From the insight and genius of Robert Hooke, we pass to the industry and precision of Johannes Hevelius.

The Polish astronomer Johannes Hevelius' initial astronomical career was crowned by success and good fortune, but later years were submersed in controversy and disaster. Born into a prosperous brewer's family, Hevelius established what was for a time the world's leading astronomical observatory. Like Jean-Dominique Cassini, Christiaan Huygens, and Pierre Petit, Hevelius had Louis XIV as a patron, and for several years he received annual research grants.

To note that the published works of Hevelius were fine examples of seventeenth century printing would be an unfair understatement. Hevelius used fine Dutch paper, drew and engraved his own magnificent plates, and printed his books on his own press. His first work, *Selenographia*, published in 1647, was on lunar topography and a marvel to all who saw it. When shown a copy sent to Italy, the Pope said it would be a book without parallel had it not been written by a Copernican heretic.

The later years of Hevelius' career were marred by controversies. The first was with Adrien Auzout, Pierre Petit, and several other contemporary astronomers over the position of the comet of 1664 in mid-February 1665, the second was with John Flamsteed (1646–1719) and Robert Hooke, who belittled Hevelius' work because he used naked-eye—rather than telescopic—sights. While Hevelius used telescopes for physical observations, he trusted only his naked-eye sights for accurate position measurements. Were it not enough to be at odds with many of his scientific contemporaries, Hevelius' famous Danzig observatory, Stellaburgum, burned to the ground on September 26, 1679. Although he immediately began rebuilding, his new instruments were fewer in number and inferior to those that had been destroyed. Grieved by the loss of his observatory and books and tired from his controversy with Flamsteed and Hooke, Hevelius died on his 76th birthday in 1687. The passing of the last great naked-eye observer in 1687 was accompanied by the printing of Newton's *Principia* the same year, an interesting transition from the old to the new astronomy.

Eager to follow his well-received *Selenographia* with another scientific masterpiece, Hevelius began working on *Cometographia*. By December 1664, when the comet of 1664 arrived, the book was three-quarters complete. Public interest and appetite for cometary treatises were not to be ignored and Hevelius set aside *Cometographia* and rushed into print a forerunner, or synopsis, of the projected work entitled *Prodromus cometicus*. Beginning with a prudent dedication to Colbert, controller general of finance to Louis XIV, this book outlined Hevelius' ideas that later appeared, in more detail, in his *Cometographia*. The *Prodromus cometicus* also detailed his observations of the comet of 1664. From only one day of observations on February 4, 1665, Hevelius found a parallax of 5000 Earth radii. This result was too small by more than a factor of six.

Hevelius stated that all comets respect the Sun as their king and center, as do the planets, making them a kind of spurious planet. He believed that without the Earth's movement, no rational account could be given of any comet's motion. However, in describing the apparent motion of the comet, Hevelius did not eliminate the Earth's annual motion and his scheme depicts the comet's path as a conic section with the Sun at one focus. Although Hevelius had stumbled upon the correct solution to the motion of comets, this result was foreign to his usual careful analysis, especially his firm belief that, as transitory objects, comets should travel on rectilinear paths. Later, in *Cometographia*, he restudied the motion of the comet of 1664, considered the Earth's annual motion, and ascribed quasirectilinear motion to the comet. Nevertheless, the notion that cometary paths were conic sections with the Sun at one focus was published as early as 1665 and was later read by Isaac Newton.

Hevelius' observations of the comet of 1664 were published in *Prodromus cometicus,* and his last observation, on February 18, 1665, placed the comet near the star alpha in the constellation of Aries, alpha Arietis. Auzout, Petit, and every other observer of the comet's motion objected. They noted that when the comet did move past a bright star in Aries three weeks later, it was beta, not alpha, Arietis. On February 18, 1665, the comet of 1664 was actually 6.5 degrees southwest of alpha Arietis and it did, in fact, pass close to beta Arietis on March 11, 1665. To point out the spurious nature of Hevelius' observation, Auzout wrote a letter to Petit, who published it along with his own at the end of his *Dissertation sur la Nature des Comètes.*

The matter should have ended there. However, far from admitting his error, Hevelius conducted a protracted defense of his *Prodromus cometicus* in letters to the secretary of the Royal Society of London, Henry Oldenburg, and in a book entitled *Descriptio cometae . . . 1665 . . . mantissa prodromi cometici.* The first part of this book presented his observations of the comet of 1665 and included a synopsis of his general cometary ideas. The second part was a defense of his *Prodromus cometicus* against Auzout, Petit, and others who had entered the fray. Hevelius finally set aside his defense only when the Royal Society of London refused to endorse his observations. The furor created by this spurious observation had delayed the publication of his cometary magnum opus and his *Cometographia* was not published until 1668.

After nearly 15 years, the *Cometographia* was published as a series of books or chapters. The first chapter presented observational data on the comet of 1652. One of the more interesting observations was the comet's expansion in size as it receded from perihelion. Later, Isaac Newton would have trouble trying to explain a similar phenomenon observed in the comet of 1680. The second and third chapters were devoted to arguments for comets existing beyond the Earth's atmosphere, and Chapters Four and Five presented a parallax determination for the comet of 1652 and a discussion of its true position and distance from the Earth. The work went on to discuss the comets of 1661, 1664, and 1665. It is interesting to note that in his discussion of the comet of 1664, Hevelius did not use his disputed observation of February 18, 1665. The book ended with a catalog of 251 cometary apparitions from the Biblical deluge to 1665.

Embedded in *Cometographia* was Hevelius' theory of comets, an interesting collection of ideas based primarily on those of Kepler. Like Kepler, Hevelius began with the premise that comets were transitory objects, and

Frontispiece from Johannes Hevelius' *Cometographia* (Danzig, 1668). Three allegorical figures showing the Aristotelian idea that comets are sublunar (left), the Keplerian notion that comets move on straight line paths (right) and the idea of Johannes Hevelius (center) that comets originate in the atmospheres of Jupiter and Saturn and move about the Sun on a curved trajectory.

hence their basic motions were rectilinear. Borrowing another idea, Hevelius considered comets to be waste matter, but instead of Kepler's spontaneous creation, Hevelius thought that comets formed in the atmospheres of the giant outer planets. Modifying slightly an idea presented 44 years earlier by Joannes C. Glorioso, Hevelius suggested that all heavenly bodies continually pour forth waste exhalations. The grosser exhalations remained bound to the parent bodies and were responsible for sunspots and clouds; the lighter, more tenuous effluvia arose into the atmosphere. Under the influence of the vertical motion—which was natural to light, tenuous effluvia—and the circular rotation of the planets, the comet effluvia spiraled upward through the atmosphere, where they were flung into space at a uniform velocity on a trajectory tangent to the atmosphere.

Although all heavenly bodies were believed to give forth exhalations, the pale color and large size of Jupiter and Saturn made them the most likely birthplace for Hevelius' comets. He classified comets by which planet produced them. Thus comets forming from Jupiter or Saturn were distinguishable by their respective colors. He noted that the color of the comet of 1531 was a beautiful yellow-gold, while the comet of 1607 was dull. While Hevelius would have attributed these two comets to different classes or birthplaces, they were successive returns of the same comet—Halley.

Once the nascent comets escaped the parent planets, the solar rays began to act upon them to form tails. The tenuous exhalations adhering to the sunward-facing side of a comet's head were rarefied by the Sun's rays and driven back in an antisolar direction. This tail formation idea was similar to Kepler's view, except that Kepler believed the tail represented the disintegration and death of the head, whereas Hevelius believed the comet's head would expand and dissipate at the end of the apparition. Hevelius' telescopic observations of various comets had convinced him that the heads were made up of discrete particles.

Although he hadn't accounted for the Earth's motion in his earlier *Prodromus cometicus,* Hevelius was careful to make this correction in his *Cometographia.* Using his accurate contemporary observations, it became apparent that comets have neither uniform motion nor rectilinear trajectories. He was forced to conclude that comets move on paths that are slightly curved toward the Sun. Due to their natural regard for the Sun and their transitory nature, Hevelius believed that comets traveled on either of the open conic sections, the hyperbola or the parabola. If a comet's path brought it quite close to the Sun, the resulting orbit would be hyperbolic; otherwise the orbit was parabolic. Unlike his earlier suggestion in the *Prodromus cometicus,* the Sun no longer occupied the focus but was now located roughly halfway between the focus and the vertex of the conic. As a device for modifying the natural uniform motion of comets, Hevelius introduced the concept of a disklike comet—an interplanetary frisbee!

As imperfect planetary effluvia, comets agglomerate, not into perfect spheres but into disklike objects. The diskcomet kept its face perpendicular to the solar rays throughout its trajectory AB. Hevelius used the analogy of the Earth's effect on a compass needle to describe the alignment of the diskcomet as it traveled along its orbit. With respect to the Sun, the orientation of the comet changed so that it traveled edge-on at perihelion and nearly face-forward in the extremities of its orbit. Hence, when traveling through

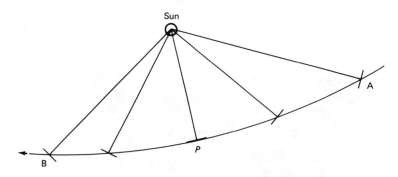

Johannes Hevelius considered comets as disk-shaped objects whose face was always maintained perpendicular to the Sun's rays. At perihelion, P, the disk-shaped comet moved edge-on through the resisting ether and so moved most rapidly there. At either end of the observational arc, AB, the disk-shaped comet was slowed by the resistance of the ether.

the ether, the disk traveled fastest at perihelion in apparent agreement with the observations. The relatively slow motion of the comet far from the Sun was due to the impedance of the ether on the disk-comet as it traveled face forward. Any irregular motion observed near the end of a comet's apparition was due to the uneven disintegration of its head.

The physical and dynamical cometary theories of Hevelius were intimately connected. By assuming that comets were transitory in nature, Hevelius was forced to assume basic, uniform, rectilinear motion, then an ad hoc physical description of comets to explain the observed deviations from uniform motion. He was fortunate that the comets he considered, from 1472 to 1665, did not show large deviations from rectilinear motion, at least during the intervals of his observations. Had he used his theory to represent the pre- and postperihelion observations of the comet of 1680, he would have failed miserably. Christiaan Huygens did try—and failed.

One of Europe's greatest scientists in the seventeenth century, Christiaan Huygens made important contributions in physics, optics, time measurement, astronomy, and mathematics. If his achievements went relatively unnoticed in the seventeenth and eighteenth centuries, it was largely due to his reluctance to publish ideas he considered incomplete or lacking great significance. He never published his ideas on comets. They have appeared only recently as successive volumes of his *Complete Works* (1888–1950) were published. A native of the Netherlands, Huygens spent most of the period from 1666 to 1681 in Paris under the patronage of Louis XIV. He observed and developed theories for the comets of 1664, 1665, and 1680.

For the comets of 1664 and 1665, Huygens began by assuming their motions were uniform and either rectilinear or curved slightly concave to-

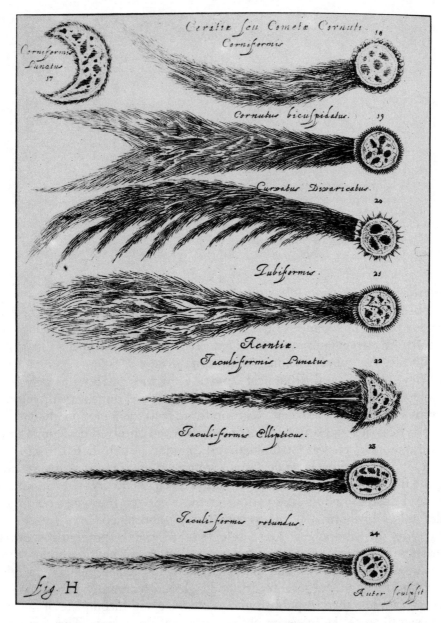

Types of cometary forms showing Hevelius' idea that the heads of comets are disklike and made up of discrete particles. From Johannes Hevelius' *Cometographia* (Danzig, 1668).

Johannes Hevelius observing with his azimuthal quadrant. From his *Machinae coelestis pars prior* (Danzig, 1673).

ward the Sun. While Kepler had introduced nonuniform motion to fit the observations of the comets of 1607 and 1618 with a straight line, Huygens maintained uniform motion, but allowed the path to curve for the comets of 1664 and 1665. To successfully represent the observations of the comet of 1664, Huygens did not have to bend the orbit significantly. However, he had great difficulty using rectilinear motion to explain the observed motion for the comet of 1665, and his final orbit was substantially curved toward the Sun.

The comet of 1680, with its incoming and outgoing paths nearly parallel, was probably the undoing of Huygens' cometary theory and may explain why he never published it. Both Cassini and Huygens reported their views on the comet of 1680 to the Royal Society of London in February 1681 and nei-

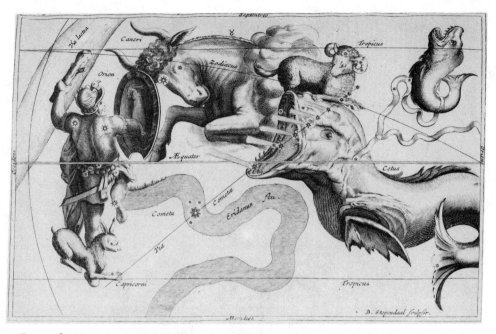

Comet of 1664 in January 1665 moving through the constellation of Cetus the whale. From Stanilaus Lubienietski's *Theatrum cometicum* (Amsterdam, 1668).

ther considered the inbound and outbound legs of the orbit to be due to the same comet. Cassini could not fit a great circle to both the inbound and outbound legs of the comet's orbit and, had he tried, Huygens could not have fit both legs with any path resembling a straight line. Even by ignoring the observations made on the inbound leg, Huygens had to devise a rather contrived theory to explain the comet's behavior. Noting that its outgoing path seemed to originate in the Sun, he suggested that it was shot from the Sun like a rocket. Solar material was ejected from time to time, and when it collected, it ignited to form a rocket-comet. A second ignition was required to form the tail and Huygens noted Robert Boyle's recent work on phosphorus as an apt analogy for a substance capable of this type of double ignition.

According to Huygens, comets differed from planets because they had an intrinsic power that gave them motion relative to the ether that was necessary to explain their deviations from strict rectilinear paths. Unlike the passive planets moved about by the solar vortex, the rocketlike thrusting of cometary nuclei gave them an intrinsic motion. Since the rocket thrust was directed toward the Sun, the deviation of the comet of 1680 from rectilinear motion was slight, since it was already traveling almost directly away from the Sun. However, the comets of 1664 and 1665 were observed traveling more transverse to the Sun-comet direction so that the sunward thrust introduced

Comet of 1665 on April 13, 1665, moving through the constellation of Pegasus, the winged horse. From Stanilaus Lubienietski's *Theatrum cometicum* (Amsterdam, 1668).

Christiaan Huygens. *(Courtesy of the Paris Observatory)*.

91

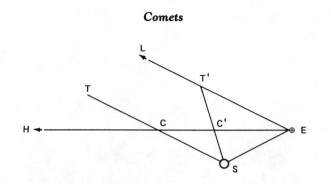

Christiaan Huygens devised a clever technique for estimating maximum Earth-comet distances. An observer's lines of sight are directed toward the end of a comet's tail, EL, and toward the comet's head, EH. A line, ST, drawn parallel to EL and through the Sun, S, establishes an upper limit to the comet's distance because at C a comet's tail would have to be infinitely long to be seen by an observer at E. Hence the comet can be no farther from the Earth than the distance EC. For the tail to have a finite length, the comet must be located at some lesser distance, EC', with the end of its tail at position T'. Using the properties of a triangle, the maximum distance of the comet from the Earth, EC, is determined in terms of the Earth-Sun distance, ES, because the angle CES is measured and the angle ECS equals the measured angle LEH.

deviations from rectilinear motion and curved the path slightly concave to the Sun.[1]

In trying to explain cometary tails, Huygens ran into additional difficulties. He rejected the traditional view that they were a result of sunlight refracted in a crystalline cometary sphere. He considered the tail to be a real entity, a kind of smoke whose density was less than that of the head but more than that of the surrounding ether. Hence, tail particles maintained the motion they shared with the head prior to ejection and the pressure of the surrounding ether imparted a slight curvature to the tail. The antisolar nature of comet tails must have been an embarrassment to Huygens because the sunward-facing rocket thrust should have sent the smokey tail particles toward—not away from—the Sun.

Although Huygens chose to ignore this problem, he was aware of the antisolar nature of a comet's tail. Indeed, he used it to crudely determine an upper limit on the distance to the comet of 1680. Assuming that the observer's lines of sight are directed to the end of the comet's tail, EL, and toward its head, EH, a line, ST, drawn parallel to EL and through the Sun, S, establishes an upper limit to the comet's distance at C. That is, a comet at point C must have a tail of infinite length in order to be seen by an observer at E. Hence, by assuming that the comet's tail is finite and antisolar, an upper limit on the Earth-comet distance, EC, can be crudely determined. Applying this technique to the comet of 1680, Huygens correctly placed the comet inside Mercury's orbit when it first emerged from the solar glare in late December 1680.

Summary

In the few decades prior to the 1687 publication of Isaac Newton's *Principia*, knowledge about the nature and motion of comets was in a confused state. Various authorities offered diverse ideas for the nature and motions of comets. However, a few common opinions were evident. For example, the supralunar position of comets was no longer seriously questioned. An increasing number of parallax determinations had convinced nearly every knowledgeable person that comets were not below the Moon. However, Aristotelian cometary ideas were still pervasive, and most contemporary savants still considered the origin of comets to be due to agglomerations or exhalations from heavenly bodies. In many contemporary theories, the all-pervasive ether took an active role in either the appearance or dynamics of comets. Hooke used the ether to dissolve a comet's head and provide the tail curvature. Hevelius suggested the ether as a resisting medium capable of explaining the observed nonuniformity of cometary motions. The ether was a common notion invoked throughout the seventeenth century. It lost some support in the eighteenth century, was resurrected in the nineteenth, and finally was banished in the twentieth century.

Even more important than the Aristotelian concepts of exhalations and the pervasive ether was the notion that transitory objects have rectilinear orbits and permanent bodies travel along circular paths. This principle, so tenaciously held, was a major obstacle to the advancement of cometary thought throughout most of the seventeenth century. The problem with this principle was its status as an axiom, rather than an hypothesis. It was not questioned and often provided a starting point for a particular theory. Hence, in the pre-Newtonian era, two schools of cometary thought were evident. The first believed, a priori, that comets are permanent celestial objects and hence had circular—or at least closed—orbits. To this school belonged Jean-Dominique Cassini, Adrien Auzout, Pierre Petit, Giovanni Borelli, and others. The second school, which included Johannes Hevelius and Christiaan Huygens, considered comets as transitory objects whose basic, intrinsic motions were uniform and rectilinear. Typically, Robert Hooke defied classification by allowing permanence and basic rectilinear motion.

Bound by the pervasive Aristotelian notion that transitory objects have rectilinear paths while permanent bodies have closed orbits, most contemporary scientists first decided whether comets were permanent or transitory, then tried to fit the observations to the appropriate path. Unfortunately, the observations of the comets of 1607, 1618, 1652, 1664, and 1665 could be crudely fit with either a closed orbit or a nearly rectilinear one. What was needed was a comet with a nearly parabolic orbit, with a large number of observations on either side of perihelion so that a definite decision could be made as to its orbital path. Also needed was a person whose genius would allow a break with Aristotelian tradition and permit a theory to be fit to the extensive cometary observations without a priori assumptions. Just such a comet arrived in 1680 and Isaac Newton was there to observe it.

NOTES

1. In 1950, Fred L. Whipple introduced a rocketlike thrusting of the cometary nucleus to successfully explain the slight nongravitational force acting on most active comets—a force that is superimposed on the comet's primary gravitational motion. Huygens incorrectly used the rocket analogy to explain a comet's primary motion.

5

The Comet of 1680

Newton's View from the Shoulders of Giants

Gottfried Kirch makes the first telescopic discovery of a comet in 1680. Georg Dörffel suggests a parabolic path for the comet of 1680. John Flamsteed and Isaac Newton consider whether one or two objects are required to explain the cometary appearances in November and December 1680. Isaac Newton establishes a method for computing the orbital paths of comets traveling on parabolic orbits.

Isaac NEWTON'S INITIAL INTEREST in comets was not altogether scientific. As a boy, on dark nights he placed candles in paper lanterns and tied them to the tails of kites to terrify the country people, who took them for comets. Later, he made a serious study of the comet of 1680 and, after a false start, developed a theory of cometary dynamics that was as elegant as it was correct. The false start resulted from his view that the comets seen in November and December of 1680 were too different objects.

The first comet to be discovered telescopically, the comet of 1680 was initially seen on the morning of November 14 in the constellation of Leo. Moving quickly toward the Sun, it developed a 20- to 30-degree tail and, for observers in mid-northern latitudes, disappeared into the Sun's glare after the first week in December. After passing perihelion on December 18, the comet was a magnificent spectacle in the late December evenings, when the tail was reported to have reached 70 degrees in length. It remained visible to the naked eye until early February 1681, and with the aid of a telescope, Newton followed it until March 19, 1681. At this time, the English were still using the Julian calendar, so contemporary English observers reported their

95

Silver medal struck in 1681 commemorating the appearance of the comet of 1680–1681. On the obverse is an illustration of the comet on a stellar background with the inscription *Ao 1680 16 Dec–1681 Jan.* The reverse is a German inscription stating "the star threatens evil things: trust in God who will turn them to good." The capitalized letters in the German legend make a chronogram when arranged into roman numerals (i.e., MDCLVVVVVVI in Arabic numerals is 1681).

observations 10 days earlier than the dates given. To them, there was a morning comet in November and an evening comet in December, a designation that will be retained here. The November comet was a morning object with a relatively modest tail, while the one in December was a bright evening object with an enormous tail. Not only did they appear physically dissimilar, but their apparent motions, when taken together, could not be represented with any figure resembling a circle or a straight line.

Kirch and Dörffel

Gottfried Kirch (1639–1710) and Georg Samuel Dörffel (1643–1688) were contemporary German astronomers who owe much of their renown to the comet of 1680. Both studied under the polymath Erhard Weigel at Jena, Germany.

After an apprenticeship with Johannes Hevelius at Danzig, Kirch earned a living computing and publishing calendars and ephemerides. His wife, Maria Margarethe Winkelmann, and two of his children, Christfried and Christine, were also active astronomers. Despite the fact that he had 14 children—or perhaps because of it—Kirch found time to systematically search the heavens with a telescope. On November 14, 1680 at Coburg, after rising early to make observations of the Moon and Mars, he discovered the comet of 1680, the first such discovery to be made with the aid of a telescope. His observations were published in a 1681 Nürnberg tract.

After studying under Weigel in Jena, Dörffel returned to Plauen, where he succeeded his father as minister in the local Lutheran church. He was married three times and had nine children. Apparently Dörffel had no telescope; most of his observations were made with a wooden radius, a device for measuring angles. In 1672, Dörffel began publishing pamphlets of his cometary observations. They were written in a coarse Gothic type and often appeared anonymously or with just the initials M.G.S.D., Master Georg Samuel Dörffel. Distributed only locally, they did not appear to have a significant contemporary impact; however—justified or not—his work has been cited as a forerunner of Newton's theory on comets.

Dörffel's original 1672 tract was entitled *Warhafftiger Bericht von dem Cometen* (True Report of the Comet). An abbreviated work was published the same year. Dörffel noted the position of the comet of 1672 with respect to fixed stars and found no parallax. He then depicted its apparent path as circular and that it moved in the same direction as the planets. Tracts by Dörffel were also published after the appearances of the comets of 1677 and 1682 (Halley), but his fame rests on the 1681 tract that presented his observations and theories on the comet of 1680. Upon receiving news of the comet from Jena and Leipzig, Dörffel observed it on three November nights, before it entered the Sun's glare. In 1680, he rushed into print a short report of these observations entitled *Neuer Comet-Stern . . .* (New Comet-Star).

Dörffel anticipated the comet's reappearance after perihelion, and when it did reemerge from the Sun's glare, he followed it with the naked eye until it faded from view in February 1681. His detailed observations and analysis of the comet's motion were then published in the form of five questions and answers in *Astronomische Betrachtung des Grossen Cometen . . .* (Astronomical Observation of the Great Comet). Dörffel began by establishing the identity of the November and December comet, then described its apparent path through the constellations. To his own question as to how far the comet was from the Earth, he responded that he could not say without more refined parallax measurements. In response to another question, he assured his readers that although he had used the Copernican system to represent the observations, he considered it only a device or useful construction and that the motions of comets could not be used to prove the Copernican system. Besides, he added, this system was contrary to Scripture. The essence of Dörffel's work was given in response to the second question concerning the shape of the comet's orbit.

Dörffel's study of the comet's orbital shape showed the influence of Hevelius' ideas as given in his *Cometographia*. Although he considered comets to be created directly by God, rather than by the planetary emanations that Hevelius suggested, Dörffel did subscribe to Hevelius' disk-comet idea. He believed their observed variable speeds and latitudes ruled out rectilinear paths and he favored the curved trajectories suggested by Hevelius. Dörffel drew a circle to represent the Earth's hypothetical orbit about the Sun. At four-day intervals, he marked off points (a, b, c) corresponding to the Earth's position while the comet was visible. He then drew lines of sight to its direction using angles between it and the Sun, as observed from Earth. (α, β, γ). The comet's actual orbital positions were fixed along the lines of sight by trial and error, choosing them (A, B, C) such that its speed increased regularly toward perihelion, then decreased regularly after perihelion. The true orbit was finally drawn by taking into account its observed latitude at selected dates. Dörffel recognized the resulting curve as a parabola, with the

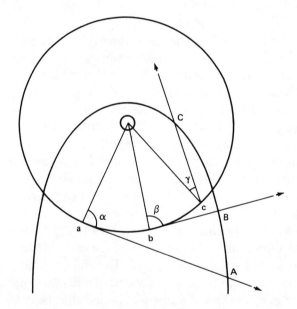

A schematic drawing showing Georg Dörffel's technique for correctly determining the parabolic motion of the comet of 1680. The points a, b, c correspond to the Earth's position when observations of the comet were made. By using the observed Sun-Earth-comet angles α, β, γ), its actual positions were fixed along the lines of sight by trial and error, choosing the positions A, B, C such that the comet's speed between AB and BC, etc., increased regularly toward perihelion and decreased regularly thereafter.

Sun at the focus, and asked whether this property might not be true of comet orbits in general.

Apart from the suggestion made in Hevelius' *Prodromus cometicus*, which was later withdrawn, Dörffel rightfully claimed priority for the idea of parabolic cometary motion with the Sun at the focus. Although his observations were crude and his orbit represented the comet's true path in shape only, his conclusions were based on his observations, and few preconceived notions entered into his discussion. Considering his observing equipment, Dörffel's efforts were quite extraordinary. However, only by gross exaggeration could his work be considered a forerunner to Isaac Newton's.

Newton and Flamsteed

Though Isaac Newton eventually realized that comets travel in highly eccentric orbits, his initial calculations were based on rectilinear paths. At first, he believed the comets seen in November and December 1680 were two separate objects, each traveling on a nearly rectilinear trajectory.

In contrast to Newton, John Flamsteed did not accept cometary rectilinear motion and predicted that November's comet would reappear after solar conjunction. Due to poor weather, Flamsteed did not observe the comet in November so his prediction was based almost entirely on his belief in comets as permanent bodies traveling on closed orbits. After it reappeared in early December, Flamsteed outlined his theory in letters to Edmond Halley, Isaac Newton, and James Crompton (1648–1694), a fellow of Jesus College, Cambridge. In a letter to Crompton dated February 12, 1681, Flamsteed noted that the comet's diffuse head might be a liquid that reflected the Sun's light less strongly than the solid nucleus. Five days later he pointed out to Halley that the comet's orbit was a curve turning not around the Sun, but before it.

Flamsteed's theory combined solar magnetic attraction and repulsion with the effect of the Cartesian solar vortex. The comet was drawn away from a straight line, AA′, by the combined effects of solar attraction and the rotation of the solar vortex. Upon reaching perihelion, P, the solar magnetic force repelled the comet and, together with the vortex motion, sent it to point B. Flamsteed likened the solar repulsion to that of a lodestone acting on a compass needle, its north pole attracting one end of the needle while repelling the other. His busy schedule prevented him from making any quantitative calculations so Flamsteed made do with some drawings on a large sheet of paper.

Also in his letter to Halley, Flamsteed outlined his ideas on the physical nature of comets. Repeating the Cartesian idea that comets were dead plan-

The Other Side of Isaac Newton

. . . dabbled with astrology and perpetual-motion machines, developed patent medicine recipes, strong interest in biblical chronology and alchemy . . .

If you were asked which famous historical figure had these interests, Sir Isaac Newton would probably not spring immediately to mind. Yet Newton's work included all of them, ranging far beyond the fields of physics and mathematics for which he was most honored.

Isaac Newton is given credit for a vast amount of original work, including the nature of colors, the particle theory of light, universal gravitation, the reflecting telescope, and calculus. These efforts are well known and documented, but his interests in other areas are not so well known. In about 1663, Newton briefly investigated judicial astrology and—sometime later—perpetual

Isaac Newton in his sixty-fourth year. Portrait by William Gandy, 1706. (*Courtesy of David Eugene Smith collection, Rare Book and Manuscript Library, Columbia University.*)

motion machines. Theology and biblical chronology occupied more of his time than physics and mathematics combined. He studied the Bible to determine and date authorship of several books in the Old Testament and discussed the prophecies given in the New Testament. His recipe for the patent medicine *Lucatello's Balsome,* which he considered effective for measles, smallpox, and the plague, contained turpentine, rosewater, beeswax, olive oil, and sack, flavored with a bit of sandalwood and a pinch of St. John's wort. The idea was to take it warm with broth—then sweat. For colic, burns, bruises and dog bites, one applied it externally.

Newton's active interest in alchemy began in approximately 1669 and extended over many years. When he died, fully one-fourth of his library consisted of alchemical works. Samples of Newton's hair tested for their content of heavy metals from some 30 years after his most active experiments in alchemy showed concentrations many times larger than normal twentieth century hair. One sample showed a mercury concentration 40 times the norm. Between liberal doses of his patent medicine and his zealous pursuit of the alchemical elixir, Newton very nearly poisoned himself. He was fortunate to have lived for 84 years.

ets, formerly from another vortex, Flamsteed thought comets differed little from the obscure large spots on the Moon, which he considered to be aqueous. The coma was due to the humid head vapors, and the tail represented the violent action of the Sun's rays carrying the stream to vast distances. As for tail curvature, he likened the effect to smoke being emitted from a chimney on a moving ship or steam falling on a moving hot iron.

In response to Flamsteed's letter to Halley, which had been passed on to him, Newton began to think seriously about the motion of this comet. In a letter to Flamsteed dated February 28, 1681, Newton wrote that he still felt that the November and December comets were two different objects and offered constructive criticisms for Flamsteed's one-comet theory. Newton objected to Flamsteed's cometary path turning before the Sun and pointed out that the comet runs contrary to the vortex motion there. In addition, the Sun cannot be magnetic because of its extreme heat. A red-hot lodestone loses magnetic properties, and even if the Sun were magnetic, the comet would never be repelled because it would simply turn its pole direction, like a small magnet floating on a walnut shell is attracted by a larger magnet nearby. Using observations on hand, Newton pointed out that if the comets of November and December were one, its motion would have accelerated and decelerated three separate times. Here Newton was misled by confused data. Flamsteed sent Newton observations made in November by the Frenchman

John Flamsteed (1646–1719).

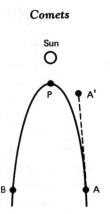

A schematic drawing showing John Flamsteed's first attempt at representing the motion of the comet of 1680. The comet is drawn away from a straight line, AA', by the combined effects of solar magnetic attraction and the rotation of the solar vortex. Upon reaching perihelion, P, the solar magnetic attraction turns to a repulsive force and repels the comet to point B. Note that the comet passes in front of, not around, the Sun.

Father Jean Charles Gallet (1637–1713). However, before passing them on, Flamsteed converted only four of the six observations from the Gregorian to the Julian calendar, which was still used in England. Hence Newton assumed there was a 12-day gap between the fourth and fifth observation times, when there was only a two-day gap.

In a March 7, 1681 response to Newton's letter, Flamsteed straightened out the confusing November observations and defended his cometary ideas against Newton's criticism. He argued that the Sun's magnetic attraction may be unlike that of a lodestone and hence unaffected by the heat. The comet could be repelled at perihelion if it were thrown violently past the Sun without the chance to reverse pole directions. Flamsteed seemed to agree with Newton's suggestion that the comet should have passed behind the Sun, rather than before it, but wondered how the extra distance could have been traveled so quickly. In his letter to Newton, Flamsteed appended a drawing of the comet's orbit showing it passing around the Sun in what appeared to be a parabolic curve.

With the corrected November observations in hand, Newton wrote a letter to Crompton dated April 1681 stating that the comets of November and December were less irreconcilable. His faith in the two-comet theory was shaken, but he still criticized Flamsteed's one-comet theory. He again argued that a hot Sun would lose its magnetic properties, and even if the comet were attracted then repelled by the Sun, it would have been continuously accelerated around the Sun and should have receded faster than it approached. This, of course, was contrary to the observed motion.

In an April 1681 letter to Flamsteed, Newton was still not convinced by Flamsteed's notion of a single comet. He thought it strange that many other comets, such as those of 1665 and 1677, had not been seen passing around the Sun and exiting the solar neighborhood in nearly the same direction from which they entered. Comets that were seen on either side of perihelion—like those of 1472, 1556, 1580, and 1664—had their incoming path in one part of the heavens and the outgoing path in another.

Flamsteed's three arguments for attributing the November and December apparitions to the same comet were all based on its apparent motion as seen from the Earth. The first argument noted its similar rate of motion on either side of perigee. Flamsteed next argued that the November and December comets each reached solar conjunction at the same time, and finally that the apparent northward motion of the November comet moved it toward the place where the December comet emerged from the Sun's glare. The principal significance of Flamsteed's one-comet explanation was not that it was a qualitatively correct scenario for the comet's motion, but rather that his ideas started Newton thinking about a quantitative solution to the problem.

By mid-1681, after corresponding with Flamsteed, Newton's insistence on two comets in November and December was not so absolute. Soon after he last saw the comet on March 19, 1681, Newton began to collect his thoughts, and by mid-1684 he was convinced that comets do travel on closed elliptical orbits. He set out to determine a method whereby these orbits could be computed, describing the problem as one of great difficulty and managing a successful solution just prior to the publication of his *Principia* in 1687. In the last of three books, Newton outlined his method for determining a parabolic cometary orbit using three observations that were nearly evenly spaced in time, then used the comet of 1680 as an example. Although his published technique was a trial-and-error, semigraphical solution, his procedure would allow as complete and exact an analytic solution as the observations would permit.

In 1925, A. N. Kriloff pointed out that it would have been impossible for Newton to achieve the accuracy he did using the technique in the *Principia*. Newton's trial-and-error construction, with compass and ruler, was drawn on a scale of 16.33 inches for the Sun-Earth distance, and his result was accurate to 0.0017 inch! Kriloff suggested that Newton either used a supplementary diagram of much larger scale or guided his construction by calculations he didn't mention. Newton may have first used his newly invented calculus to work the problem rigorously, then laid out the solution in his *Principia* using a simplified, semigraphical form. He also made remarks about how his method could be generalized to elliptic orbits if the comet's period could be established from the interval between returns to perihelion. He had previously shown in Book I that a parabolic path would result from an inverse square force law if the Sun was located at the focus and the object's velocity was at any time the square root of two multiplied by the velocity required to establish a circular orbit through the same point. For example, a comet located at the Earth's distance from the Sun and moving on a parabolic orbit would have an orbital velocity, 42 kilometers per second, equal to the Earth's circular velocity about the Sun, 30 kilometers per second, multiplied by the square root of two, 1.4.

The triumph of Newton's method for the unknown paths of comets was a significant step forward in the history of science. Although a few of his predecessors had correctly suggested qualitative descriptions of cometary orbits, it was Newton who developed a mathematical model for the comet's motion then successfully tried the numbers. Using his orbit determination technique on the comet of 1680, he successfully fit the observations of Flamsteed, Gallet, and others. Where his predecessors used analogies, preconceived ideas, and guesswork, Newton fit his dynamical theory of comets into the framework of his universal gravitational attraction. One of the most influential scientific works ever written, Newton's *Principia* laid a firm mathematical foundation for all subsequent work on the dynamics of the heavenly bodies.

In the same period that Newton developed his cometary orbit determination technique, he established his views on the physical nature of comets. Apparently prompted by Descartes' suggestion that comets were located beyond Saturn, Newton argued that they were actually seen in a region interior to Saturn's orbit. If comets were located beyond it, they would appear most frequently when their angular distances from the Sun were largest and their Earth-comet distances smallest; that is, they would be seen

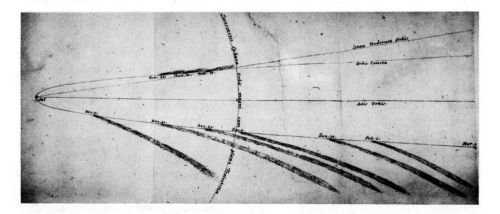

Engraving from Isaac Newton's *Principia* showing the parabolic path of the comet of 1680. (*Photograph courtesy of Owen Gingerich.*)

most frequently when on the opposite side of the Earth from the Sun, in opposition. However, the observed evidence showed four or five times as many comets in the hemisphere facing the Sun, Newton used this evidence to support his belief that the region where comets were observed was interior to Saturn's orbit. He envisioned a cometary nucleus that was a durable, solid, and compact object whose light was derived from the Sun. To support this latter notion, Newton pointed out that they appeared brightest, not at perigee, but at perihelion.

Concerning the visibility of cometary atmospheres, Newton dismissed the notion of sunlight passing through a diaphanous nucleus. Like Kepler, he argued that there was no matter in space to reflect the sunlight. Because no spectral colors were observable, he also dismissed Descartes' idea that the cometary head was visible because of the lenslike refraction of sunlight in the nucleus. Newton argued that the extreme tenuity of the cometary atmosphere and tail was apparent by analogy with the Earth's atmosphere. Only a few miles thick, it is dense enough to render the stars invisible when illuminated by sunlight, but sunlight illuminating the comet's atmosphere does not appreciably dim the background stars.

The least successful of Newton's works were his views on cometary tails. He acknowledged, and thought possible, Kepler's notion of tail particles carried along by the action of solar rays. This mechanism could operate in free space because the tenuous ether could be expected to yield to the action of the rays, whereas on Earth, this action was not apparent on more dense materials. Although he did not dismiss Kepler's tail formation theory, he offered one of his own. For the comet of 1680, he noted that the postperihelion tail was largest and suggested that it was a very fine vapor, which the head emitted after being heated near perihelion. Analogous to smoke rising in a chimney, cometary tail particles were heated by the Sun's rays, then—in turn—heated the surrounding ether. The subsequent rarefaction and diminished specific gravity of the ether caused it to ascend away from the Sun, carrying along the tail particles. To explain its curvature, Newton used Flamsteed's suggestion that it was due to the motion of the head from which the smoke ascends. The leading convex edge of the tail appeared brightest because the emitted tail particles were more dense there.

Physical observations had shown the atmosphere to shrink, not expand, as the comet approached perihelion. This was troublesome, and Newton suggested that as the comet approached perihelion, the intense heating of the nucleus caused a denser, blacker smoke to be apparent just above the heated surface. The observed shrinking of a cometary atmosphere as it approaches perihelion is currently thought to result from the increasing strength of solar radiation. The excited molecules responsible for light emission in a comet's atmosphere survive only until they are dissociated by solar radiation. As the Sun-comet distance decreases, these molecules encounter more and more intense solar radiation. Very close to the Sun, they dissociate too quickly to travel great distances from the nucleus. Thus, the visible coma can shrink as it approaches the Sun.

Newton's ideas on cometary tails were not well developed. One glaring inconsistency was his use of the ether for his tail theory; he had previously rejected it because there was no evidence of its resistive effect on the planetary and cometary motion. Newton may have grown tired of the topic or he may not have had enough time to better develop his ideas on cometary tails. There is some evidence to indicate that they were a last minute insertion in the *Principia.* After reading a first draft of Book III, Halley wrote to Newton in April 1687:

The Comet of 1680
and Its Fowl Deeds

The appearance of the comet of 1680 marked the zenith of cometary superstition. This credulity would be reversed somewhat after Isaac Newton's work on the comet of 1680; however, before his efforts could stem the tide of superstition, the comet of 1680 would witness a tidal wave of foolishness. Of the approximately 208 known broadsides on all comets, 62 refer to the comet of 1680. At least a dozen different medals were struck, and in Germany alone there were nearly 100 tracts published. Of these, only four were written to quiet superstitious fears. That these four skeptical authors were writing unpopular views is evident; three were anonymous and the other was signed only by the author's initials. Even the significant tracts by Gottfried Kirch and Georg Samuel Dörffel noted that while comets may be natural phenomena, they were nevertheless signs from God.

In Rome, even barnyard fowl responded to the comet's appearance. A letter to the prestigious Academy of Sciences in Paris announced that on the evening of December 2, 1680, a Roman chicken that had never laid an egg began to cluck in a loud and extraordinary fashion. She succeeded in laying an enormous egg with natural markings resembling a comet on a stellar background. A representative of the academy, after apologizing for taking notice of the occurrence, noted that the egg was not marked by a comet as many believed, but rather with several stars. In scientific circles, the comet egg was not taken seriously, but the very fact that the Academy of Sciences felt compelled to

I do not find that you have touched that notable appearance of Comet tayles, and their opposition to the Sunn; which seems rather to argue an efflux from the Sunn than a gravitation towards him. I doubt not that this may follow from your principles with the like ease as all other phenomena; but a proposition or two concerning these will add much to the beauty and perfection of your Theory of Comets.

In one final speculation in Book III of the *Principia*, Newton suggested that the tail vapors, continually rarefied and dispersed, would spread throughout the heavens and be gradually accreted by the planets, resupplying them with vital fluids spent on "vegetation and putrefaction." Thus, cometary vapors became essential for life on Earth.

comment on it implies that many took it as a portent. Comet eggs were also identified during comet Halley's return two years later and again in 1910. During the 1986 return, the hens of the world seemed devoid of cometary influences.

German broadside by Friedrich Madeweiss showing the path of the comet of 1680 through the constellations. Note the Roman comet egg in the lower right corner. (*Courtesy Adler Planetarium, Chicago.*)

. . . for all vegetables entirely derive their growths from fluids, and afterwards, in great measure, are turned into dry earth by putrefaction; and a sort of slime is always found to settle at the bottom of putrefied fluids; and hence it is that the bulk of the solid earth is continually increased; and the fluids, if they are not supplied from without, must be in a continual decrease, and quite fail at last. I suspect, moreover, that it is chiefly from the comets that spirit comes, which is indeed the smallest but the most subtle and useful part of our air, and so much required to sustain the life of all things with us.

Newton's 1687 speculation of earthly fluids being replenished by cometary encounters was without observational support, yet nearly three centuries later, John Oró would speculate that cometary encounters may have, in fact, provided the building blocks of life by depositing a layer of organic molecules and water on the early Earth (see Chapter 11).

Summary

From the first telescopic discovery by Gottfried Kirch on November 14, 1680, to Isaac Newton's final telescopic observation on March 19, 1681, the comet of 1680 ushered in a new era of cometary thought. First Georg Dörffel used his own observations to correctly determine that the shape of its orbit was a parabola with the Sun at the focus. Although Dörffel correctly suggested the shape of the comet's orbit, he could not use his theory to accurately define its true orbital path. His work represented the best pre-Newtonian cometary ideas; using his somewhat crude observations and a trial-and-error technique, he achieved a qualitatively correct solution. Although Dörffel's work preceded Newton's by six years, it was unknown to Newton.

John Flamsteed, believing that comets were permanent bodies, supposed that the comets seen in November and December of 1680 were one and the same object moving on a greatly curved trajectory. He suggested that the comet seen in November 1680 would reappear after perihelion. When his prediction was borne out by the December observations, Flamsteed gloried in his successful theory. Apart from the correct connection of the November and December comets, his ideas were largely untenable; however, his unyielding position forced Newton to seriously consider cometary motion.

Using the comet of 1680 as an example, Newton finally developed a technique for determining the parabolic orbit of a comet from three observations. Though his successful solution was the inspired work of just one man in a rather short period of time, Newton built upon the wisdom of his predecessors. In an oft-quoted letter to Robert Hooke dated February 5, 1676, Newton stated "if I have seen further, it is by standing on the shoulders of giants." With respect to Newton's cometary theory, the giants included

Johannes Kepler and his three laws of planetary motion; Johannes Hevelius, who first suggested parabolic motion with the Sun at the focus; Robert Hooke, who touched upon universal gravitational attraction; and John Flamsteed, who not only provided the most accurate contemporary observations of the November and December comet, but also argued for a single object.

Although the physical nature of comets was still largely unknown, Newton had essentially solved the problem of their dynamical behavior. Subsequent research would improve and refine his theory of cometary motion, but all subsequent work would be built on the foundation he laid. After he had successfully tested his theory using the comet of 1680, the technique was available for additional comets and in a prophetic letter to Newton, Halley wrote in April 1687:

> Your method of determining the Orb of a Comet deserves to be practised upon more of them, as far as may ascertain whether any of those that have passed in former times, may have returned again.

Taking up his own suggestion, Edmond Halley would use Newton's method to discover the first periodic comet. Fittingly, it would bear his name thereafter.

6
The Return of Comet Halley

Using Isaac Newton's method, Edmond Halley computes the orbits for 24 comets, finding those of 1531, 1607, and 1682 very similar. Halley predicts his comet's return in late 1758 or early 1759. Astronomers race to recover comet Halley. Clairaut refines Halley's prediction. Palitzsch, a German amateur astronomer, finds Halley's comet Christmas evening 1758—several weeks before the French professional astronomer, Charles Messier.

The Predicted Return of Comet Halley

In 1687, Edmond Halley wrote to Isaac Newton suggesting that Newton's orbit determination technique be tried on comets other than that of 1680. It was not until eight years later, however, that Halley heeded his own suggestion and determined an orbit for another comet. In a letter to Newton dated September 7, 1695, Halley noted that a parabolic curve most closely represented his own and John Flamsteed's observations of the comet of 1683, and that the agreement between these observations and the places predicted by Newton's method was always within one arc minute. Three weeks later in another letter to Newton, Halley mentioned his reexamination of the orbit for the comet of 1680 and indicated that Flamsteed's observations might better be represented by an elliptic orbit, rather than a parabolic one. Halley also requested that Newton procure Flamsteed's observations of the comet of 1682, and in a rather prophetic statement, Halley wrote:

> I must entreat you to procure for me of Mr. Flamsteed what he has observed of the Comett of 1682 particularly in the month of September, for I am more and

111

more confirmed that we have seen that Comett now three times, since ye Yeare 1531, . . .

By 1695, Halley clearly suspected the periodicity of the comet that was to bear his name. He had tentatively identified, as the same object, the comets of 1531, 1607, and 1682. He was troubled, however, by the two unequal periods between these three apparitions. In another letter in early October 1695, Halley thanked Newton for procuring Flamsteed's observations and requested that he think about the perturbative effects of Saturn and Jupiter on a comet's period. Halley also presented his completed work on the comet of 1680 and again noted that the observations might be made more compatible with the theory if an elliptic orbit was assumed. On October 15, 1695, Halley wrote to Newton stating that he had finished an orbit for the comet of 1664 and suggested that Johannes Hevelius "to help his calculations to agree with the hevens, added 8 or 9 minutes to the places observed, on the 4th, 5th and 8th of December . . . ." After observing that Jean-Dominique Cassini's geocentric theory could not represent the observations to within two degrees, Halley mentioned that his next orbit determination would be for the comet of 1682.

In a letter to Newton dated October 21, 1695, Halley wrote that he had almost finished his computations for the comet of 1682 and reasserted his

German broadside of comet of 1607 (Halley). The German title reads, "About the comet or tailed new star which made itself seen in September of this year 1607." The approximate path of the comet is traced on the constellations from September 16 through October 4 (Julian calendar). (*Courtesy of the Bibliothèque Nationale, Paris.*)

German broadside showing comets of 1680, 1682 (Halley), and 1683. The illustration shows a view of Augsburg, Germany with the comets of 1680, 1682, and 1683 in the sky. Three horsemen of the Apocalypse are in the foreground. The scene is bordered by a clock face, the numerals of which are made of bones, weapons, and instruments of torture. Each of the four corners outside the dial contains an allegorical figure with an appropriate biblical text. (*Courtesy, Adler Planetarium, Chicago.*)

suspicion that those of 1607 and 1682 were one and the same. Since he was becoming rather adept at orbit determination, Halley wrote that he planned to compute orbits for all well-observed comets. By 1696, he was convinced that the comets of 1607 and 1682 were one and according to an entry in the *Journal Book of the Royal Society of London* dated June 3, 1696, Halley demonstrated to the society that these comets' orbits were similar. In another presentation to the society a month later, Halley gave parabolic orbital elements for the comet of 1618. This orbit successfully represented most of the observations and correctly brought the comet inside Mercury's orbit at perihelion.

Shortly after reading his results on the comet of 1618 to the Royal Society, Halley left London to take up his duties at Chester as deputy comptroller of the English mint. Between 1698 and 1700, Halley was commissioned as a naval captain and sailed his ship, the *Paramore Pink*, across the Atlantic

113

Ocean twice, charting the variations, or *declinations*, between magnetic north measured with a compass and true north determined from celestial positions. He hoped that once these local magnetic variations were charted at various longitudes, future sailors could use the charts and their own variation measurements to determine longitude at sea. Unfortunately, magnetic variation measurements do not show systematic changes with longitude, nor

Edmond Halley, Renaissance Man

There are widely divergent opinions as to Edmond Halley's place in history. The nineteenth century mathematician and biographer Augustus De-Morgan noted:

> The period during which he (Halley) held the post of Astronomer Royal, compared with those of his predecessor Flamsteed, and his successor Bradley, is hardly entitled, if we look at its effect upon the progress of science, to be called more than strong twilight night between two bright summer days.

Portrait of Edmond Halley* at about age 30 when he was seeing Newton's *Principia* through press. The inscription above the portrait was added later. (*Reproduced by permission of the president and council of the Royal Society.*)

DeMorgan's unfair judgment of Halley is balanced by the equally unfair assessment by his two surviving daughters, who had the following inscription placed on their father's tombstone at the churchyard of St. Margaret, Lee near London:

> Under this marble, together with his beloved wife, rest Edmund Halley, LL.D. unquestionably the greatest astronomer of his age.

Born on November 8, 1656, Edmond Halley's early interest in science was encouraged by his father, a prosperous London soapmaker and salter. At the age of 20, Halley published his first scientific paper, dropped out of Oxford, sailed south to the island of St. Helena, and over 13 months made the observations that allowed him to compile the first systematic catalog of stars in the southern hemisphere. His *Catalogue of the Southern Stars* was useful for navigation at sea and was partly responsible for Halley being awarded a Master of Arts degree at Oxford without having to take the usual examinations.

*Although Halley's first name is often given as Edmund, he always wrote it as Edmond.

do they remain constant with time. While Halley's voyages failed to solve the problem of determining longitude at sea, he did publish the first map showing regions of equal magnetic variations. He also charted the positions of coasts and ports encountered during his voyages. In 1701, he made a third voyage to undertake a detailed study of the English Channel, which resulted in the first published tidal chart.

Edmond Halley's genius was diverse. His ideas ranged from the design of an underwater diving bell to the possible infinity of space. During his 85-year lifetime, Halley served as the first corresponding secretary to the Royal Society of London and published the scientific works of its members. Halley himself published significant works in the fields of meteorology, mathematics, and navigation, as well as astronomy. He virtually founded the sciences of geomagnetism and physical oceanography and initiated work in life insurance statistics by publishing the first tables of life expectancies. He was unselfish in recognizing the genius of others, including Isaac Newton. Halley edited Newton's classic work, the *Principia* and, when the Royal Society could not afford to publish it, provided the necessary funds from his own pocket.

In computing the orbit of the comet of 1682 that would later bear his name, Halley used a modified version of Newton's method and the observations of John Flamsteed, wisely ignoring his own less accurate observations. Halley's work was not equal to Newton's profound accomplishments in mathematics, nor did he possess the precise observing techniques of Flamsteed. He was what we might today call an "idea man." His intellect was too lively for him to focus on a single issue for very long, yet he made remarkable contributions. He was the first to recognize that stars have their own motions with respect to each other, and he correctly suggested that some stars appear to vary in intensity because they change their light output over time. He pointed out that observations of the planet Venus during its occasional transits in front of the Sun would allow the distance between the Sun and Earth to be determined; his technique later served as the basis for observing the Venus transits in 1761 and 1769.

Halley's wide range of interests and warm, gregarious manner contrasted sharply with his contemporary colleagues, Isaac Newton and John Flamsteed. One can imagine that Newton was at his best when left alone in his study to focus his prodigious intellect on mathematical constructs. The stern Flamsteed was at his best in the Greenwich observatory making precise and careful position measurements of celestial bodies. No doubt Halley was at his best discussing new scientific ideas with friends at a favorite coffee house.

Two pages from one of Halley's notebooks are shown in this illustration. Edmond Halley recorded his observations of the comet that was to bear his name in the late summer of 1682. On the left-hand page Halley has recorded, in Latin, cometary position measurements for September 4, 1682, measurements that were made with respect to neighboring stars. The right-hand page details, in English, some unrelated notes on the properties of a parabola. (*Courtesy of the Royal Greenwich Observatory.*)

Halley's three voyages are regarded as the first sea journeys undertaken for purely scientific purposes. In 1702 and 1703 Queen Anne sent Halley on diplomatic missions to Europe to advise Emperor Leopold of Austria about the fortifications of seaports on the northern shores of the Adriatic. With Halley's busy schedule, he did not have enough time to bring together and publish his work on cometary orbits. We learn from a letter than Flamsteed wrote on May 30, 1702, to the astronomer Abraham Sharp that Halley had a treatise on comets finished by mid-1702, but this work was not published until 1705.

Halley's famous treatise on comets was first published in Latin as a six-page folio pamphlet. The first edition was printed at Oxford, probably in a very limited quantity as less than a half dozen copies are now extant. Later in the same year, an English translation was published in London and a longer and slightly modified Latin version appeared in the *Philosophical Transactions of the Royal Society.*

	MOTUUM COMETARUM IN ORBE PARABOLICO ELEMENTA ASTRONOMICA.								
Cometæ Anni.	Nodus Afcend. (° ' ")	Inclin. Orbitæ. (° ' ")	Perihelium. (° ' ")	Diftantia Perihelia à Sole.	Logarithmus Diftantiæ Perihelia à Sole.	Logarithmus Medii Motus Diurni.	Temp. Æquat. Perihel. (D. H. ')	Perihel. à Nodo. (° ' ")	
1337	♊ 24 21 0	32 11 0	♉ 7 59 0	40666	9 609236	0 546274	Junii 2 6 25	46 22 0	Retrog.
1472	♑ 11 46 20	5 20 0	♉ 15 33 30	54273	9 734584	0 358252	Feb. 28 22 23	123 47 10	Retrog.
1531	♉ 19 25 0	17 56 0	♒ 1 39 0	56700	9 753583	0 329754	Aug. 24 21 18	107 46 0	Retrog.
1532	♊ 20 27 0	32 36 0	♋ 21 7 0	50910	9 706803	0 399924	Oct. 19 22 12	30 40 0	Direct.
1556	♍ 25 42 0	32 6 30	♑ 8 50 0	46390	9 666424	0 460492	Apr. 21 20 3	103 8 0	Direct.
1577	♈ 25 52 0	74 32 45	♌ 9 22 0	18342	9 263447	1 064958	Oct. 26 18 45	103 30 0	Retrog.
1580	♈ 18 57 20	64 40 0	♋ 19 5 50	59628	9 775450	0 296953	Nov. 28 15 0	90 8 30	Direct.
1585	♉ 7 42 30	6 4 0	♈ 8 51 0	109358	0 038850	9 901853	Sept. 27 19 20	28 51 30	Direct.
1590	♍ 15 30 40	29 40 40	♏ 6 54 30	57661	9 760882	0 318805	Jan. 29 3 45	51 23 50	Retrog.
1596	♒ 12 12 30	55 12 0	♏ 18 16 0	51293	9 710058	0 395041	Julii 31 19 55	83 56 30	Retrog.
1607	♉ 20 21 0	17 2 0	♒ 2 16 0	58690	9 768490	0 307393	Oct. 16 3 50	108 5 0	Retrog.
1618	♊ 16 1 0	37 34 0	♈ 2 14 0	37975	9 579498	0 590881	Oct. 29 12 23	73 47 0	Direct.
1652	♊ 28 10 0	79 28 0	♈ 28 18 40	84750	9 928140	0 067918	Nov. 2 15 40	59 51 20	Direct.
1661	♊ 22 30 30	32 35 50	♋ 25 58 40	44851	9 651772	0 482470	Jan. 16 23 41	33 28 10	Direct.
1664	♊ 21 14 0	21 18 30	♌ 10 41 25	102575½	0 011044	9 943562	Nov. 24 11 52	49 27 25	Retrog.
1665	♏ 18 2 0	76 5 0	♊ 11 54 30	10649	9 027309	1 419164	Apr. 14 5 15	156 7 30	Retrog.
1672	♑ 27 30 30	83 22 10	♉ 16 59 30	69739	9 843476	0 194914	Feb. 20 8 37	109 29 0	Direct.
1677	♑ 26 49 10	79 3 15	♌ 17 37 5	28059	9 448072	0 788020	Apr. 26 0 37	99 12 5	Retrog.
1680	♑ 2 2 0	60 56 0	♐ 22 39 30	00612½	7 787106	3 279469	Dec. 8 0 6	9 22 30	Direct.
1682	♉ 21 16 30	17 56 0	♒ 2 52 45	58328	9 765877	0 311313	Sept. 4 7 39	108 23 45	Retrog.
1683	♍ 23 23 0	83 11 0	♊ 25 29 30	56020	9 748343	0 337614	Julii 3 2 50	87 53 30	Retrog.
1684	♐ 28 15 0	65 48 40	♏ 28 52 0	96015	9 982339	9 986620	Maii 29 10 16	29 23 0	Direct.
1686	♓ 20 34 40	31 21 40	♊ 17 0 30	32500	9 511883	0 692304	Sept. 6 14 33	86 25 50	Direct.
1698	♐ 27 44 15	11 46 0	♑ 0 51 15	69129	9 839660	0 200638	Oct. 8 16 57	3 7 0	Retrog.

Edmond Halley's table of 24 comets with parabolic orbits as represented in the 1752 edition of his *Astronomical Tables*. Note that the comets of 1531, 1607, and 1682 have similar orbital elements. Halley used this similarity to correctly suggest that they were one and the same object.

For such an important work, Halley's "Synopsis of the Astronomy of Comets" was rather brief. Much of the important information is contained in one table giving parabolic orbital elements for 24 comets observed from 1337 through 1698. According to Halley, the table was framed as a result of a "prodigious deal of calculation." It was obvious that the orbits showed no preferred inclination or orientation, and the differences in the perihelion distances made him suspect that many more of them moved unseen in regions more remote from the Sun. Halley also mentioned that none of the 24 comets had hyperbolic orbits. Though he used parabolas in his orbit determination work, Halley believed that the true paths of comets were very eccentric ellipses. After computing a parabolic orbit for a particular comet, he compared its predicted and observed positions at the same time, formed the differences, or *residuals*, between the two angles, then noted whether a slightly different orbital eccentricity might reduce these residuals. In this manner, he showed that the 24 computed orbits could not be hyperbolic, hence each comet must be bound to the Sun.

Although the list of cometary apparitions for which orbital elements are available has grown from Halley's 24 to more than 50 times that, his astute observations are still valid. The most famous conclusion of his treatise of 1705 was that the comets seen in 1531, 1607, and 1682 were the same object. Halley noted that their orbital elements were similar, except for unequal periods between their perihelion passages. The respective periods were over 76 years between 1531 and 1607, but just under 75 years between 1607 and 1682. He suggested that Jupiter's gravitational pulls, or perturbations, on the comet were responsible for the unequal periods, adding that the comet of 1456 was seen passing retrograde between Earth and the Sun:

> tho' no Body made Observations upon it, yet from its Period, and the Manner of its Transit, I cannot think different from those I have just now mention'd. Hence I dare venture to foretell, That it will return again in the Year 1758. And, if it should then return, we shall have no Reason to doubt but the rest must return too.

Using an orbit determination technique developed from Newtonian dynamics, Halley was the first to correctly predict the return of a comet. He was rewarded by having this most famous comet named after him.[1] In his

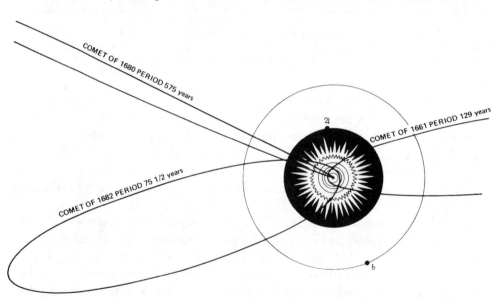

Diagram showing the three periodic comets suggested by Edmond Halley. The comet of 1682 is Halley's comet and its return was predicted by his comparing the similar orbital elements for the comets of 1531, 1607, and 1682. The possible periodicity for the comet of 1661 was noted in Halley's 1705 work on comets and that of the comet of 1680 was first mentioned in the second edition, 1713, of Isaac Newton's *Principia*. The periodic nature of comets 1661 and 1680 was suggested only because of similar time periods between returns; neither one is actually periodic.

1705 treatise, Halley also suggested a possible identity for the comets of 1532 and 1661 but considered Peter Apian's observations of the 1532 comet, which were the only ones available to him, too crude to allow certainty. Inadvertently, Halley also sowed a seed for the post-Newtonian fear of comet-Earth collisions in his 1705 work. Discussing the comet of 1680, he noted—incorrectly—that on November 11 (November 21 in the Gregorian calendar), at 1:06 P.M., this comet passed less than one solar radius above the Earth's orbit.

> This is spoken to Astronomers: But what might be the Consequences of so near an Appulse; or of a Contact; or, lastly, of a Shock of the Coelestial Bodies, (which is by no means impossible to come to pass) I leave to be discuss'd by the Studious of Physical Matters.

Unfortunately cometary collisions were discussed by the not-so-studious as well. Newton's successor in the Lucasian professorship at Cambridge, William Whiston (1667–1752), would later use the comet of 1680 to explain the Biblical flood and predict the end of the world (see Chapter 7).

Although Halley's 1705 treatise predicted the 1758 return of his comet, it was not until his posthumous *Astronomical Tables* that it became generally known that he had revised his prediction to late 1758 or early 1759. Halley's *Tables* contained the necessary information for computing the positions of the Sun, Moon, planets, and comets. Although first published in Latin in 1749, English precepts were added in a 1752 edition. Most of the book, including a revised edition of his "Synopsis of the Astronomy of Comets," was completed in 1717 and printed in 1719. In 1720, Halley succeeded John Flamsteed as Astronomer Royal and set aside the publishing of *Tables* until he could check the lunar tables with his own observations.

Much of his revised "Synopsis of the Astronomy of Comets" was identical with the 1705 edition, but with important exceptions. Halley worked out an extension to his orbit determination procedure to include elliptic orbits and applied this new technique to the comets of 1607 and 1682. He assumed an average period P of 75.5 years for the comet of 1682 and 76 years for that of 1607, then used Kepler's third law ($a^3 = P^2$) to determine each comet's semimajor axis, a. The eccentricity was then determined from the properties of an ellipse, whereby the known perihelion distance q is related to the semimajor axis a and the eccentricity e by $q = a(1 - e)$.

For the first time, Halley estimated the perturbative effect of Jupiter on the comet's orbit. He correctly gave it as the primary cause of the increase, with time, of the comet's longitude of the ascending node. More importantly, Halley recognized the effect of Jupiter on the comet's orbital period. In preparing *Astronomical Tables* for publication, he became aware of the sizable ef-

119

Defining the Orbit of a Comet

As the Earth circles the Sun once each year, it does so in a plane called the *ecliptic*. To a rough approximation, other planets in the solar system also move about the Sun in this ecliptic plane. On or about March 21 each year, the Sun appears in the same region of the heavens, and the length of day and night are equal everywhere on Earth. At this time the Sun is said to have reached the *vernal equinox* and Spring begins for the northern hemisphere. The ecliptic plane and the vernal equinox provide convenient references for describing the paths of comets.

Unlike planets, most comets do not confine their paths to the ecliptic plane, nor do they travel in the same direction around the Sun. The Sun's position at one focus of the ellipse is denoted by S, and the point on its orbit where a comet can come closest, perihelion, is P. The comet's distance from the Sun reaches a maximum at aphelion, denoted as A. The direction of the vernal equinox is denoted by the sign of the ram's horns (γ) because this point was once the beginning of the constellation of the ram, Aries. Only three angles are required to completely describe the orientation of the cometary orbit with respect to the ecliptic plane and the vernal equinox. The longitude of the *ascending node*, Ω, is an angular measure of the distance between the vernal equinox and the point (ascending node) where the comet crosses the ecliptic plane going from south to north. The opposite point, where the comet crosses the ecliptic plane going south, is called the *descending node*. At these nodal crossing points the comet is briefly in the plane of the Earth's orbit. If its path passes near enough to Earth's orbit, a meteor shower may result as the Earth runs smack into cometary debris. The two other angles used to orient the comet's orbit are the argu-

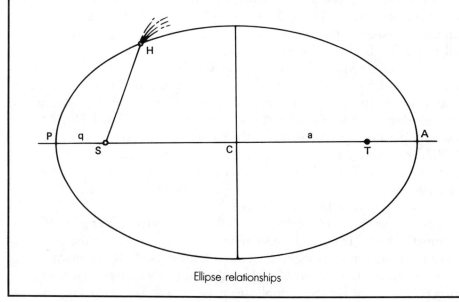

Ellipse relationships

ment of perihelion, ω, the angular separation between the ascending node and perihelion, and the orbital inclination, i. If the inclination is less than 90 degrees, the comet's motion is called *direct*. If the inclination is greater than 90 degrees, the comet's motion is termed *retrograde,* since it would then move in opposition to the eastward motion of the planets about the Sun.

The size and shape of an elliptical orbit are described by the eccentricity and semimajor axis. The eccentricity is given by the ratio of the distance CS to CP. Since the Sun is located at S, the distance CS = 0 for a circular orbit. The eccentricity is also zero in this case. As the eccentricity of an orbit increases, the elliptical path becomes more and more elongated. A parabolic orbit has an eccentricity of one, and a hyperbolic orbit greater than one. Comets moving on either parabolic or hyperbolic orbits are not periodic; they will never return to the Sun. The major axis of an ellipse is the distance AP and, not surprisingly, the semimajor axis a is one-half this distance. This distance is usually expressed in terms of an astronomical unit, AU, the mean distance between the Sun and Earth, approximately 150 million kilometers, or 93 million miles. The perihelion distance in AU, q, between the comet at perihelion and the Sun, distance SP, is given by the expression $q = a(1 - e)$.

The final piece of information that allows a comet's motion to be described is the time, T, that it passes the perihelion point at P. The comet's position in its orbit is often specified using the true anomaly, ν, the Sun-centered angle between the comet and the perihelion point, angle HSP. For any comet moving about the Sun, its orbit can be defined uniquely by the six orbital elements, T, e, q, ω, Ω, and i. However, the introduction of a perturbing planet or planets causes slight changes in these orbital elements with time. Thus when the perturbing effects of the planets are taken into account during orbit determination computations, the six orbital elements are strictly valid only for a particular instant of time, or *epoch.*

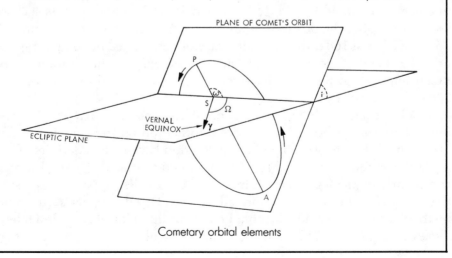

Cometary orbital elements

fect that Jupiter's perturbations had on Saturn's period; by neglecting these perturbations, Saturn's period could be in error by as much as a month. Halley wondered, How much greater would Jupiter's effect be on a comet traveling in a highly elliptical orbit nearly four times farther from the Sun than Saturn? He noted that in the summer of 1681, the comet of 1682 approached Jupiter close enough that it experienced a perturbative force one-fiftieth of the primary Sun-comet force. This effect would increase the comet's orbit size and period so that[2]

> it is possible that its return will not be untill after the period of 76 years or more, about the end of the year 1758, or the beginning of the next.

Halley discussed Apian's observations of the comet of 1531 and again concluded, as he had in 1705, that—though crude—they were adequate to identify this comet with those of 1607 and 1682. After searching through an unspecified catalog of ancient comets, he concluded that the comets of 1305, 1380, and 1456 appeared at 75- to 76-year intervals.[3] Although the comets of 1305 and 1380 subsequently turned out not to be previous apparitions of his comet, Halley had gathered an impressive amount of data to substantiate his comet's periodic returns in 1531, 1607, and 1682. He expertly presented his case:

> Wherefore if according to what we have already said it should return again about the year 1758, candid posterity will not refuse to acknowledge that this was first discovered by an Englishman.

Hereby acknowledged Dr. Halley!

In his 1749 *Astronomical Tables,* Halley also speculated on the identity of comets seen in 44 B.C., 531, 1106, and 1680. The comet of 44 B.C. was seen shortly after the death of Julius Caesar. However, unlike the comets of 1531, 1607, and 1682, their identity was not determined by comparing their orbital elements. Like previous attempts based only on similar intervals between returns and physical characteristics, the suggested identity of these comets was incorrect.

With regard to Halley's prediction of his comet's return, it is interesting to note that as he gathered more information on the comets of 1531, 1607, and 1682, the wording of his prediction became less definite. Thus, while the statement in the 1705 first Latin edition reads, "I shall venture confidently to predict its return in 1758," the English version in the same year states, "I dare venture to foretell . . . ." In David Gregory's astronomy textbook of 1715 we read, "I think, I may venture to fortel . . . ." In Halley's *Tables* of 1749 and 1752 we read, "if . . . it should return again about the year 1758 . . . ." Halley's decreasing confidence in his own prediction was likely due to his increasing awareness of the problem's complexity and, in particular, the role of planetary perturbations. Halley was aware that an accurate prediction would require perturbation computations that he was unable to make. Though he had done extraordinary, pioneering work in unraveling the mysteries of cometary motions, he had gone as far as he could with the mathematical tools available to him.

The Race for Recovery

After Halley initially predicted his comet's return in late 1758 or early 1759, additional attempts were forthcoming from the Swiss astronomers Jean Philippe Loys de Cheseaux (1718–1751) and Leonhard Euler (1707–1783) as well as from Thomas Stevenson (d. 1764), a plantation owner on the Caribbean island of Barbados. Citing the unequal 76- and 75-year periods between the returns of the comets seen in 1531, 1607, and 1682, Cheseaux thought it probable that there were two comets, each traveling on identical orbits such that when one comet was at perihelion, the other was close to its aphelion. Their orbital periods were constant and identical. Calculating the interval between the perihelion passages in 1531 and 1682 to be 151 years and 10 days, he predicted that the comet last seen in 1607 could be expected to reach perihelion on November 7, 1758, by the Gregorian calendar. This prediction was given in his 1744 work on the comet of that year. Though not discovered by Cheseaux, this comet is often referred to as Cheseaux's because he computed its orbit and ephemeris and described its impressive, multiple tails.

Probably without the knowledge of Cheseaux's work, Stevenson also presented a two-comet theory which found its way into several London newspapers and magazines in October 1758. He suggested a perihelion passage of April 1 for the comet of 1305 and noted that an interval of 302 years and 198 days occurred between this date and the October 16 perihelion that Halley had computed for the comet of 1607. Taking half of this interval and adding it to the 1607 date, Stevenson derived a perihelion passage of January 23, 1759 for the expected return, or February 3, 1759 on the Gregorian calendar. This work is a fine example of how perverse nature can be. Despite using the wrong data—the comet of 1305 was not Halley's—and an incorrect theory, Stevenson arrived at a nearly correct result; the actual perihelion passage time for the 1759 return was March 13 using the Gregorian calendar.

In 1746, the mathematician Leonhard Euler published his prediction for the return of comet Halley. Euler, whose prolific scientific output included some 560 books and articles, worked or corresponded with the science academies at Berlin and St. Petersburg for most of his career. Having been asked by his colleague Joseph-Nicolas Delisle (1688–1768) to determine paths for the comets of 1742 and 1744, Euler was familiar with the technique for computing cometary orbits. Like Halley, Euler was disturbed with the different time intervals between the comet's returns in 1531, 1607, and 1682. Rather than attribute the difference to planetary perturbations, Euler assumed that the comet's period decreased from 76 to 75 years because of the drag effects of the interplanetary ether. Hence Euler concluded that the next return would be in 1757.

Without a date for the expected comet's return to perihelion, accurate predictions for its specific celestial locations at various times, an ephemeris, could not be determined. However, a number of more general ephemerides were computed in an attempt to assist the expected recovery. In 1754, Thomas Barker (1722–1809), a grandson of William Whiston, constructed 12 different ephemeris tables, each one based on the supposition that the comet would reach perihelion during a different month. From these tables, he constructed an additional table indicating, for each month, where in the sky the comet was likely to be first visible.

In 1755, the Dutch astronomer Dirk Klinkenberg (1709–1799) produced a set of ephemerides similar to Barker's. Under several different suppositions for the comet's perihelion passage times from June 14, 1757 to May 15, 1758, he plotted the comet's resultant positions as seen from Earth. In February 1757, the English instrument maker and science popularizer Benjamin Martin (1705–1782) published a broadside showing the comet's orbit overlying that of the Earth. Its path was represented from its ascending node, 115 days before perihelion, to its descending node, 30 days past perihelion. In 1757, T. J. Jamard, a French cleric and colleague of comet cataloger Alexandre Pingré, charted the positions of the comets of 1531, 1607, and 1682 on a celestial chart and presented it to King Louis XV and the Royal Academy of Sciences in Paris. In an accompanying memoir, Jamard made 14 assumptions as to the time of year the comet would reach perihelion, then described viewing conditions, probable brightness, and tail lengths that might be expected for each time.

In 1757, the French astronomer Joseph-Jérôme Lefrançais de Lalande suggested that the comet would be most easily seen during the month of November, when it and the Sun would be on opposite sides of the Earth. Assuming that the comet would first be sighted 0.4 times farther from Earth than the distance between the Sun and Earth, or 0.4 AU, he gave the most likely positions for first sighting it on various November dates.

After spending 22 years as an astronomer in St. Petersburg, where he helped establish an observatory, Joseph-Nicolas Delisle returned to Paris in 1746. He was made astronomer to the Navy and put in charge of its new observatory at the Hotel de Cluny, near the Sorbonne. Delisle hoped to facilitate the comet's recovery by charting regions of the sky where it might first be visible. He noted that the comet in 1531 was first seen 18 days prior to perihelion, in 1607 it was sighted 33 days prior to perihelion, and in 1682 the interval between perihelion and discovery was 24 days. Hence, he expected the comet might be first sighted some 35 days prior to perihelion. Using this assumption, Delisle then calculated, at 10-day intervals, where the comet would be expected to appear on the sky and plotted these positions on a celestial map. The resultant plots were two ovals, the smaller one for sighting the comet 35 days prior to perihelion and the larger one for recovering the comet some 25 days prior to perihelion. It was along arcs connecting identical days on these two ovals that Delisle instructed his assistant Charles Messier (1730–1817) to look for the comet.

Using a 4.5-foot focal length, Newtonian reflector telescope, Messier began searching for the expected comet in mid-1757. The comet's path was fairly well known from the orbits computed from the 1607 and 1682 observa-

Hotel de Cluny as it looked in the eighteenth century showing Messier's observatory atop the tower. (*Courtesy of Owen Gingerich.*)

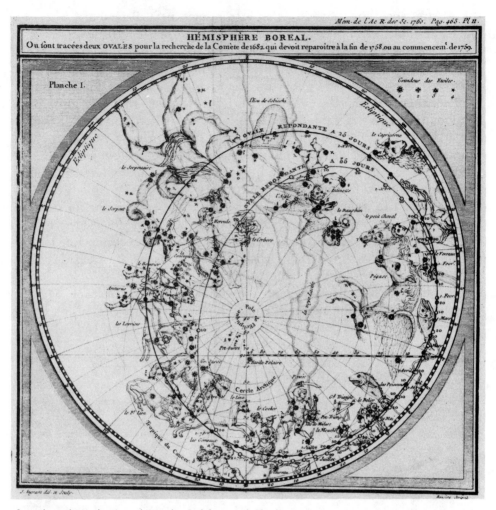

Star chart drawn by Joseph-Nicolas Delisle to aid Charles Messier's search for comet Halley. The ovals drawn on the chart represented the expected positions of the comet for various dates, assuming that the comet would be discovered 25 days prior to perihelion, large oval, and 35 days prior to perihelion, small oval. (*Courtesy Crawford Library, Royal Observatory, Edinburgh.*)

tions, but its actual position along this path at any given time was still very uncertain. What was needed was an improvement in the comet's predicted time of perihelion passage. Enter Alexis-Claude Clairaut (1713–1765).

A child prodigy, the French mathematician Clairaut read his first published scientific paper at the age of 12. In the late 1740s, Euler, Clairaut, and the French mathematician Jean LeRond D'Alembert published ap-

Alexis-Claude Clairaut

proximate solutions to the three-body problem whereby the motion of one body about another is determined taking into consideration the perturbative effects of a third body. They applied their techniques to the theory of the Moon's motion. These efforts are considered the first approximate solutions of the three-body problem in celestial mechanics. In this case, the three-body problem requires the solution of the Moon's motion about the Earth, taking into consideration the perturbations of the Sun. According to Lalande, it was he who first suggested that Clairaut apply his theory of three bodies to the motion of comet Halley. When Clairaut began his computations in June 1757, he used a modified version of the analytic technique developed for his lunar theory to refine the predictions for comet Halley's return. Since the return was imminent, Clairaut was racing against time. His refined prediction would have to precede the comet's recovery. To assist him in the lengthy computations, Clairaut enlisted the aid of his

127

Charles Messier, the Ferret of Comets

Charles Messier had such single-minded devotion to the discovery of comets that King Louis XV of France gave him the nickname *birdnester*, or ferret of comets. Born in Badonviller, Lorraine, France on June 26, 1730, Messier was the tenth of 12 children. When 21 years of age, Messier arrived in Paris and was hired by Joseph-Nicolas Delisle as a clerk assistant at the Marine Observatory at the Hotel de Cluny. Messier's neat handwriting and drafting skill were his only job qualifications; his consummate skill in making astronomical observations would require on-the-job-training. Even by contemporary standards, the telescopes he employed were crude and inefficient.

Charles Messier

None had a light-gathering power greater than a refracting telescope of 3.5 inches aperture.

Although possessed with keen eyesight and extraordinary zeal for observing, Messier had little mathematical ability and left the computation of cometary orbits to colleagues such as Pierre-François André Méchain (1744–1804), a rival comet hunter and calculator in the Department of the Navy, and Jean-Baptiste-Gaspard De Saron, an eminent lawyer and president of the Paris parliament. During the day, Messier would often make meteorological and sunspot observations, and during moonlit nights he would observe the occultation of stars by the Moon. However, the time when the Moon was down and the skies were dark was reserved for comet searches. Messier is credited with 13 comet discoveries; his first, comet 1759 II, and his last, comet 1798 I.

young colleague Lalande, who in turn enlisted the aid of Madame Nicole-Reine Étable de la Brière Lepaute, wife of the clock maker to King Louis XV, Jean André Lepaute.

Initially the plan was to compute the comet's motion around the Sun over the 1607 to 1759 interval taking into account Jupiter's perturbative effects. Not only does Jupiter influence the comet's motion by its direct gravi-

On September 12, 1758, Messier noted a little cometlike cloud or nebula in the constellation of Taurus. So not to confuse this type of nebula with comets just beginning to appear, Messier began to systematically note their celestial positions. Once cataloged, Messier would no longer confuse these objects with the comets he so eagerly sought. What appeared to Messier as cometlike patches of light were actually galaxies, star clusters and nebulae of glowing gases. Descriptions and positions of some 45 *Messier objects* were published in 1774; an additional 58 objects were added in publications in 1780 and 1781. Although Messier did not originally find many of the objects in his catalogs, he was the first to clearly describe them, note their positions in the sky, and publish them collectively.

Today, his cometary discoveries are less well known than the so-called *Messier objects,* though he considered them only annoying distractions in his continuing search for comets. Messier was ambitious and sought recognition of his observing skills through membership in and correspondence with various scientific societies. Though he was made a member of the Royal Society of London in 1764 and of the academies at Berlin and St. Petersburg, he was denied membership in the Academy of Sciences in Paris until 1770 because the French were reluctant to add a mere observer to their ranks. In 1806, he was awarded the Cross of the Legion of Honor from the emperor Napoleon.

One contemporary anecdote, though perhaps apocryphal, demonstrates Messier's single-minded and zealous pursuit of cometary discoveries. From J.-F. La Harpe, a correspondent in St. Petersburg, we learn that Messier's attendance at his wife's deathbed cost him his thirteenth comet discovery—French amateur astronomer and apothecary of Limoges, Jacques Leibax Montaigne, found it first. When a visitor offered condolences for his recent loss, Messier—thinking only of the comet—answered, "I had discovered twelve; alas, to be robbed of the thirteenth by that Montaigne!" His eyes filled with tears. Then, realizing that the visitor was commenting on his wife's death, Messier added "Ah! the poor woman."

tational attraction, it also indirectly affects the comet's motion by slightly altering the Sun's position. Both these direct and indirect perturbative effects of Jupiter were taken into account. Lalande was charged with providing Clairaut with tables of elongations (Jupiter-Sun-comet angles) and distances between Jupiter and the comet for hundreds of points around the comet's orbital path. The effects were so large that Clairaut decided to compute the effects of Saturn as well, so Lalande and Lepaute calculated the tables for Saturn as they had for Jupiter. According to Lalande, the three of them made calculations from morning to night—sometimes even at dinner—for over six months. He also claimed the arduous work left him with an unspecified malady that changed his temperament for the rest of his life. Lalande, who can always be counted upon to supply juicy bits of astronomical biography, relayed that Madame Lepaute's work was not properly acknowledged by Clairaut because of the demands by an envious but unnamed female friend. Madame Lepaute was first given proper credit for her work in the 1759 edition of *Tables Astronomiques de M. Halley* edited by Lalande.

Computing only the perturbative effect on the comet's orbital period, Clairaut established that the interval between the 1682 and 1759 perihelion passage would be 618 days longer than the corresponding interval between 1607 and 1682; 518 days due to Jupiter's influence and the remaining 100 days due to Saturn's. To test his method, he computed the perturbative effects of Jupiter and Saturn over the 1531 to 1607 interval and found that the computed, or predicted, time of perihelion passage in 1682 differed from the actual time by 27 days.[4] Clairaut felt that his prediction for the comet's upcoming return to perihelion in mid-April 1759 would also be accurate to within a month. Although not explicitly stated, his computations indicated April 15, 1759, as the predicted perihelion passage time—nearly 33 days too late. This is only a modest error considering the uncertainty in the planetary masses, the perturbations from neglected or undiscovered planets, and the approximations that had to be made in the method itself. Clairaut emphasized that his calculations could not take into account unknown forces acting in the distant regions of the solar system—forces due to other comets or possibly an undiscovered planet.

Clairaut's first paper on the predicted return of comet Halley was read to the Academy of Sciences in Paris on November 14, 1758. After an exhausting computational effort, Clairaut, Lalande, and Lepaute must have been very anxious before the paper's presentation. If the comet was recovered prior to their announced result, their work might have been perceived as a mere footnote in astronomical history rather than the classic work it turned out to be. They could not afford to wait until the paper was published to announce their result. Clairaut had to appear in person before the academy to win the race between himself and the comet. His strategy succeeded. The comet was recovered less than six weeks after his verbal announcement. The published version of his prediction did not appear until January 1759—well after the first sighting.

In 1760, after the comet was recovered, Clairaut corrected some errors in the earlier work, made more comprehensive perturbation calculations for

Saturn, and suggested a perihelion passage of April 4, 1759. His essay two years later moved this date back further, to March 31, 1759, which Clairaut considered to be within 19 days of the observed perihelion passage. A comparison by Peter Broughton in 1985 between Clairaut's work and the 1971 work by Tao Kiang identified six days of this remaining error as due to the planets Uranus and Neptune, which hadn't yet been discovered; another six days due to the neglected effects of Mercury, Venus, Earth, and Mars; and four days from errors in the masses of Jupiter and Saturn that Clairaut adopted.

Clairaut shared with Leonhard Euler's son, Johann-Albrecht Euler, the award offered in 1759 by the St. Petersburg Imperial Academy of Sciences for the best essay on comet Halley's motion. Both works were published in 1762. In his essay on highly eccentric orbits in the three-body problem, Euler observed that the comet's mass must be very small relative to the Earth since no effects on Earth's motion were apparent during the comet's close approach in late April 1759, when it actually came within 0.15 AU of the Earth. In his own prize essay, Clairaut refined his computations for Jupiter's perturbations on the comet's motion for a period in 1681 and, in response to the contest's guidelines, added a discussion of a resisting medium to his 1762 essay. He considered the effect of ether drag and decided that it was negligible.

Recovery at Last

During the mid-eighteenth century, the dominance of French astronomy was clear. In the post-Newtonian era, French astronomers abandoned the ideas of René Descartes and embraced Newtonian mechanics. The preeminent astronomical discipline was celestial mechanics, and the most visible problem was the approaching return of comet Halley. By the end of 1758, Delisle's search had been carried out by Messier for over a year and Clairaut had refined Halley's prediction for the comet's perihelion passage. It seemed likely that when comet Halley was recovered, the Frenchman Messier would find it first.

As part of his search for comet Halley, Messier discovered another comet on August 15, 1758, which he followed until November 2. An earlier, independent discovery of the comet of 1758 was made on May 26, 1758 by De la Nux, observing on what is now Reunion Island in the Indian Ocean. De la Nux was an advisor to the superior counsel for the island and a correspondent to the French academy. Clairaut was given a fright by this comet, but soon realized that this was not the one he had been laboring over for several months.

For the months of November and December 1758, the skies over Paris were frequently cloudy and Messier managed to continue his search with only a few hurried observations. Finally, on the cloudless evening of January 21, 1759, he noticed a hazy patch in the constellation Pisces that physically resembled the comet he had observed a few months earlier. Due to some neighboring, hazy nebulae he was uncertain as to his discovery but the comet's motion soon gave it away. Messier's observations were made using a 4.5-foot Newtonian reflecting telescope, and it is clear that he had extended his search beyond the confines of the search ovals drawn on Delisle's chart. Messier later speculated that if Delisle had made less restrictive assumptions on the number of days before perihelion that the comet would first be discovered, he might have found it much earlier. Whereas Delisle had assumed the comet would not be sighted more than 35 days prior to perihelion, Messier's recovery was actually some 51 days before. There is no evidence to indicate that Messier's search took advantage of Clairaut's mid-April prediction for the comet's perihelion passage. Delisle's search chart did not assume when the comet would reach perihelion, only the time interval between perihelion and its first appearance. After his recovery of comet Halley on January 21, Messier continued to observe it on January 22, 23, 25, 27, 28, 31 and February 1, 3, 4, 13, and 14, until it was lost in the evening twilight.

With what must have been great excitement, Messier immediately announced his discovery to Delisle, who urged him to carefully follow the comet but not to reveal its presence to anyone else. As Delisle's assistant, Messier had to suppress his historic observations, and Delisle refused to announce his assistant's recovery for more than two months. The motive for this suppression is not clear. Perhaps Delisle considered the discovery his own personal property, to be revealed only at the proper moment in history, or perhaps he wished to be absolutely sure that the comet Messier had discovered was the returning comet Halley. In any case, Delisle's announcement of the comet's return was made only after the comet was observed by Messier on April 1, 1759 as it exited from solar conjunction in the morning sky. On the same day, word was received in Paris that the triumphal recovery by Messier on January 21 had been preceded by nearly a month—by a German amateur!

Johann Georg Palitzsch (1723–1788) was a farmer and amateur astronomer living in Prohlis, a small town near Dresden in Saxony. At 6 P.M. on Christmas evening, 1758, Palitzsch directed his eight-foot telescope toward the heavens. He was well aware that comet Halley was expected, but there is no clear evidence that he was using some sort of search ephemeris. Upon directing his telescope toward the constellations of Cetus and Pisces, he noted a nebulous star near the variable star Mira Ceti between the stars ε and δ Piscium. His observations on the next two nights indicated the object's

Johann Georg Palitzsch

movement and he recognized it as a comet. Palitzsch's observations were communicated to Christian Gotthold Hoffman, who in turn published them in the second 1759 issue of Dresden's newspaper. Hoffman, who was chief commissioner of the excise in Dresden, observed the comet himself on December 27, 1758, but neither Hoffman nor Palitzsch identified it with comet Halley.

The discovery of comet Halley was announced in an anonymous, 15-page tract published in Leipzig dated January 24, 1759. Soon after the author learned of Palitzsch's observations, he realized that the comet was the one Halley had predicted. Using a geometrical construction, which probably relied on the 1682 orbit with an adjusted perihelion passage, the author computed the circumstances of the comet's future path from January 28 through May 13, 1759. This ephemeris gave not only the comet's longitude and latitude for various dates, but also when its position, brightness, and tail length would make for advantageous viewing. The author checked his computations by observing the comet on January 18 and 19 with a three-foot telescope to make sure his predictions were in accord with his observations. The ephemeris was in error by only about 0.3 to 0.8 degrees in longitude through mid-April and grew to approximately 1.5 to 2.0 degrees thereafter. The date of the

133

comet's closest approach to Earth was correctly predicted to be April 26, and the predicted time of perihelion passage, March 14, was only one day too late.

Overall, this pamphlet served as a complete and accurate observing guide for the comet's apparition. This remarkable work is so rare that outside Germany, only the copies at the Crawford Library in Edinburgh, Scotland and at the Stanford Library in California have been located. The title of the tract begins *Anzeige dass der im Jahre 1682 erschienene und von Halley nach der Newtonianischen Theorie . . .* (announcement of the reappearance of Halley's comet, which appeared in the year 1682 and according to Newtonian theory . . .). The author was probably Gottfried Heinsius, professor of mathematics at the University of Leipzig. His identity is evident from the obvious astronomical knowledge required to write the tract and from a letter he wrote to Leonhard Euler dated April 21, 1759, reporting that he observed the comet with a three-foot telescope on January 18, 19, 22, and 27 and that he authored a tract on it.

Word of this tract reached Paris on April 1, 1759. In a letter dated February 20, probably from Christian Mayer, professor of mathematics at Heidelberg, Delisle and Messier learned of the German observations. They must have been shocked to hear that several prior German observations and an ephemeris had been published in late January 1759. On this same day, Messier had already told several astronomers at the Paris academy that he had observed the comet from January 21 through February 14, when it entered solar conjunction, and on that very morning, April 1, 1759, he had recovered the comet as it emerged.[5]

There was a good deal of confusion at the Paris academy in early April 1759 when the German observations became known and Messier's observations of January and February were finally made public. Some members of the academy refused to believe that Messier's observations were genuine and suggested they be omitted from any orbit computations. Messier's observations were soon admitted to be genuine, however, and he traced the comet's motion on a large celestial map. He was then allowed to accompany Delisle to Versailles on April 5th to announce the comet's appearance to King Louis XV and to present the king with the celestial map showing its motion through the heavens. Having had the rug pulled out from under him, Delisle formally published Messier's recovery in the first volume of *Mercure de France* in July 1759. Seven years later, Messier's observations were published in the *Mémoires de l'Académie Royale des Sciences*.

The collective pride of the Paris academy must have been deeply wounded, since the recovery of comet Halley was made by a German amateur more than three months before they knew anything about it. This embarrassment may have still been evident at comet Halley's next return in 1835. In

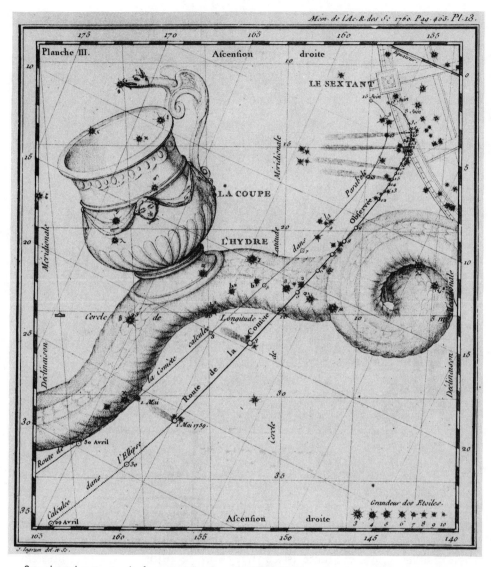

Star chart showing track of comet Halley from April 29, 1759, until June 15, 1759. This chart was similar to the one presented by Joseph-Nicolas Delisle and Charles Messier to King Louis XV on April 5, 1759. (*Courtesy Crawford Library, Royal Observatory, Edinburgh.*)

what appears to be an attempt to shift the blame to the Leipzig astronomer, who did in fact publish his observations, the French celestial mechanician, Philippe G. de Pontécoulant, stated in 1835 that

> An astronomer of Leipzig (Palitzsch) observed it soon after , but as a lover of his mistress, and a miser of his treasures, jealous of his discovery, which he

German broadside illustration of comet Halley in 1682. Nürnberg, 1682. The engraving shows comet Halley just before dawn in late August. It is being observed by a number of people, some using telescopes, at the private observatory of Georg Christoph Eimmert, located on the western part of the city wall of Nürnberg, Germany. (*Courtesy Adler Planetarium, Chicago.*)

would participate with no one, the German astronomer abstained from divulging the secret . . .

Summary

Using a modified version of Isaac Newton's method, Edmond Halley computed parabolic orbits for 24 well-observed comets and concluded from their similar orbital characteristics that those seen in 1531, 1607, and the one he and Newton observed in 1682 were one and the same. He boldly predicted that this comet would return again in 1758. Although he had verbally announced his results to the Royal Society of London in 1696, it was not until 1705 that Halley first published his prediction. In 1717 Halley made a revised prediction for the comet's return in late 1758 or early 1759, but it was not published until 1749, seven years after his death.

Alexis-Claude Clairaut's classic refinement of Halley's prediction was completed only six weeks prior to the comet's actual discovery. His prediction was in error by less than 33 days.

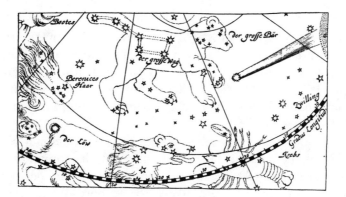

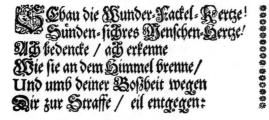

Franckfurt am Mayn / zufinden bey Johann-Georg Walthern / den Laden auff dem Pfarneysen / habend.

German broadside of comet Halley in 1682. The Julian calendar was in use in Frankfurt at this time. The German text, translated by Ruth Freitag, follows:

True picture and position of the comet as it appeared to astronomical observation over Frankfurt am Main on the 15th, 17th, and 21st of August, in this year of 1682, from nine o'clock at night until 3:30 in the morning, to the northeast, in the sign of Cancer, under the forefeet of the Great Bear.

Look at the wonderful torch-candle!
O human heart, so prone to sin!
Take heed and recognize
How it burns in the sky
And because of your wickedness
Hastens to bring you to punishment;
Assemble then with penitence,
Extinguish these flames of wrath,
That on the soil of the German land
God's fury will be abated
Who threatens us with the comet
And requires our repentance and prayer.

(*Courtesy of Martin Luther University, Universitäts- und Landesbibliothek, Halle-Wittenberg, Germany.*)

In mid-1757 Joseph-Nicolas Delisle and his assistant, Charles Messier, began searching for the comet from their observatory in the tower of the Hotel de Cluny in Paris. Delisle mapped out an area on a celestial chart where he felt the comet would first be recovered and instructed Messier to begin the search. Finally, on January 21, 1759, Messier found the comet, but Delisle insisted on keeping his assistant's precious discovery a secret. The announcement of Messier's discovery of comet Halley was delayed 10 weeks, until the comet passed perihelion and emerged from the Sun's glare. Unknown to Delisle and Messier, Halley's comet had already been discovered by a German amateur astronomer on Christmas evening 1758 in the constellation of Pisces.

The bold prediction and successful recovery of comet Halley in 1758 and 1759 was the most visible confirmation of Newtonian dynamics in the eighteenth century. At least this one comet was periodic, and—by inference—many others were likely to be periodic as well. Moreover, Clairaut's technique for accurately predicting future returns of periodic comets was used as a model for 150 years, including the 1835 and 1910 returns of comet Halley. Especially in France, celestial mechanics was the preeminent astronomical discipline during the eighteenth century and the successful prediction of comet Halley's return prompted many enthusiastic studies of cometary orbits. Efforts to discover additional periodic comets continued throughout the eighteenth century but with only two successes, several failures, and at least one outright fraud.

NOTES

1. The term *Halley's comet* appeared as early as 1765, when Nicolas-Louis de Lacaille (1713–1762) used it in *Mémoires de mathématique et de physique, tirés des registres de l'Académie Royale des Sciences, de l'année 1759*, 1765, p. 522–524.

2. In a memoir published in 1765, Joseph-Jérôme Lefrançais de Lalande (1732–1807) pointed out that in 1683, the comet was again near Jupiter and experienced a perturbative acceleration nearly opposite to that suffered in the summer of 1681. Hence, Lalande implied that Halley's delaying the perihelion time into early 1759 may have been fortuitous.

3. In David Gregory's 1715 textbook, *The Elements of Astronomy, Physical and Geometrical*, Halley commented that on looking over some histories of comets, he found one seen about Easter 1305 that may have been a previous apparition of the comet seen in 1456, 1531, 1607, and 1682. Sometime before 1717, when he had finished "Synopsis of the Astronomy of Comets" for inclusion in *Astronomical Tables*, Halley added the comet of 1380 as a possible return of the same comet.

4. Clairaut actually gave 37 days as the difference between the actual and predicted times of perihelion passage in 1682, but it is clear from his work that he meant 27 days.

5. For mid-European observers, the comet's apparition was broken into three phases. The first phase, from December 25, 1758, through February 14, 1759, ended when the comet disappeared into the evening twilight. Rounding perihelion on March 13, 1759, the comet again became visible in early April before it sank below the local horizon. Its third period of visibility was from early May to when it was last seen on June 22. The comet passed within 0.12 AU of the Earth on April 26, 1759, and became a rather impressive naked-eye object.

7
The Eighteenth Century
The Orbit Computers

Due to the complexity of Newton's method, few cometary orbits are determined in the first half of the eighteenth century. Methods of Olbers, Laplace, and Gauss facilitate computational procedures. Comet Lexell passes close to Earth in 1770 but hasn't been seen since. Eighteenth century orbit computers search unsuccessfully for the return of a second periodic comet. Cometary thought evolves in the eighteenth century.

Theoretical Work

Isaac Newton's method for computing parabolic cometary orbits required that the three utilized observations be fairly close to one another and far enough from perihelion that the comet's velocity could be assumed uniform. Subsequent orbit determination methods slowly improved on Newton's semigraphical procedure, but it would be nearly a century after Newton's work was published in 1687 before they were improved significantly.

In 1746, 1749, and 1774, Yugoslavian-born astronomer Rudjer J. Boščović (1711–1787) published his ideas on comets, including an orbit determination technique similar to Newton's. It used repeated trials to compute a parabolic orbit from three observations sufficiently close to one another that the comet could be considered traveling on a nearly rectilinear path between them. However the method was complex and difficult, so apart from his own orbit for the comet of 1774, it was not often employed.

The first purely analytic orbit determination method that did not require graphical techniques was Leonhard Euler's. The most prolific mathematician in history, Euler's memory was photographic. He knew Virgil's *Aenid* by heart and although he seldom looked at the book since his childhood, he could recite the first and last lines on any given page. He was also capable of extraordinary mental calculations. Euler's genius is suggested by his attacking the extremely complex problem of the Moon's motion, a problem that gave Newton headaches, when he was 59 years old and blind. He performed much of the complicated analysis in his head.

Euler left his native Switzerland in 1727 to work at St. Petersburg, now Leningrad, and in 1740 he began working in Berlin. In 1744, he published his classic work in celestial mechanics, *Theoria motuum planetarum et cometarum,* in which he established a purely analytic technique for computing the general orbit of a comet (parabolic, elliptic, or hyperbolic) directly from observations. Euler's technique employed three observations close to one another and, for each trial orbit based on these observations, the heliocentric distance of the comet at the middle observation was assumed known approximately. The optimum orbit was selected by examining which trial orbit best represented a fourth observation a good distance from the other three. While rigorously correct, Euler's method did not find wide use because the required calculations were quite lengthy. He published the results of his method applied to the comet of 1744, at first determining a hyperbolic orbit based on observations made at the Berlin observatory and then modifying the orbit to a parabola after receiving additional observations from Jacques Cassini (1677–1756) in Paris. Euler's initial orbit for the comet of 1744 was

Leonhard Euler as he appeared on a Swiss 10 Franc note.

probably the first legitimate hyperbolic orbit computed, although in 1735, Pierre Bouguer (1698–1758) published a crude one for the comet of 1729. Bouguer is best remembered for his pioneering work in astronomical photometry.

Euler's 1744 book also gave, for a general orbit, an infinite series for representing the time interval, t−T, elapsed since a body passed the time of perihelion, T. For a parabolic orbit, the expression reduces to a cubic equation in terms of the true anomaly (v),

$$\frac{k\,(t-T)}{\sqrt{2}\,q^{3/2}} = \tan \frac{v}{2} + \frac{1}{3}\,\tan^3 \frac{v}{2}$$

where k is the so-called Gaussian constant defined as the square root of the product of the gravitational constant and the solar mass, and q is the perihelion distance in astronomical units, AU. This expression is used to determine the comet's position, or true anomaly (v), on its parabolic orbit for any given time, t. It was derived by Euler using the calculus and was extremely useful for computing ephemerides for comets traveling on parabolic orbits. Although he derived it geometrically, it is apparent from his work in 1705 that Edmond Halley was aware of this relationship. Thomas Barker published a work on comets in 1757 in which he gave a table of values for

$$\tan \frac{v}{2} + \frac{1}{3}\,\tan^3 \frac{v}{2}$$

for various values of v and these types of tables became known as *Barker's tables* even though both Halley in 1705 and Euler in 1744 had previously published them.

In 1743, Euler published a relationship, Euler's theorem, that eventually led to improved orbit determination techniques. Euler's theorem states that the time interval required, $t_2 - t_1$, for a comet to traverse an arc between two radius vectors (r_1, r_2) of a parabola is a function of the corresponding lengths of the radius vectors and the chord, s, joining them. Mathematically, this is written as

$$6k\,(t_2 - t_1) = (r_1 + r_2 + s)^{3/2} \pm (r_1 + r_2 - s)^{3/2}$$

where k is the Gaussian constant and the "−" sign applies if the arc between r_1 and r_2 is less than 180 degrees and the "+" sign applies if it is greater then 180 degrees. Although not expressed analytically, the same relationship was give in Book 3 of Isaac Newton's 1687 *Principia mathematica*. The German mathematician Johann Heinrich Lambert (1728–1777) gave a generalized formulation of Euler's theorem in 1761 that was good for elliptical, para-

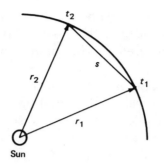

Euler's theorem states that the time required for a body to traverse an arc between two heliocentric radius vectors, r_1 and r_2, of a parabola is a function of the corresponding lengths of the radius vectors and the chord, s, joining them.

bolic, and hyperbolic orbits. However, Lambert did not utilize his formulation for a complete method of orbit determination. Several complete methods were available in the second half of the eighteenth century, but they all required lengthy computations. What was needed was an easy-to-use method, and just such a technique was finally supplied by Heinrich Wilhelm Matthias Olbers (1758–1840).

Olbers was a physician practicing the then new field of ophthalmology near Bremen, Germany. Requiring only four hours of sleep each night, he was free to devote his extra hours to astronomy. Olbers installed an observatory above his house and equipped it at various times with two achromatic Dollond refracting telescopes, a reflecting telescope by Johann H. Schröter, a heliometer, refracting telescope, and comet seeker from the shop of Joseph Fraunhofer, in addition to other instruments. Though he was an amateur astronomer, his observatory was one of the best contemporary facilities. Olbers amassed a large collection of astronomical books and literature, which was particularly rich in cometary works. After his death, Friedrich G.W. Struve purchased the library for the Pulkovo observatory near St. Petersburg.

Comet searching was Olbers' primary astronomical activity, and he is given credit for the first discovery of two, in 1796 and 1815, and a joint discovery with Jacques Leibax Montaigne for a third, 1780 II. For the comet of 1796, Olbers computed a parabolic orbit with a new method he developed. His orbit for this comet and a description of his new method were sent for comments to the Hungarian-born astronomer Baron Franz von Zach. Zach was then the director of the Seeberg observatory near Gotha, Germany. After studying Olbers' method and testing it on the comet of 1779, Zach personally saw the paper through publication. Olbers' paper was aptly titled "An Essay on the Easiest and Most Convenient Method of Calculating the Orbit

Heinrich Matthias Olbers

of a Comet." It is a credit to his 1797 work that this technique, modified only slightly, is still in use today.

By constraining the comet's motion to a parabolic path with the Sun at one focus and employing Euler's theorem together with the properties of a plane surface, Olbers worked out a technique whereby the geocentric distance of a comet at the time of its first and third observations were determined using repeated trials. Once these two quantities were known, it was relatively straightforward to calculate the comet's heliocentric coordinates and then the five unknown orbital elements. The value of the sixth orbital element, the eccentricity, was assumed equal to one. Olbers' method, which immediately established him as one of the finest contemporary astronomers, owes much to Euler's work in 1743, Lambert's 1761 work, and perhaps also to the works of Achille-Pierre-Dionis du Séjour (1734–1794). As pointed out by Wilhelm Fredrik Fabritius in 1883, Séjour's 1782 work contains a technique for computing orbits that is basically the same as Olbers'.

An entirely new approach to orbit determination was published in 1784 by Pierre-Simon Marquis de Laplace (1749–1827). Laplace was an outstanding French mathematician, physicist, and celestial mechanician; he actually coined the term *celestial mechanics*. Previous methods of computing

145

orbits from three observations required that they be fairly close. Hence, small errors in the observations could seriously affect the results. Laplace's method can employ observations separated by as much as 30 to 40 degrees, and the method becomes more precise with the use of many observations. When only three observations of right ascension and declination are used, the first and second time derivatives of the middle observation are determined from the first and third observations. From the right ascension and declination of the middle observation, and their first and second time derivatives, Laplace demonstrated that it was possible to compute the heliocentric position and velocity vectors of the object and hence all the orbital elements for any general orbit about the Sun.

Although mathematically elegant and straightforward, Laplace's method suffers from the difficulty in obtaining sufficiently accurate derivatives. In addition, only the middle observation is represented exactly; the first and third observations have differences, or residuals, from the method's position predictions. The method can be unsatisfactory when only three observations are available. Even with improvements, principally by Armin O. Leuschner in 1913, Laplace's method is limited.

Pierre-Simon Marquis de Laplace

Jean Louis Pons, the Champion Comet Hunter

Jean Louis Pons

With at least 34 comet discoveries to his credit, Jean Louis Pons (1761–1831) will likely remain one of the most successful comet hunters in history. Like his rival, Charles Messier, Pons rose from humble beginnings and used his tenacity and keen eyesight to make a name for himself by discovering comets. In 1789, he was hired as the doorkeeper for the observatory in Marseilles, France. Astronomers at the observatory instructed him, and by 1813 he had the position of adjunct astronomer. Five years later, he was appointed assistant director of the observatory. The next year, with the recommendation of his friend Baron von Zach, Pons became director of an observatory in the Royal Park La Marlia near the city of Lucca in Tuscany, now western Italy, and subsequently, in 1825, he became observatory director of the museum for physics and natural history in Florence, Italy.

Pons' fame was almost entirely due to his extraordinary success in discovering comets. Using telescopes he constructed himself, Pons discovered or codiscovered three-quarters of all comets found during the period 1801 to 1827. He ended his extraordinary career by finding five comets in one year, between August 7, 1826 and August 3, 1827. Pons discovered periodic comet Encke at both its 1805 and 1818–1819 apparitions. After the German astronomer Johann Franz Encke (1791–1865) identified it as a short-period comet and successfully predicted its return in 1822, the comet was named after him; however, Encke himself always referred to it as Pons' comet.

Once when Pons had gone some time without a comet discovery, he wrote to Baron von Zach complaining of the paucity of comets in the sky. Zach, a man of sly wit, responded that few spots had been seen on the Sun lately and when the sunspots came back, the comets would surely follow. A short time later, Zach received another letter from Pons informing him that very large spots had appeared on the Sun and he had found a fine comet shortly thereafter.

147

In the Autumn of 1800, Baron von Zach initiated a campaign to discover a suspected planet between Mars and Jupiter. The zodiac was divided into 24 zones and 24 astronomers were each assigned a zone to search for the possible planet. Soon after the campaign was underway, a new planet was discovered, but not by any of the astronomers engaged in looking for it. Before the Italian astronomer Giuseppe Piazzi (1746–1826) received word of the proposed search, he discovered the first minor planet, Ceres, on the first day of the nineteenth century, January 1, 1801. Piazzi was the director of the Palermo observatory in Sicily and was measuring star positions when he detected an uncharted object. The following evening, he noticed that this new star had moved four arc minutes: the first minor planet had been detected. He observed the new object for 41 days, but before other observers received word of his discovery, it disappeared into solar conjunction. Without a search ephemeris based on a reliable orbit, the minor planet would be extremely difficult to recover when it exited from the Sun's glare.

Rising to the task, the 24-year-old German genius, Carl Friedrich Gauss (1777–1855), developed a technique for determining an orbit from three observations, which he used for the object Piazzi had discovered. He

Carl Friedrich Gauss

148

then generated a search ephemeris for several months to facilitate the recovery attempts. Using this ephemeris, Baron von Zach was the first to confirm the minor planet's recovery on January 1, 1802, exactly one year after its original discovery. A few hours after Zach, Heinrich Olbers also confirmed the recovery.

Actually Zach had first observed Ceres on December 7, 1801. Lacking adequate star maps of the area, however, he could not be sure of his observation. Fighting cloudy weather, Zach finally determined on December 18 that one of the stars seen on December 7 was no longer there. However, it was not until January 1, 1802 that he was convinced of his recovery. In testimony to the accuracy of Gauss' new orbit determination method, Ceres was found only 15 to 20 arc minutes from Gauss' predicted position.

In contrast to Laplace's method, which computes position and velocity vectors at the middle observation time, Gauss' procedure determines position vectors for the first and third observation times. These two vectors can also be used to compute orbital elements, which then represent the first and third observations exactly. Gauss' method leaves only the middle observation with a position residual, whereas with Laplace's method, two of the three observations are left with residuals. Gauss' general orbit theory was based to some extent on the earlier work of Leonhard Euler, and one of the few significant improvements to it was published in 1889 by the American mathematician Josiah Willard Gibbs (1839–1903). The technique Gauss used to ensure the recovery of Ceres underwent many improvements before it was finally published, along with his method for a least squares adjustment of the preliminary orbit, in his 1809 classic work *Theoria motus corporum coelestium*. Laplace, who had a fair amount of self-esteem, freely admitted Gauss' mathematical superiority. In an age of extraordinary mathematical geniuses—Euler, Lambert, Laplace, Lagrange—Gauss was acknowledged as the prince of mathematics.

The orbit determination methods of Boscovic, Euler, Laplace, Olbers, and Gauss allowed a cometary orbit to be fit to a set of observed positions at one apparition. However, to successfully follow the motion of a periodic comet from one apparition to another, the perturbative effects of the planets must be taken into account. Clairaut's modification of his lunar perturbation theory allowed him to successfully follow the motion of comet Halley from 1607 to 1759. Drawing on this work and related works by Jean Le Rond D'Alembert (1717–1783) and Leonhard Euler, Joseph Louis Lagrange (1736–1813) published a perturbation technique in 1785 that is conceptually similar to methods used today. Lagrange's method is termed the *variation of arbitrary constants,* or *variation of orbital elements* technique. Using this method, it is possible to satisfy the perturbed equations of a comet's motion using formulae for the unperturbed motion if its orbital elements are as-

The Bogus Comets of 1784, 1793, and 1798

For amateur and professional astronomers alike, the desire to discover a comet is a strong incentive for observing the heavens. Typically a comet observer must scan the skies for hundreds of hours before discovering one, and usually only the most persistent observers are rewarded with a discovery. The instant fame and recognition brought to a comet discoverer is a great honor and that person is assured a place in history by having his or her name forever associated with the comet.

The rewards of a comet discovery were apparently tempting to the Chevalier Jean Auguste D'Angos (1744–1833). D'Angos was a captain in the prerevolutionary French army and became a member of the Knights of Malta. He was also a physician, chemist, and astronomer. In 1783, the Grand Master of the Knights of Malta invited D'Angos to Malta, and a fine observatory was built for him in the Palace of Valletta. The observatory lasted only six years before D'Angos' experiments with chemistry, particularly phosphorus, resulted in a disastrous fire that destroyed the observatory and its records.

One year after occupying the observatory, D'Angos sent Messier a letter dated April 15, 1784, announcing the discovery of a comet in Vulpecula. D'Angos enclosed two positions of his comet and a subsequent letter to Messier included an orbit allegedly computed from his own observations. Despite repeated attempts, neither Messier nor others could find D'Angos' comet. In 1806, the French celestial mechanician, Johann K. Burckhardt, wished to recalculate this comet's orbit and requested the observations from D'Angos. However, D'Angos reported that all his records of the comet were lost in the observatory's fire 17 years earlier. Under various assumptions, Burckhardt could find no orbit that resembled that of D'Angos. Subsequently a German periodical of 1786 was discovered that listed 14 observations of the comet as reported by D'Angos.

The German astronomer, Johann Encke, then began an investigation. In 1820, Encke found that the orbit computed by D'Angos was consistent with his 14 observations provided that Encke, in each case, used an Earth-comet distance that was exactly 10 times the correct value. D'Angos had fabricated his observations by incorrectly using his own bogus orbit! As Encke noted, "D'Angos had the audacity to forge observations that he never made, of a comet that he had never seen, based upon an orbit he had gratuitously invented, all to give himself the glory of having discovered a comet."

D'Angos was also the only observer who was successful in observing a comet he had allegedly discovered in 1793 and his discovery of a 14-year-period comet in 1798 was also fabricated. This latter comet D'Angos claimed to have observed transiting the Sun on January 18, 1798. According to Lalande's textbook *Astronomie,* the Sun was at the descending node of the comet of 1672 on this date. Stating that he had observed a comet transit the Sun in 1784 and again in 1798, D'Angos claimed that the comet of 1672 had a 14-year period. When it was pointed out that the value given in Lalande's textbook for the node of the comet of 1672 was in error by 60 degrees, D'Angos' fraud was discovered again.

In the short space of 15 years, D'Angos had managed to discover three bogus comets and burn his observatory to the ground. He lived to be 89 years old, and for his patients' sake, we can only hope that his career as a physician was more successful than his work in astronomy.

Lithograph by French satirist Honoré Daumier. (*Photograph courtesy of P. Veron.*)

sumed to vary with time. One finds analytical expressions for the time deriv-atives of the orbital elements, which are then integrated to find the instantaneous, or *osculating*, orbital elements for any particular time. From a set of osculating orbital elements, the object's positions on the celestial sphere are obtained in a straightforward manner.

The Orbit Computers

In 1705 Halley published 24 cometary orbits that he computed using a procedure similar to the one published by Isaac Newton in 1687. However, Newton's method was not easily followed, and in the first half of the eight-eenth century there were very few attempts to compute cometary orbits. The third Astronomer Royal, James Bradley (1693–1762), was one of the few who could make the necessary computations after Halley's death in 1742. Using Newton's method and many of his own observations, Bradley com-puted parabolic orbits for the comets of 1723 and 1737 I and never found the residuals between the observations and his predicted positions to be larger than one arc minute.

James Bradley, England's third Astronomer Royal.

In 1744, the Italian astronomer Eustachio Zanotti (1709–1782) published a parabolic orbit for the comet of 1739, and eight years later the Dutch astronomer Nicholaas Struyck (1687–1769) published a list of 19 parabolic orbits for comets seen between 1533 and 1748. The various orbits were computed by Nicolas-Louis de Lacaille, comets of 1699 I, 1702, and 1739; James Bradley, 1723 and 1737 I; Dirk Klinkenberg, 1743 II; Giovanni Domenico Maraldi, 1748 I; Joseph Betts, 1744; Jean Philippe Loys de Cheseaux, 1747; Struyck himself, 1706, 1707, 1742, 1743 I, and 1748 II; his assistant Cornelius Douwes, 1533, 1678, 1718, and 1729; and Isaac Newton, 1680. In his publications of 1740 and 1753, Struyck gave a history of comets and a compilation of computed orbits to that time.

By the 1750s there was an increase in cometary orbit determination that was due in large part to a work by Nicolas-Louis de Lacaille. In 1751 Lacaille published a clear and straightforward orbit determination technique. Pierre Charles Lemonnier (1715–1799) filled in a few details in 1743, but the necessary computations in this latter work were unclear. The historian Jean-Baptiste Joseph Delambre (1749–1822) suggested that those few astronomers who could compute orbits in the early eighteenth century had purposely kept the method a secret. However, Lacaille's work laid out a modified version of Newton's method in enough detail to allow contemporary astronomers to understand and employ it. Delambre noted that nearly all Lacaille's colleagues—including Lalande, Méchain and Pingré—followed his procedures.

As well as reviewing older, well-observed comets, contemporary astronomers seized on each new cometary apparition as a chance to compute its orbit. Often, each new comet would have several different orbits computed by various astronomers. The enthusiasm for computing new comet orbits was prompted by the possibility of discovering a periodic comet. Halley had shown the way with his successful prediction of a comet's return, and the chances of finding more periodic comets seemed good. However, imprecise observations, overzealousness, and circumstance would allow only a few periodic comets to be successfully identified in the eighteenth century.

In the second volume of his collected works, published in 1753, Struyck compared time intervals among several cometary apparitions and suggested periods for the comets of 1580, 1585, 1596, 1652, 1664, 1665, 1677, and 1686. He regarded two cases as likely, three cases as very probable, and three more cases as certain. In fact, none were correct. In a manuscript sent to Alexandre Pingré in 1759, Struyck suggested periods for several additional comets. Once again, none were correct. In a more conservative vein, Joseph-Jérôme de Lalande, in 1771, suggested three comets that were likely to be periodic. According to Lalande, the comet seen in 1532 and 1661 could be expected back in 1789–90, the comet of 1264 and 1556 should return in

This eighteenth century cometarium was designed to exhibit the nature of cometary motions. On turning a handcrank, the radial rod rotated, while the comet, free to move along the rod, was constrained to move within the elliptical grove. Keplerian motion was achieved by two elliptical pulleys within the device that rotated upon each other. (*Photograph courtesy of George III collection, Science Museum, London.*)

1848, and the comet of 1106 and 1680 was due back in 2254. In his publications of 1705 and 1749, Edmond Halley had already suggested the periodic nature of the comets seen in 1532/1661 and 1106/1680. None of Lalande's three identifications were correct.

The supposed identity of the comets of 1532 and 1661 was the subject of an intensive study by Pierre Méchain. Méchain calculated orbits for both comets and proved they were not one and the same object; hence the comet's return should not be expected in 1789–1790. In 1782, his research on these two comets won him a prize offered by the Academy of Sciences in Paris.

Pierre-François-André Méchain

Méchain was also an active comet hunter with nine first cometary discoveries to his credit and, unlike Messier, he computed orbits for all his own comet discoveries and others as well.

Eighteenth century orbit computers not only incorrectly identified several nearly parabolic cometary orbits as being periodic, they missed identifying several actual short periodic comets. Johann Evangelist Helfenzrieder (1724–1803) discovered the comet of 1766 II on April 1, 1766, and Pingré was unable to satisfactorily compute a parabolic orbit using both the pre- and postperihelion observations. He concluded that the orbit was uncertain. It was not until 1915 that Carl W. Wirtz recognized this comet as having a short periodic orbit, and in 1985 Donald K. Yeomans determined that it had a period of 4.4 years, giving it the shortest period of any comet except comet Encke. Unfortunately this comet's orbit is poorly known and it remains a one-apparition, short-period comet.

Comet Encke, with the shortest period of any comet, 3.3 years, made two observed returns in the eighteenth century without its periodic nature being detected. It was first discovered on January 17, 1786, by Méchain and rediscovered on November 7, 1795, by Caroline Lucretia Herschel (1750–1848) at Slough, England using a 16-centimeter aperture reflecting telescope built by her brother, William Herschel. There were only a few nights of observations for the 1786 I apparition of comet Encke, so an orbit could not be determined with any certainty. The 1795 apparition, however, covered a period of three weeks and its periodic nature might have been detected.

In a similar fashion, short periodic comets 1783 Pigott, 5.9 years, and 1790 II Tuttle, 13.9 years, appeared in the eighteenth century but their periodic nature went undetected. Caroline Herschel's second comet discovery was made on December 21, 1788; her first was comet 1786 II. Herschel's comet 1788 II later turned out to have a period of approximately 150 years. It is now called periodic comet Herschel-Rigollet since the variable star observer, Roger Rigollet, rediscovered it at its second apparition in 1939. This comet's periodic nature, finally recognized by Leland E. Cunningham in 1939, also escaped detection in the eighteenth century.

The second comet after Halley's to be correctly recognized as periodic was not a short-period comet. Elliptic orbits were computed for the comet of 1769 by Anders Johan Lexell (1740–1784) and Leonhard Euler in 1770, and

Caroline Lucretia Herschel

by Pingré, who included these elliptic orbits in his *Cométographie*. The accepted orbit for the comet of 1769 was computed by the German astronomer Friedrich Wilhelm Bessel (1784–1846) in 1807. Lexell and Euler determined the period to be between 449 and 519 years, Pingré computed 1231 years, and Bessel 2100 years. Though its period was quite uncertain, it was clear to contemporary astronomers that the observations of the 1769 comet were best represented by an elliptical orbit.

The most interesting orbital behavior of an eighteenth century comet was that due to comet 1770 I Lexell. Charles Messier discovered it on June 14, 1770, as a nebulosity in Sagittarius. It moved rapidly toward the Earth and, on the evening of July 1, approached to within 0.015 AU, making this the closest recorded approach of any comet to the Earth. During its close passage, the apparent coma diameter reached 2 degrees 23 minutes, nearly five times the diameter of the full moon. On July 4, it was lost in the solar glare but using Pingré's ephemeris, Messier recovered it on August 4 and observed it until October 2. It remained visible to the naked eye until August 26. Parabolic orbits were tried, without success, by several computers including Pingré, Lambert, and the finest contemporary American astronomer David Rittenhouse (1732–1796).

Alexandre Guy Pingré

After unsuccessfully trying to fit a parabolic orbit to the observations, Lexell, at St. Petersburg, succeeded with an elliptical orbit having a period of only 5.6 years. In 1772 and 1779, Lexell published his research on the comet that was to bear his name and concluded the observations would not allow a period much larger than the 5.6-year period he had computed. To answer why the comet was not seen prior to 1770, Lexell pointed out that it had approached close to Jupiter in the early summer of 1767 when the separation distance was only $^{1}/_{58}$th that of the comet-Sun distance. Jupiter's attraction on the comet was three times that of the Sun, hence the strong perturbation

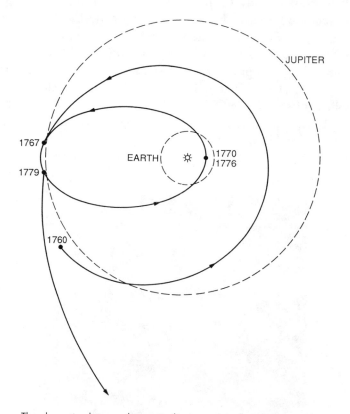

The dramatic changes that periodic comet Lexell underwent in the late eighteenth century are represented in this diagram. Prior to 1767, comet Lexell's perihelion distance was approximately 2.9 AU, too distant to be observed. A Jupiter close approach in 1767 kicked the comet into a path that reached perihelion in 1770 and 1776 inside Earth's orbit. The perihelion distance was then 0.7 AU and the comet was well observed in 1770. A subsequent Jupiter close approach in 1779 transformed the comet's orbit, such that its perihelion distance was then near Jupiter's orbit. Once again it became unobservable.

from its close approach threw it into a different orbit. Due to its unfavorable position with respect to the Sun, the comet was not observed in 1776. Lexell pointed out that another, even closer, approach to Jupiter in the summer of 1779 could have retarded the comet's motion so that its next perihelion passage might well be later than 1781. Astronomers looked in vain for the return of comet Lexell in 1781 and 1782.

The French celestial mechanician Urbain Jean-Joseph Le Verrier (1811–1877) gave a complete discussion of comet Lexell's motion in 1857 and pointed out that it was possible the comet passed within the satellite system of Jupiter in 1779. However, Le Verrier was careful not to predict its precise orbit prior to the 1767 Jupiter encounter nor subsequent to the 1779 event. He knew that the comet's motion, following its close Jupiter encounters, was critically dependent on the uncertain circumstances of the encounters themselves. Le Verrier gave 33 different sets of orbital elements that might represent the comet's path as a result of the large Jupiter perturbation in 1779. This extensive collection of orbital elements could then be matched with the elements of newly discovered comets in an effort to rediscover comet Lexell at some future date. The second short-period comet to be recognized, comet Lexell made only one observed return to the Sun; severe per-

Urbain Jean-Joseph LeVerrier. (*Yerkes Observatory photograph.*)

159

turbations by Jupiter in 1767 kicked it into the Earth's neighborhood, then kicked it out again in 1779.

Anders Lexell was also instrumental in resolving the question of a comet reported by William Herschel (1738–1822), father of John and brother of Caroline Herschel. In 1781, Herschel reported that he had discovered a new comet on March 13 near the constellation Taurus. Herschel's object was observed by Nevil Maskelyne (1732–1811) at Greenwich from April 1 and by Messier at Paris from April 16. Various attempts to fit an orbit to the observations were made by Méchain, Laplace, Boscović, and others. Initial orbit attempts assumed that the perihelion distance was less than 4 AU, the largest value computed for a comet to that date. In 1782, Jean-Baptiste-Gaspard de Saron (1730–1794) was the first to suggest that the perihelion distance was not only larger than 4 AU but likely to be larger than 12 AU. Once Saron pointed out that Herschel's comet was perhaps a new planet instead, Lexell suggested a year later that the observations could be satisfied with a circular orbit whose radius was 18.9 AU. The new planet was ultimately named Uranus.

The last premier president of the parliament of Paris, Saron was also a magistrate, an authority on jurisprudence, a skilled instrument maker, and a mathematician. A close friend of Messier, Saron made his last astronomical contribution while imprisoned and awaiting the guillotine during the French revolution. Saron was able to compute a postperihelion ephemeris for a comet that Messier had discovered on September 27, 1793. Using Saron's ephemeris, Messier was able to recover the comet after perihelion and to secretly inform Saron of his success just prior to his execution.

After comet Halley, the next short-period comet to be recognized, at more than one return to perihelion, was comet Biela. It was first seen on March 8, 1772, by Jacques Montaigne at Limoges, France. Approximate parabolic orbital elements were given by Pingré and when Gauss' elliptic orbit for the comet of 1806 I showed elements similar to the comet of 1772, it was strongly suspected that the comets seen in 1772 and 1806 were the same object. Periodic comet Biela was destined to split, then disappear, in the middle of the nineteenth century (see Chapter 8).

Cometary Thought in the Eighteenth Century

The eighteenth century witnessed several unusually bright and peculiar comets. The comet of 1729 was perhaps intrinsically the brightest comet for which an orbit has been computed. It was observed from August 1, 1729, through January 21, 1730. Although its perihelion distance was 4 AU, it was a naked-eye object at discovery on August 1, 1729, when the comet-Earth

Jean-Philippe Loys de Cheseaux

and comet-Sun distances were 3.1 and 4.1 AU respectively. Assuming this comet's brightness varied as the traditional inverse fourth power of heliocentric distance, its absolute magnitude—defined as its apparent magnitude when both the comet-Earth and comet-Sun distances are 1 AU—would have been a rather extraordinary −3. For comparison, the planet Mars at its brightest is slightly fainter, reaching a magnitude of −2.8.

The magnificent comet of 1744 was both bright and unusual in that it was reliably reported that it had multiple tails spread out like a fan. The Swiss astronomer Jean Philippe Loys de Cheseaux, after whom the comet is often named, began his observations on December 13, 1743, and computed a parabolic orbit based on his own observations through March 1, 1744. The comet rapidly brightened during December and its tail became more and more impressive. By mid-January, the apparent tail length was 7 degrees and on February 1 its apparent brightness was reported to be brighter than the brightest star, Sirius, with a curved tail some 15 degrees long. On January 18, it matched Venus in brightness and possessed a double tail with one branch extending 7 degrees in length and the other some 25 degrees. On February 25, in bright twilight, the comet was seen 3 degrees above the horizon. Two days later, it appeared brighter than Venus and was still visible to the naked eye 6 minutes before sunrise. Later that same day, the comet was

161

reported to be plainly visible only 12 degrees from the bright Sun. Before morning twilight on March 7 and 8, 1744, Cheseaux reported seeing a multiple-tail system, with 6 distinct rays extending above the horizon. These observations were confirmed through similar reports by Delisle at Paris and by Margaretha Kirch in Berlin, the daughter of Gottfried Kirch, who discovered the comet of 1680. On March 18, 1744, the comet's tail was reported as being some 90 degrees long. One month later, on April 18, it was fading rapidly, but was still a naked-eye object.

Discovered by Messier on August 8, 1769, the comet of 1769 also displayed an impressive tail, which grew to a maximum of 90 to 98 degrees on September 11 and still showed telescopic images of a tail into December 1769. The tail exhibited wavelike motions similar to those observed in aurorae. Pingré noted this motion on September 4, 1769. Earlier, Johann Cysat reported a similar tail behavior for the comet of 1618 while Johannes Hevelius noted this phenomenon in the comets of 1652 and 1661.

The eighteenth century had an unusually large number of comet-Earth close approaches, with eight comets coming within 2500 Earth radii. These comets include that of 1702 (minimum separation = 1025 Earth radii), and the comets of 1718 (2418); 1723 (2430); 1743 I (915); 1759 III (1598); 1763 (2190); and 1797 (2062). Of course the record Earth-comet close approach was that of comet Lexell, with a minimum separation distance of 354 Earth radii. In 1805, Laplace pointed out that the close approach of comet Lexell to the Earth did not sensibly alter the length of the year and hence the comet could not have been more massive than $\frac{1}{5000}$ of the Earth's mass. This result contributed to the notion that, despite an enormous atmosphere, the solid portion of a comet was very tiny indeed.

The eighteenth century was the age of Newton, and most contemporary scientists adopted his ideas for cometary tail formation, just as they adopted his theory for their motion. A notable exception was the prolific Leonard Euler. In 1748, Euler, arguing for the wave theory of light, claimed that the vibratory nature of light could affect cometary tail material in a manner similar to the effect of sound waves on tiny airborne particles. The French Cartesian astronomer, Jean-Jacques Dortous de Mairan (1678–1771), pointed out that although Euler's theory might lead to vibrations in cometary tail particles, it could not explain a net motion directed away from the Sun. According to the account given in Mairan's 1752 work on the aurora, the Sun was surrounded by an ellipsoidal atmosphere, flattened at the solar equator, which could occasionally reach the Earth's neighborhood. When this material fell into the atmosphere, the Earth's rotation tended to scatter it, but the effect was diminished at the poles. The Earth's polar atmosphere was then illuminated by the solar material and the aurora borealis resulted. Either the light was due to the faint glow of the solar material or it was illuminated by solar light. In a

Cheseaux's comet of 1744 showing the six separate tail structures observed on March 7–8, 1744. The head of the comet was below the horizon. Illustration from Cheseaux's *Traité de la comète qui a paru en Décembre 1743 & Janvier, Fevrier & Mars 1744.* (Lausanne and Geneva, 1744.) (*Photograph courtesy of the Crawford Library, Royal Observatory, Edinburgh, Scotland.*)

similar fashion, Mairan speculated that when a comet entered the solar neighborhood, it captured some of the solar material, which was repelled by the impulsion of the solar rays and pushed behind the comet. The tail phenomenon was then due to the resulting cometary aurora borealis.

163

An alternate explanation for the formation of comet tails was due to the American savant Andrew Oliver (1731–1799). In 1772, Oliver explained the cometary tail phenomenon by supposing that the atmospheres of the Sun, Earth, and comets were similar and mutually repellent. Although Oliver used the analogy of charged pith balls to demonstrate mutual repulsion, he stated that comet tails cannot be electrically charged because close cometary approaches to planetary atmospheres would cancel each other's charge—the arc of lightning being fatal to both inhabited worlds. Oliver was not the first to suggest that comets were inhabited. Bernard le Bovier de Fontenelle (1657–1757), in his enormously popular 1686 work *Entretiens sur la pluralité des mondes*, regarded comets as habitable, and Johann Lambert, in his 1761 work *Cosmologische briefe*, devoted an entire chapter to this idea. Even as late as 1828, the Scottish astronomer, David Milne, expressed his belief in the habitability of comets in his *Essay on Comets*.

One of the first scientific attempts to explain cosmology was due to William Whiston (1667–1752) in his 1696 book *A New History of the Earth*. Perhaps motivated by Edmond Halley's 1694 suggestion that the Biblical deluge may have been due to an Earth-comet encounter, Whiston's book was an impressive manipulation of Biblical chronology and Newtonian mechanics. To Whiston, the six days of creation actually meant six years because, at the time, the Earth's day was equal to one year. That is, the periods of the Earth's rotation about its axis and revolution about the Sun were both equal to one year. According to Whiston, on Friday, November 28, 2349 B.C., a comet passed very close to the Earth. The near collision caused a tidal breakup of the Earth's crust. The subsequent release of subterranean waters, together with precipitation from the comet's atmosphere and tail, caused the Biblical deluge by raising a tide several miles high. In addition to this explanation of Noah's flood, Whiston went on to predict the end of the world by supposing another close approach of the same comet on its next return to the Earth's neighborhood. This time the comet would alter the Earth's present orbit, causing it to approach the Sun and be consumed by the solar inferno. Alternatively, Whiston mentioned the end of the world might occur if the comet collided with the Earth.

Whiston originally had no particular comet in mind, but when Edmond Halley's erroneous suggestion that the comet of 1680 had a period of 575 years was first published in the 1713 edition of Newton's *Principia*, Whiston appended later editions of his book to specify this comet as the cause of the Biblical flood and the future end of the world. The end would come at the comet's next return to the Earth's neighborhood in about A.D. 2255. Whiston's book was well received, passing into its sixth edition in 1755, and helped establish the notion that a cometary collision with the Earth was a distinct possibility.

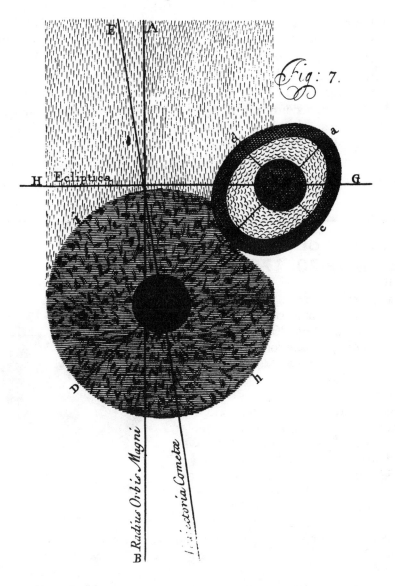

William Whiston supposed the biblical deluge was due to the close Earth approach of a comet that broke the Earth's crust and released subterranean waters. With the help of water released from the comet's heavy atmosphere and tail, an enormous tide was created. Illustration from Whiston's book entitled *A New Theory of the Earth.* (London, 1696.)

Even before Whiston predicted the comet of 1680 would return in the twenty-third century, the Swiss mathematician Jakob Bernoulli (1654–1705) had announced this comet would return on May 17,1719. Although Bernoulli's work in analytic geometry and the theory of probability is of fundamental importance, his 1682 treatise on comets was absurd. He suggested that comets were satellites of a single invisible planet located some 250 times more distant than Saturn. Noting that the comet of 1680 traveled 1.28 degrees in 50 days, Bernoulli reasoned that the comet would complete one revolution in 38 years and 147 days, bringing it back to perihelion on May 17, 1719. He failed to account for the nonuniform motion of the comet around its highly elliptical orbit. In memory of his father's opposition to his study of mathematics and astronomy, Bernoulli adopted the motto "against my father's will, I study the stars." As far as his work on comets is concerned, Bernoulli should have listened to his father.

The End of the World on May 20, 1773

Many people in Paris were convinced that a comet would destroy the Earth on May 20, 1773. On April 21, 1773, Joseph-Jérôme Lefrançais de Lalande had scheduled a lecture before the Paris Academy of Sciences to discuss his computations on 61 comets whose orbits brought them close to Earth's orbit. There was no time for Lalande's complete paper to be given and in summarizing the work, he noted that planetary perturbations might make a collision possible. While he knew the probability of a comet-Earth collision

Joseph-Jérome Lefrançais de Lalande

was extremely small, he failed to emphasize this point. The result was panic, based on the rumor that Lalande had withdrawn his complete paper to avoid widespread panic over a comet that was to destroy the Earth. The concern reached such a fever pitch that the Archbishop of Paris was asked by concerned citizens to proclaim 40 hours of prayer to avert the disaster. He was on the point of ordering the prayers when some academicians advised him of the situation's absurdity. In an attempt to quiet the fears, Lalande inserted an announcement in the May 7 issue of the *Gazette de France* stating that it was not possible to fix the dates of comet-Earth encounters.

Lalande's newspaper announcement did not completely quiet the public's fears and his efforts to suppress the furor may have been less than strenuous. Throughout his life, Lalande drew attention to himself

Another eighteenth century cosmological work invoked cometary near collisions to explain the origin of the solar system. Strongly influenced by Whiston's earlier work, Georges-Louis Leclerc (1707–1788) embedded his ideas in his 1749 multivolume work entitled *Histoire Naturelle.* Leclerc, or Comte de Buffon as he was to be remembered, suggested that the planets were formed in a filament detached from the Sun by a passing comet. The lighter parts of the Sun's mass necessarily received the greater velocity during the cometary encounter, so the outer planets would be less dense than those closer in. In the 1778 edition, Buffon speculated that the origin of comets was due to random stellar explosions. Prior to these explosions, the stars were

through his controversies with other astronomers, his prolific writings, and occasional bizarre episodes. One such episode was his campaign to lessen the popular fear of spiders—showing his devotion to the cause by eating a number of the little creatures.

The foolishness of the comet scare of 1773 was immortalized by the sarcastic pen of Voltaire. Then 79 years old, he wrote in his *Letter on the Alleged Comet:*

Grenoble, May 17, 1773

Some Parisians, who are no philosophers—and, if they are to be believed, will not have time to become so—have informed us that the end of the world is approaching, and that it will infallibly take place on the 20th of this present month of May.

On that day they expect a comet which is to overturn our little globe and reduce it to impalpable powder, according to a certain prediction of the Academy of Sciences which has not been made.

Nothing is more probable than this event; for Jacob Bernoulli, in his "Treatise upon the comet," expressly predicted that the famous comet of 1680 would return with a terrible crash on the 17th of May, 1719. He assured us that its wig would signify nothing dangerous, but that its tail would be an infallible sign of the wrath of heaven. If Jacob Bernoulli has made a mistake in the date, it is probably by no more than fifty-four years and three days.

Now, an error so inconsiderable being looked upon by all mathematicians as of no account in the immensity of ages, it is clear that nothing is more reasonable than to expect the end of the world on the 20th of the present month of May, 1773, or in some other year. If the event should not happen, what is deferred is by no means lost.

Voltaire's letter was published anonymously in the June 1, 1773, issue of the French *Journal Encyclopédique.*

similar to glass and received their heat from the pressure of comets orbiting about them. While Buffon's ideas were popular in the eighteenth century, contemporary astronomers never fully accepted his rather qualitative notions.

Summary

Cometary thought in the eighteenth century focused on the dynamics of comets. Initiated by Isaac Newton and Edmond Halley, orbit determination techniques were successfully developed by Rudjer Boščović, Leonhard Euler, Achille du Séjour, Pierre Laplace, Heinrich Olbers, and Carl Gauss; the techniques of the last three are still used. The number and quality of mathematicians in this era were remarkable, and many of them brought their talents to bear on cometary orbit determination.

By the end of the century, orbit computers thought they had identified several periodic comets to follow Halley's triumph. In fact, they had found only two—neither was observed a second time. Most comets identified as periodic were not, and the few comets that actually were went unrecognized. The identification of one periodic comet was an outright fraud. Despite the questionable success record for identifying periodic comets, the techniques and skills developed for computing cometary orbits helped usher in the new field of celestial mechanics. The understanding of cometary motion is one of the principal achievements of eighteenth century astronomy.

The techniques for computing cometary orbits were well in hand by the end of the century and contemporary astronomers were content in the knowledge that comets were solid, permanent, celestial bodies, whose motions obeyed the laws of Newtonian mechanics. To contemporary astronomers, the appearance of a comet was still a mystery, but at least they were permanent objects whose motions could be followed by straightforward application of the same Newtonian techniques used to predict the motion of planets. It was only a matter of time before more periodic comets would be successfully predicted. Indeed, in the following century, comets Encke and Biela were identified as the next periodic comets, after Halley, whose returns were successfully predicted. However, comet Encke was to reveal a dynamic behavior that was not consistent with Newtonian mechanics and comet Biela was to split, disintegrate, and disappear, showing rather dramatically that comets are not permanent celestial objects.

8
The Nineteenth Century

The Shrinking Orbit of Comet Encke, the Disintegration of Comet Biela, and the Meteor Showers

The orbit of comet Encke is shrinking in size: Johann Encke suggests a resisting medium, while Wilhelm Bessel favors rocketlike thrusting of the comet itself. Biela's comet disintegrates in 1846 and 1852. A connection is established between the debris of comets and meteor showers, or shooting stars.

Comet Encke and the Resisting Medium

On January 17, 1786, Pierre Méchain discovered a faint comet in the region of Aquarius. After notifying Charles Messier of his discovery, the comet was observed two nights later by Méchain, Messier, and Jean-Dominique Cassini (1748–1845). Because it was only observed on these two nights, an orbit was not attempted. Nearly 10 years later, on November 7, 1795, Caroline Herschel, at Slough, England, discovered her sixth comet and observed it long enough to determine its orbit. In another 10 years, a faint comet was discovered on October 20, 1805, by Jean Louis Pons at Marseille, Johann Sigismund Huth at Frankfurt an der Oder, and Alexis Bouvard (1767–1843) at Paris. When Pons discovered another comet on November 26, 1818, it was well observed for nearly seven weeks and Heinrich Olbers suspected that this comet, 1819 I, was identical with the comets of 1805, 1795, and 1786 I.

However it was the German astronomer and mathematician Johann Encke who provided the necessary computations in 1819 to show that comets 1819 I, 1805, 1795, and 1786 I were the same comet returning at intervals of 3.3 years. Encke identified the four apparitions by mathematically integrating the comet's motion backward in time, including approximate perturbations by all the known planets except Uranus. His computed positions for the comet agreed with the observed positions at each of the four apparitions. Encke also integrated its motion forward to correctly predict the next perihelion passage time as approximately May 24, 1822. This was the second confirmed prediction for the return of a periodic comet, and—as with the honor bestowed on Edmond Halley—Encke's name would be attached to this comet thereafter, though Encke himself modestly referred to it as Pons'

Johann Franz Encke. (*Courtesy of Special Collections, San Diego State University Library.*)

comet. At the time, Encke was studying under Carl Gauss and with Gauss' help, he later procured a position at the small Seeberg observatory near Gotha, Germany in 1816 and soon became its director. Nine years later he became the director of the Berlin observatory. For the better part of his career, Encke studied the motion of the comet that bears his name.

Although Encke's comet was poorly placed for northern hemisphere observations in 1822, it was observed in June 1822 for three weeks by Karl Rümker at the private observatory of Sir T.M. Brisbane at Paramatta, New South Wales, Australia. Using Rümker's observations, Encke noticed that the observed perihelion passage was a few hours before his prediction. More certain of a suspicion alluded to in his previous work, Encke postulated in 1823 that the comet moved under the influence of a resisting medium. The resisting force was supposed proportional to the density of the medium through which it traveled and to the square of the comet's orbital velocity. The density of the resisting medium was assumed to vary as the inverse square of the distance from the Sun.

Using his mathematical model of the resisting medium to decrease the comet's computed period by approximately 2.5 hours at each return, Encke successfully predicted the perihelion passage times for the next returns from 1825 to 1858. In his work published in 1859, Encke related that from 1819 to 1848, planetary perturbations were taken rigorously into account, but during the 1848 to 1858 period only Jupiter's perturbations were considered. The effects of the remaining planets were ignored because of a condition that is still a problem today—the computers were overloaded. While modern electronic computers were a century away, nineteenth century computers were young men and women who were hired to figure particular computations in assembly line fashion. Because the many newly discovered minor planets occupied so much of the Berlin observatory's computer force, Encke had to neglect planetary perturbations other than Jupiter's. However, we can surmise that Encke got the most out of his computers because he was a master of rapid, accurate, and efficient computation. Although he was anticipated by the American astronomer George P. Bond (1825–1865) in 1849, three years later Encke independently developed a numerical integration technique still used to compute the motions of celestial bodies. The Bond-Encke method computes perturbations, step-by-step, in an object's rectangular, heliocentric coordinates, rather than its orbital elements.

Encke's work on his comet included efforts to represent its motion over the interval from 1786 to 1858. His success in representing its dynamic behavior with his resisting medium hypothesis was strong evidence for the medium he envisioned as either an extension of the solar atmosphere or the debris of planetary and cometary atmospheres left in space. Because of Encke's research on this comet's anomalous, nongravitational motion, many

171

contemporary astronomers favored the resisting medium hypothesis; one notable exception was Friedrich Bessel.

While working as an apprentice in a mercantile establishment in Bremen, Germany, young Bessel attended lectures by the astronomer Heinrich Olbers. His interest awakened by Olbers, Bessel purchased Joseph-Jérôme de Lalande's popular textbook on astronomy and with this reference computed an orbit for the 1607 apparition of comet Halley. Bessel analyzed a set of observations that had been recently discovered by Baron Franz von Zach in 1793. After dismissing the observations of a William Lower (mistakenly attributed to a Nathaniel Torporley) and a man named Standish, Bessel based his final orbit on one observation each by Johannes Kepler and Longomontanus and eight by Thomas Harriot.

On July 28, 1804, Bessel chanced upon Olbers on a Bremen street and the 20-year-old Bessel inquired if Olbers would permit him to bring to his house a short manuscript on the comet of 1607. Olbers consented to review Bessel's first scientific work and the next day Bessel received a letter and some books from Olbers. The letter stated that Olbers had read the manuscript and that it showed remarkable mathematical and astronomical knowledge. In 1804, Bessel's work was submitted by Olbers himself for publication in Baron von Zach's *Monatliche Correspondenz*. Thus Bessel, who is best remembered for his later discovery of stellar parallax, began his extraordinary career computing orbits for comets. Olbers would later quip that his own greatest contribution to astronomy had been to encourage the young Bessel to become a professional astronomer.

During the 1835 apparition of comet Halley, Bessel observed its nuclear region carefully and noticed that some emanations facing the Sun looked like a burning rocket. Analogous to a rocket, these sunward emanations should produce a nearly radial thrust away from the Sun. Arguing that a resisting medium was not evident in the motion of the planets, Bessel suggested that the orbital periods of comets could decrease with time as a result of reactive forces acting on the comets themselves. As an example, Bessel calculated how much comet Halley's period would be shortened if its emanations took place before perihelion and lasted 23 days beyond October 2, 1835. If the daily mass loss from the nucleus was assumed to be 1/1000 of the total, the result was a shortening of the comet's orbital period by 1,107 days.

Bessel's example was quite an overestimate as the actual change in comet Halley's period is an increase—not a decrease—of approximately four days per revolution about the Sun. However the idea of a rocket effect to explain the anomalous motions of comets was perceptive and anticipated the currently accepted explanation for these effects advanced in 1950 by Fred L. Whipple.

Friedrich Wilhelm Bessel

In 1878, Friedrich Emil von Asten (1842–1878) published his investigation into the motion of comet Encke over the observed period from 1818 to 1875. He tried to determine whether Encke's or Bessel's hypothesis better explained the comet's anomalous motion. Asten thought that Encke's resisting medium would cause variations with time in both the orbit's semimajor axis, a, and eccentricity, e, while Bessel's reactive model would cause only a determinate variation in the semimajor axis. Hence Asten made the time variation of the eccentricity one of the quantities he solved for in his orbit determination. The remaining parameters in his solution set were the six orbital elements, the change with time in the comet's mean motion, and the masses of Mercury, Jupiter, and the Earth-Moon system. The change in the mean motion would determine the variation with time in the semimajor axis, because the mean motion, n, and the semimajor axis, a, are related by the expression

$$n \text{ (degrees/day)} = \frac{k}{a^{\frac{3}{2}}}$$

173

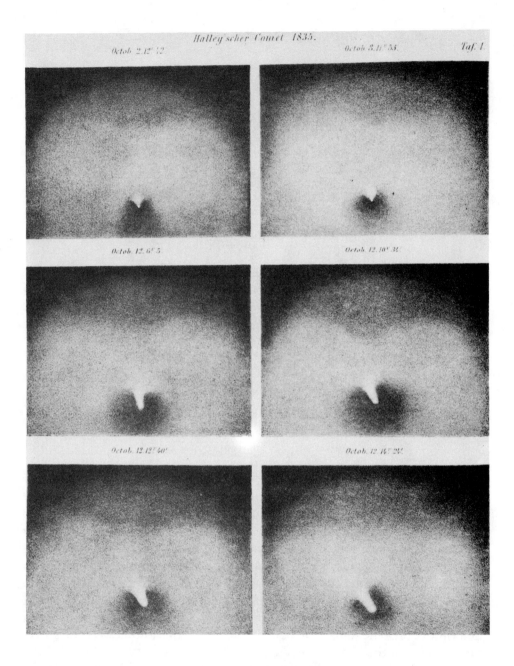

Drawings of comet Halley as seen by Friedrich Bessel from October 13 through October 25, 1835. The Sun's position is toward the top of the page. Bessel used his observations of the sunward emanations to suggest that the comet's motion could be affected by rocketlike thrusting.

Bessel's drawings of comet Halley's outgassing in 1835.

Top:	1835 Oct. 2 23h 42m	Oct. 3 22h 53m
Middle:	1835 Oct. 12 17h 5m	Oct. 12 21h 34m
Bottom:	1835 Oct. 12 23h 40m	Oct. 13 1h 24m

To Bessel's times we add 12 hours to put beginning of day at midnight and subtract 1 hour to get from Berlin mean time to GMT.

Friedrich Emil von Asten. (*Photograph courtesy of N.A. Belyaev.*)

where a is the semimajor axis expressed in astronomical units, AU, and k is the so-called *Gaussian constant* having a value of 0.9856 for the units used here. From this relationship, an increase with time of a comet's mean motion implies the orbital semimajor axis and the period are decreasing with time. In the absence of planetary perturbations, a comet's mean motion remains constant, with the value depending only on the orbital semimajor axis. However, forces acting on a comet as a result of a resisting medium or reactive rocket-like thrusts from the nucleus itself would introduce time variations in a comet's mean motion. These latter effects are termed *nongravitational* because they are not due to gravitational interactions with other solar system bodies.

Over the 50-year period from 1818 to 1868, Asten determined that the computed decrease in eccentricity with time was very close to the value predicted by the resisting medium hypothesis, and he concluded Encke's concept was valid. Since other periodic comets, with larger perihelion distances, were apparently not affected by nongravitational forces, Asten suggested that the dense region of the resisting medium did not extend much beyond the orbit of Mercury.

While Asten's work on comet Encke over the 1818 to 1868 interval confirmed Encke's previous work, the 1871 return of the comet revealed no time variation in the mean motion. Asten speculated that the apparent disappearance of these nongravitational effects may have been caused by a collision with a minor planet in June 1869. He pointed out that though comet Encke passed within six million miles of minor planet (78) Diana[1] in late May 1869, this close approach could have accounted for only 1/100 part of the observed effect. Asten finished his investigations by noting that nothing short of an actual collision with some undiscovered minor planet was capable of explaining the orbital motion of comet Encke in 1871.

The Swedish astronomer Jöns Oskar Backlund (1846–1916), who became director of the Pulkovo observatory in 1895, also devoted a major portion of his career to the study of comet Encke. For his work, Backlund received scientific recognition and the Royal Astronomical Society of London's gold medal for 1909. In spite of the honors and recognition, Backlund had only limited success with the troublesome comet. Backlund rejected Asten's suggestion of a collision with a minor planet and noted that the observations of comet Encke from 1858 to 1871 were compatible with a constant time variation in its mean motion, but the magnitude of this effect was smaller after 1858. He decided to recalculate all planetary perturbations on comet Encke since 1819. By 1886 he had obtained the necessary financing and hired enough computers to carry out this enormous task.

Ultimately, Backlund concluded that, although the change of comet Encke's mean motion with time was nearly uniform from 1819 to 1858, it was impossible to fit the observations over the longer 1819 to 1891 interval using a constant rate of change for the mean motion. He suggested that discrete changes in the mean motion had occurred in 1858, 1868, 1895, and probably also in 1904. In other words, Backlund concluded that the comet's semimajor axis and orbital period were decreasing with time but the rate of decrease was not as fast in the late nineteenth century as it was earlier. To explain comet Encke's peculiar nongravitational motion and its continued departures from his predictions, Backlund considered two possibilities:

1. After noting the coincidence between the times of change in the comet's mean motion and maximum solar surface activity, Backlund suggested a possible connection with tangential electric forces.

2. The resisting medium took the form of a meteoric ring near the comet's perihelion. The changes in the comet's mean motion with time were then a measure of the reduction in the resisting medium's density with time.

176

Jöns Oskar Backlund. (*Photograph courtesy of N.A. Belyaev.*)

Backlund ruled out Bessel's hypothesis, whereby reactive forces acted on the comet itself, because there was no obvious change in its appearance after each change in its mean motion. Backlund clearly preferred the meteoric ring hypothesis for explaining comet Encke's anomalous motion.

The French astronomer, Hervé Faye (1814–1902), developed an hypothesis whereby the time variation of comet Encke's mean motion was considered due to the repulsive force of solar rays. In 1858, Faye argued that the Sun's influence introduced the required force contrary to the comet's motion via an aberration effect. This solar repulsive force also formed the tail phenomenon. A periodic comet, discovered by Faye in 1843, also played a role in the resisting medium controversy.

In two scientific papers published in 1861, the Swedish astronomer Didrik Magnus Axel Möller (1830–1896) used the observations of comet Faye over the 1843 to 1858 interval to determine the time variations in its orbital mean motion and eccentricity. These variations apparently agreed with Encke's resisting medium and also appeared consistent with Faye's solar radiation hypothesis. However, in papers published in 1865 and 1872, Möller

177

The Millerites, the Great Comet of 1843, and the End of the World

The great comet of 1843 witnessed a return to the notion that comets were indications of impending disaster—in this case the end of the world. William Miller and his followers, the Millerites, sincerely believed the world would come to an end in 1843.

Miller lived most of his life as a simple farmer in Hampton, New York and Poultney, Vermont. During the war of 1812, Captain Miller fell from the back of a horse-drawn wagon onto his head, which some people suspected was a factor in his later delusions. After the war, Miller began an intensive 14-year study of the Bible. Based upon a rather strange interpretation of the eighth chapter of Daniel, Miller worked out his prophetic view. The Earth would be destroyed by fire sometime in 1843. Before the destruction, Christ would appear a second time and raise the righteous, and the Earth below would be purified by fire. After the purification, the millennium would begin.

Driven by an intense missionary zeal and aided by an electrifying speaking manner, Miller began lecturing in 1831. In the autumn of 1836 he delivered some 82 lectures and Prophet Miller, as he was now called, began to collect followers. In 1840, Miller and the Millerites moved their lecture circuit from rural New England to the limelight of the larger cities. It made little difference to them that opposing clergy refuted his theories.

As the fateful year approached, people who previously regarded the Millerites only with curiosity were now a bit on edge. What if Miller was right? By early 1843, there were more than 50,000 true believers and whenever Prophet Miller and his followers held meetings, crowds gathered. During these meetings, some of the more ardent Millerites expected to rise up into the air to fulfill the prophecy. Although Miller himself had stated no specific date as the end of the world, his followers fixed on some dates of their own. In early 1843, the *New York Herald* newspaper announced the Millerites had selected April 3. As the date approached, excitement grew and some Millerites abandoned their crops, donned white ascension robes, and waited on their rooftops for their skyward trip. In New England's villages and cities, the Millerites gazed heavenward looking for a sign that Prophet Miller said would announce the Earth's destruction and the beginning of the millennium. Enter the great comet of 1843.

The comet of 1843 was first seen by the New England populace in late February in broad daylight only a few degrees from the noonday Sun! To easily see a comet so close to the Sun is an extraordinary event and when the comet had rounded the Sun and moved far enough away from the solar glare, it appeared as an impressive object in early March 1843, low in the southwest evening sky. As bright as the brightest stars and with a tail stretching upward for 50 degrees, the comet must have made quite an impression. The exaltation of the Millerites knew no bounds. Here at last was their celestial sign.

However, the appearance of the comet in March 1843 was the peak of the Millerite movement. The comet itself faded quickly from view in late March. The world did not end on April 3—nor did it end on subsequent dates that the retreating Millerites suggested. Continually revising the date for the end of the world, the Millerites were able to sustain the movement well into 1844. Finally, Prophet Miller himself endorsed October 22, 1844, as the world's end, but when that date came and went without incident, the Millerites began to disband. The humiliation of repeated failures was too much to sustain even the most dedicated believers. Miller himself admitted his prophetic Biblical interpretations were in error—what else could he do?

Engraving of the great comet of 1843 seen over the city of Paris.

Hervé Faye

found his previous result to be in error so that no time variations in the mean motion or eccentricity of comet Faye were evident.[2]

Initial support for Encke's resisting medium hypothesis also came from the work on periodic comet Pons-Winnecke by the Austrian astronomer Theodor Ritter von Oppolzer (1841–1886). In 1880, Oppolzer announced that he had found a uniform increase in the mean motion of comet Pons-Winnecke that was consistent with Encke's resisting medium hypothesis. However, a more thorough study of this comet by Eduard F. von Haerdtl nine years later did not indicate any such increase. In the late nineteenth century, this comet actually had a slight decrease in its mean motion with time, but the effect was probably too small to have been detected by methods then employed.

The final blow to the resisting medium came when Michael Kamienski (1880–1973) in 1933 and Albert William Recht in 1940 found uniform decreases in the mean motion of periodic comets Wolf and d'Arrest respectively. With the discovery of mean motion that decreased with time, as well as increased, a successful hypothesis had to explain both phenomena. A resisting medium could only cause the latter phenomenon.

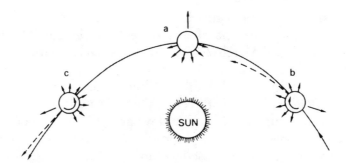

Schematic diagram showing Fred Whipple's model for explaining the nongravitational forces acting on comets. In case a, the comet's nucleus is not rotating, so the outgassing of the cometary ices is toward the Sun, with the resultant rocketlike thrust acting in a radial, antisolar direction. In case b, the comet is rotating in a retrograde fashion so that the outgassing is offset from the purely radial case, and a transverse force component acts counter to the comet's motion about the Sun. In case c, the direct rotation of the comet's nucleus produces a force in the same direction as the comet's motion about the Sun.

In 1950, Fred Whipple solved the problem by suggesting that the vaporization of a dirty-ice nucleus, as it approached the Sun, provided the mechanism for explaining the increases and decreases observed in a comet's mean motion. Whipple's rocket effect was reminiscent of that suggested by Bessel, except that Whipple explained the physical cause of the effect and allowed the nucleus to rotate in either a direct or retrograde sense. This provided an outgassing, rocketlike thrust either acting along, or contrary to, the comet's orbital motion. Whipple's model relied on the nucleus rotation to create rocketlike acceleration components acting both in a radial, Sun-comet direction and in a transverse direction, the latter being in the comet's orbit plane and at a right angle to the radial direction. Bessel's mechanism accomplished the same effect using only a radial acceleration of short duration that occurred primarily before or after perihelion.[3] Bessel's hypothesis was originally suggested by the emanations seen near the nucleus of comet Halley in 1835. It was obvious to Bessel that the comet's motion must be affected by this rocketlike phenomenon. It was also obvious that comet Halley was losing mass, or disintegrating, with time. However, it would be the third confirmed periodic comet, Biela, that would provide the most dramatic evidence for cometary disintegration.

The Short Life of Comet Biela

For a comet of such notoriety, periodic comet Biela's observed lifetime began in a rather unimpressive fashion. Jacques Leibax Montaigne, at

Limoges, France, discovered this faint comet on March 8, 1772, using a Dolland achromatic telescope of one meter focal length. Montaigne observed the comet until March 20. Montaigne's rival, Charles Messier, had been searching for it since March 15 but only managed to observe it four times between March 26 and April 3. While no astronomers suspected its periodic nature in the eighteenth century, a few suggested its periodicity at its next observed return in 1805. On November 10, 1805, Jean Louis Pons rediscovered the comet and it was observed until December 15. Near its perigee on December 10, when it passed within 0.04 AU of the Earth, it was observed with the naked eye by Heinrich Olbers, Friedrich Bessel and Johann H. Schröter (1745–1816) at Lilienthal, near Bremen, Germany. Bessel computed an orbit for this comet and suggested its identity with the comet of 1772. Carl Gauss also computed its orbit using the technique he had recently developed for the orbit of Ceres. Gauss found the comet's orbital elements similar to those for the comet seen in 1772 and suggested this was the same comet returning every 4.7 years.

The third observed return of comet Biela was in 1826, when it was discovered by an Austrian infantry captain, Wilhelm von Biela (1782–1856). With the Napoleonic wars concluded, Biela began a study of astronomy in 1815 at Prague's Charles University. He attended the astronomical lectures of Alois David, director of the Prague observatory, and became an avid amateur astronomer. Biela also became acquainted with Josef Morstadt, an Austrian civil servant and amateur astronomer with a private observatory in Prague. In 1826, Biela was stationed at his garrison in Josefov, a small fortress town in eastern Bohemia. It was there, on February 28, 1826, that Biela, armed with a small telescope, discovered the comet that was to bear his name.

Although some historians have suggested that Morstadt and Biela concluded the comet's period was 6.75 years and both expected its return in 1826, neither the discovery announcement nor the subsequent orbital computations by Biela mention his anticipation of its arrival. Thus it is not altogether clear whether Biela's discovery was by chance or by design. On March 9, the same comet was independently discovered by Jean Félix Adolphe Gambart (1800–1836) at Marseille, France. Gambart, Biela, and Thomas Clausen (1801–1885) computed orbits for this comet and recognized that the comets seen in 1772, 1805, and 1826 were the same object. All three suggested orbital periods of approximately 6.7 years and communicated their results in letters to the editor of the *Astronomische Nachrichten* in 1826. Although claims were made on behalf of Gambart's priority, Biela made the first discovery and the honor of the comet's name belonged to him.

After the comet's 1826 apparition, its next perihelion passage was predicted to be late November 1832 by both Marie Charles Théodor Damoiseau

Wilhelm von Biela. (*Photograph courtesy of Karel Hujer.*)

(1768–1846) at Paris and Giovanni Santini (1787–1877) at Padua, Italy. Both computers were within two days of the correct result. Using Santini's ephemeris, John F.W. Herschel, at Slough, England, first recovered comet Biela on September 24, 1832. It was followed through the months of October, November, and December, and was last seen by Thomas Henderson at the Cape of Good Hope on January 4, 1833.

Although comet Biela was too poorly placed for any observations in 1839, the 1846 apparition was well observed. Santini continued his work on its orbit and again predicted an accurate perihelion passage of 9 P.M. on February 1846, within 10 hours of the correct time. The comet was recovered on November 26, 1845, by Francesco P. DiVico (1805–1848) at Rome and two days later by Johann G. Galle (1812–1910) at Berlin. Its routine recovery and faint appearance in November 1845 gave no clue to the startling sequence of events that followed. The comet was observed on December 21 by Encke at Berlin and on Christmas night by Jean Élix Benjamin Valz at Marseille. Both observers reported it to be a rather lackluster 8- to 9-magnitude, diffuse object.

The first indication of the comet's peculiar behavior came on December 29, 1845, when Yale College librarian Edward C. Herrick (1811–1862), and a New Haven, Connecticut banker, Francis Bradley, used Yale's five-inch Clark refracting telescope to observe a faint secondary comet more than one arc minute northwest of the primary comet. On January 13, 1846, the first director of the newly founded U.S. Naval Observatory in Washington, D.C., Matthew Fontaine Maury (1806–1873), also observed a faint sec-

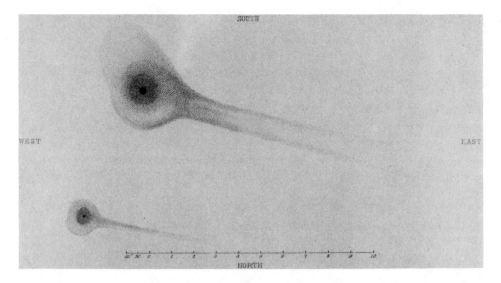

Double comet Biela as noted by Otto Struve at Pulkovo on February 19, 1846. The primary comet is shown to be southeast and approximately 6.5 arc minutes from the secondary. In this illustration, north is down and east is to the right.

ondary object 1 to 2 arc minutes north of the primary comet. Two nights later, the comet was observed as a double object by Moritz Ludwig Georg Wichmann at Konigsberg, Germany and by James Challis (1803–1882) at Cambridge, England. Challis only noted the phenomenon briefly because he was anxious to observe the newly discovered asteroid (5) Astraea. When Challis returned to his observations of comet Biela on January 23, 1846, the comet again appeared double, but he did not publish his observations until March 1846.[4] On January 27, the comet was observed to be double by both Encke and d'Arrest at Berlin and by Valz at Marseille. By the end of January, its duplicate nature was apparent to all observers and both comets exhibited nuclei, comae, and short tails parallel in direction and nearly perpendicular to an imaginary line joining them.

In mid-January, the new secondary comet to the northwest was extremely small and faint compared with the primary. By February 10, 1846, they were equally bright and on February 10 to 16 the secondary comet was larger and brighter. However, by February 18, the secondary had faded to half as bright as the primary. Maury observed a fine luminous arc between the two comets toward the end of February, and a month later the secondary component had faded from sight. The primary component remained visible until April 27, 1846.

The last apparition of comet Biela began on August 26, 1852, when Angelo Secchi (1818–1878), at Rome, recovered the comet at a considerable

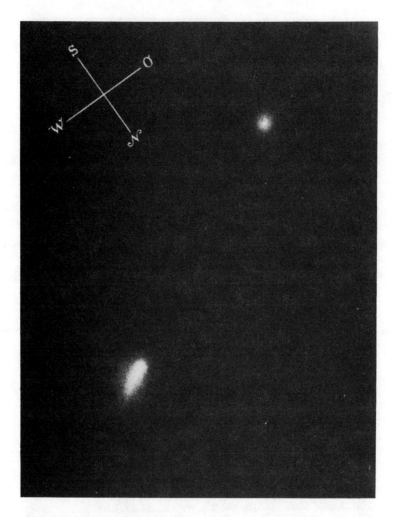

On September 25, 1852, Otto Struve sketched double comet Biela as seen from the Pulkovo Observatory (near Leningrad, Soviet Union). North is down and east is to the right.

distance from its predicted position. When the secondary component was re-covered by Secchi on September 16, it was still northwest of the primary. The double comet was extensively observed by Otto W. Struve (1819–1905) with the Pulkovo 15-inch refractor from September 18 to 28. On the 18th, Struve recorded that the secondary component was of magnitude 8 to 9, with no nu-cleus, but with a slight central condensation. On the 20th, the secondary component had a starlike nucleus with a coma diameter of approximately 40 arc seconds. In addition, Struve observed a jet emanating from the secondary directed toward the approximate position of the primary. On September 23,

The End of the World on June 13, 1857

For some Parisians in 1857, there was a frightening comet scare. Their concern was all the more unfounded because no real comet was involved. In 1751 the English astronomer Richard Dunthorne studied the observations of an impressive comet seen in 1264 and determined its orbit. Comparing this orbit with one computed by Edmond Halley for the comet of 1556, he was struck by the similarities and suggested the comet's period of 292 years would bring it back to the Sun's neighborhood in 1848. The French astronomer Alexandre Pingré arrived at a similar conclusion around 1760. In the 1840's, the Englishman John Russell Hind obtained a somewhat different orbit for the comet of 1556 than Edmond Halley. Depending on whether one used Halley's orbit or Hind's, the comet's return was predicted for August 1858 or August

_La voyez-vous, la comète ? ... là-haut ... au bout de mon doigt ... ne perdez pas de vue mon doigt ! ...

An enduring legacy of the 1857 comet scare was a series of contemporary cartoons drawn by the French satirist Honoré Daumier. The first cartoon shows a Parisian pickpocket relieving a citizen of his pocket watch while diverting his attention by saying "See the comet? . . . up there . . . at the end of my finger . . . don't lose sight of my finger!"

1860. Hind concluded that with all the uncertainties, the comet might be expected sometime within two years on either side of 1858.

Based on these calculations and his own imagination, a German astrologer inferred that the comet would not only arrive in 1857, but would strike the Earth on June 13. This disturbing news was published in an almanac and the bogus prediction spread rapidly throughout Europe, particularly Paris. A foreign correspondent for *Harper's* weekly reported that parts of Paris were panic-stricken. Women miscarried, crops were neglected, wills were drawn, comet-proof suits were manufactured, and a cometary life insurance company was established—premiums payable in advance. It was reported that the churches and confessionals were crowded for days. June 13, 1857, came and went. Again the Earth escaped destruction at the hands of a marauding comet. It was later shown by Martin Hoek, a Dutch astronomer at Utrecht, that the comets of 1264 and 1556 were two different comets and should not have been expected to return at all.

_ J'en voulons point d'vot' billet ......l'échéance est au quinze juin et l'monde finissiont l'treize !....

The second cartoon shows a Parisian refusing the note of another by pointing out that the due date of the note is June 15 while the world ends on the thirteenth!

the secondary coma was circular, but two days later it was slightly elongated and 50 to 60 arc seconds in diameter. On September 28, the secondary component could just be seen in bright moonlight and strong twilight. On the same evening, Struve could not discern the southeast primary component.

The primary component started out brighter than the secondary in 1852, but on September 20 they were comparable. By September 25, the primary was considerably fainter than the secondary and was not seen at all on the 28th of the month when the secondary was last seen. In a similar fashion, the primary component started out with a greater coma diameter, 1 minute 40 seconds compared with 40 seconds on September 20, but finished with a smaller angular extent, 30 seconds compared with 50 to 60 seconds on September 25. Similar relative changes in the apparent extent and brightness of both comets were noted in 1846. Those astronomers fortunate enough to observe comet Biela in either 1846 or 1852 were witnessing the death throes of a periodic comet. Comet Biela was poorly placed for observation in 1859, but an extensive search for the double comet, during what should have been an easily observed 1866 apparition, proved futile.

Brian G. Marsden and Zdenek Sekanina concluded in 1971 that comet Biela split between early 1842 and mid-1843 with a relative velocity between the two portions of only one meter per second. They identified the southeastern component in both 1846 and 1852 as the primary one and suggested that, if it had not dissipated completely, the primary comet should have passed within 0.05 AU of the Earth in November 1971. However, no trace of the comet was found. The dramatic splitting of comet Biela and its failure to reappear after 1852 strongly suggested to contemporary astronomers that comets disintegrate with time. This concept was reinforced with the appearance of another double comet, 1860 I, observed by the French astronomer Emmanuel Liais in Olinda, Brazil from February 27 until March 14, 1860. The debris from the cometary disintegration was soon recognized as the answer to another impressive celestial phenomenon—meteor showers.

Comets and Meteor Showers

The modern definition of the term *meteor* refers to the track of light, or shooting star, in the night sky that occurs when an interplanetary particle of rock or dust burns while falling through the Earth's atmosphere. Exceptionally bright meteors, called *fireballs,* occur when larger particles enter the atmosphere. A meteor shower results when the Earth collides with a group of interplanetary particles traveling together as a meteor stream. An interplanetary chunk of rock, with sufficient size and cohesiveness to survive its fiery trip through the Earth's atmosphere, is called a *meteorite.* Meteor stream particles are thought to be the remnants of cometary disintegration, while fireballs and meteorites are thought to originate as asteroidal fragments.

As early as 1714, Edmond Halley wrote that meteors might be of extraterrestrial origin. Noting their extraordinary velocities in the Earth's atmosphere, Halley suggested that they may be caused when matter formed in the ether "by some fortuitous concourse of atoms" collides with Earth in its motion about the Sun. Halley's idea, together with similar notions put forward by English scientist John Pringle (1707–1782) in 1759 and American astronomer David Rittenhouse in 1786, were among the few arguments for the cosmic origin of meteors prior to Ernst Florenz Friedrich Chladni's (1756–1827) work of 1794. Before the early 1800s, meteors and meteorites were generally considered terrestrial in origin.

On July 24, 1790, a number of meteorites fell near Agen, in southwestern France, and the phenomenon was seen and documented by no less than 300 people. Meteorite fragments were exhibited but Pierre Bertholon, editor of the *Journal des Sciences Utiles,* after publishing the account in 1791, dismissed the reports as groundless and physically impossible.

In the early morning hours of December 14, 1807, a huge fireball swept over New England and crashed into the Earth near the town of Weston, Connecticut. Benjamin Silliman (1779–1864), then professor of chemistry at Yale College, and college librarian James L. Kingsley (1778–1852) collected many samples of the meteorite, including a piece weighing approximately 200 pounds. When the evidence for the Weston meteorite shower was brought to the attention of President Thomas Jefferson, he remained skeptical but the remark often attributed to him is probably apocryphal: "it is easier to believe that two Yankee professors would lie than that stones would fall from heaven."

The father of acoustic science, Ernst Chladni, published a work in 1794 that argued for the cosmic origin of fireballs and the generic connection between fireballs, meteorites, and meteors. Chladni systematically analyzed a large amount of published information on meteorite falls and proposed that certain metal-rich masses, such as iron meteorites, could not have been produced by either natural or artificial processes on Earth. Chladni considered the likely source of fireball material was the debris remaining in interplanetary space after the formation of the planets. Although this idea was not well received by his contemporaries, it is remarkably in step with current ideas.

Between September 11 and November 4, 1798, two students at Göttingen University, Johann F. Benzenberg (1777–1846) and Heinrich W. Brandes (1777–1834), observed 22 meteors simultaneously from different locations a few kilometers apart. Their parallax measurements were designed to determine the height at which these meteors were extinguished, or disappeared, in the Earth's atmosphere. For 17 of the meteors, they determined

Ernst F.F. Chladni. (*Courtesy of N. Sperling.*)

heights that varied from 10 to 214 kilometers above the Earth's surface. Two meteors appeared to descend in the Earth's atmosphere, while two more appeared to ascend. From their data, the mean height of disappearance was 89 kilometers, and although they did not conclude that their observed meteors were extraterrestrial, at least it was clear that these were very high atmospheric phenomena.

At the beginning of the nineteenth century, the connection between meteors and cometary debris was suspected by some scientists, but widespread belief would come only after the particles that caused the meteor shower of November 27, 1872, were recognized as debris from the disintegrated comet Biela. However the pioneering work on meteor astronomy that took place in the nineteenth century was done, not on the late November Bielid shower but rather on the August Perseid and early November Leonid showers.[5] The historical sequence of events for both the Perseid and Leonid showers was, first, the establishment of their celestial origins; second, their periodicities; and finally, their identification with specific comets.

The study of modern meteor astronomy was initiated during the spectacular Leonid meteor shower witnessed in eastern North America during the early morning hours of November 13, 1833. Observers were stunned by the virtual storm of meteors. Contemporary reports were gathered and pub-

Denison Olmsted. (*Courtesy of N. Sperling.*)

lished in 1834 by Denison Olmsted (1791–1859), a professor of natural philosophy at Yale College. Earlier reports had been made of a similar display seen on the morning on November 12, 1799, by Friedrich Wilhelm Heinrich Alexander von Humboldt (1769–1859) and the French natural historian Goujaud Aimé J.A. Bonpland (1773–1858) in Cumana, South America, and by Andrew Ellicott (1754–1820) aboard a ship off Cape Florida near the edge of the Gulf stream.

From Olmsted's report on the November 1833 event, it was clear that several observers noted the point from which the meteors seemed to emerge was stationary in the neck of the constellation Leo. Those who first identified this radiant position included Olmsted, Alexander C. Twining (1801–1884), a civil engineer at West Point, W.E. Aikin, professor of chemistry at Mount St. Mary's College in Maryland, Virgil H. Barber in Frederick, Maryland, and a J.L.R. from Worthington, Ohio who Charles P. Olivier (1884–1975) identified as J.L. Riddell. Olmsted himself drew a number of conclusions from the data he gathered. He pointed out that the meteors originated above the atmosphere, approximately 2238 miles above the Earth's surface. They were attracted to the Earth by gravity, fell in nearly par-

191

Woodcut of November 12, 1799, Leonid meteor shower event.

allel lines with a velocity of approximately four miles per second, and were
made up of a light, transparent, and combustible material that ignited in the
atmosphere. The nebulous, cometlike body that supplied the meteors was
supposed revolving about the Sun on a 182-day period orbit inside the
Earth's orbit that was little inclined to the ecliptic plane and had an aphelion distance near the Earth's orbit. At the end of his published work,
Olmsted pointed out that Twining independently reached some of these
same conclusions.

Twining published his own results shortly after Olmsted's work appeared. Like Olmsted, Twining concluded that the fixed radiant point in Leo proved the celestial nature of the meteors and that their measured velocities, of at least 14 miles per second, implied an orbit interior to Earth's. The systematically low velocities measured by Olmsted, Twining, and others led to the notion of a very short-period comet as the source of the Leonid meteor stream. This erroneous concept persisted until 1866. However, Olmsted and Twining both noted the celestial origin and periodicity of the November Leonid meteor showers.

The first person to point out the fixed radiant position of the August Perseid showers was apparently professor John Locke (1792–1856) of the Ohio Medical College. After outlining his observations of August 8, 1834, Locke noted that the meteors seemed to radiate from a point near the star Algol in Perseus. During a December 3, 1836, session of the Royal Academy of Brussels, the mathematician and pioneer in the field of statistics, Lambert-Adolphe-Jacques Quetelet (1796–1874), stated that the mid-August meteors returned annually and predicted a shower on about August 9, 1838. In 1838, Edward C. Herrick also deduced the annual nature of the Perseid showers from reports of August showers in 1781, 1798, 1823, 1833, and 1836. He mentioned that Irish peasants apparently knew of this shower's annual nature for some time because they referred to these meteors as the burning tears of St. Lawrence, whose festival is August 10.

By 1838, the celestial origin and annual nature of the Perseid and Leonid showers were known. The annual nature of each shower is due to the Earth returning to nearly the same spot in space each year, where its orbit and those of the meteor stream intersect. The celestial nature of the meteor showers is evident because, while the Earth rotates, the radiant point of a meteor shower remains fixed with respect to the stellar background. What was not clear was the actual period that meteor stream particles travel about their orbits. Based on the November Leonid showers of 1799 and 1833, Olbers, in 1837, suggested a period of 3, 6, or 34 years for the stream particles and noted the possibility of another great storm in 1867. Six years later, Sears Cooke Walker (1805–1853) used radiant observations and meteor velocities to directly compute orbits for the Leonid and Perseid streams. Walker used poorly determined velocity information and arrived at incorrect orbital periods of 0.356 and 0.568 years respectively. Walker also suggested that the variation in the mean motion of comet Encke may have been due to the comet encountering an aggregation of small asteroids within the orbit of Mercury. A similar idea was later used by Oskar Backlund for his work on comet Encke.

The work of Hubert Anson Newton (1830–1896) is fundamentally important to the understanding of meteor streams. At age 25, Newton was al-

Hubert Anson Newton. (*Courtesy of N. Sperling.*)

ready a full professor at Yale College and chairman of their mathematics department. In a series of scientific papers in 1863 to 1865, Newton offered additional proof for the cosmic origin of meteors, noted the cometlike orbits of meteor stream particles, provided evidence that would later allow the orbital period of the Leonid particles to be determined, and gave the first successful prediction for a meteor shower that was not an annual event.

Aided by the brilliant young Yale student Josiah Willard Gibbs, Newton collected historical data on meteor streams and demonstrated their celestial nature by showing that the April Lyrid, August Perseid, and November Leonid showers return in periods of a sidereal, rather than a tropical, year. The time the Earth takes to revolve about the Sun with respect to the fixed stars (*sidereal* year) is about 0.014 day longer than the Earth takes to complete one revolution about the Sun with respect to the moving equinox (*tropical* year). Newton correctly pointed out that rings of particles, revolving about the Sun in orbits that intersect Earth's, would collide with the Earth in cycles of a sidereal year. Using the collected historical accounts of the Leonid showers from 902 to 1833, Newton established a time of 33.25 years between major shower events. In his view, the Leonid stream particles were not uniformly spread around their orbit, but rather bunched together in groups so the most likely time of the next major event would be November 1866.

Newton also determined that the nodal positions of the Leonid stream orbits advance with respect to the equinox at a rate of 1.711 arc minutes each year. Under the influence of planetary perturbations, the longitude of the nodes for a solar system object will regress for a direct orbit or advance in the direction of increasing longitude for a retrograde orbit.[6] Because the equinox itself regresses at a rate of 50.26 arc seconds per year, Newton's nodal advancement rate with respect to a fixed point in space was 52.4 arc seconds per year, or 29 arc minutes per 33.25 years. From the historical data, Newton computed that the Leonid stream particles must have an orbital period of either 0.49, 0.51, 0.97, 1.03, or 33.25 years. Newton preferred the 0.97-year period but suggested that theoretical perturbation calculations on a particle having each of these periods would settle the question when the computed nodal rate in each case was compared with the observed rate of 29 arc minutes per 33.25 years.

John Couch Adams took Newton's suggestion and found that only the 33.25-year orbit could produce the observed nodal rate. Adams computed elliptical elements for the Leonid stream particles and found a theoretical nodal advancement of 20 arc minutes due to Jupiter, 7 due to Saturn, and 1 due to Uranus; the remaining planets producing no sensible effect. The close agreement between the predicted nodal advance computed by Adams of 28 arc minutes per 33.25 years and the observed nodal advance determined by Newton of 29 arc minutes finally settled the orbital period of the Leonid stream. The period at which the main showers occur was found equal to the orbital period of the Leonid stream particles themselves, 33.25 years.

By analyzing sporadic meteors that did not return at regular intervals, Newton concluded that the average number of meteors traversing Earth's atmosphere daily is greater than 7.5 million and that a large number of meteors have absolute velocities upon encountering the Earth that are larger than the Earth's orbital velocity of 18.5 miles per second. Hence, the sporadic meteors cannot all belong to one narrow heliocentric ring with a diameter nearly equal to the Earth's orbit. Their orbits must resemble the eccentric ellipses of comets rather than Earth's near circular path.

Newton was not the first to mention a connection between meteors and comets. In 1834, W.B. Clarke mentioned the Leonid and Perseid meteors of 1833 as possible cometary fragments and Olmsted compared the material of meteors to that which forms the tails of comets. Walker, in 1843, noted the great eccentricity of meteor orbits and drew an analogy with cometary orbits. In 1861, Daniel Kirkwood (1814–1895) suggested that periodic meteors were the debris of ancient, disintegrated comets whose matter had become distributed about the orbit, but his ideas were not widely known until the publication of his popular *Meteoric Astronomy* six years later. Kirkwood's idea is the currently accepted explanation for the origin of meteor streams.

Daniel Kirkwood. (*Courtesy of the Mary Lea Shane Archives of the Lick Observatory.*)

As others had speculated, and Newton had predicted, an impressive Leonid meteor shower occurred on November 13, 1866. Although the display did not compare with the great storms of 1799 and 1833, it was an impressive spectacle—all the more so because it was successfully predicted, and hence anticipated.

During 1866, Giovanni Virginio Schiaparelli (1835–1910), director of the Brera observatory in Milan, wrote an important series of letters to Angelo Secchi about meteor streams. Secchi wisely had these letters published. From his studies, Schiaparelli established a definite connection between comets and meteors. He compared the observed hourly meteor frequency from evening until dawn throughout the year with a mathematical model of the expected frequency variation. He found that meteors occur more frequently in the morning than in the evening. In the morning hours, the leading face of the Earth runs into cosmic dust as the Earth moves about the Sun, whereas the same dust particles must have sufficient velocity to catch up to the Earth in order to be visible on the Earth's trailing, or evening, face. Within his mathematical model, Schiaparelli adjusted meteor velocities until the observed rates were simulated. The theoretical velocities required to fit the observations did not significantly exceed those of comets moving on parabolic orbits.

Giovanni Virginio Schiaparelli. (*Yerkes Observatory photograph.*)

Schiaparelli correctly concluded that the orbits of meteor streams resemble those of comets, rather than planetary bodies. He provided the first identification of a meteor stream with a specific comet by computing the orbit of the Perseid stream and demonstrating its similarity with an orbit of comet Swift-Tuttle, or 1862 III, computed in 1862 by Theodor von Oppolzer. Schiaparelli also computed orbital elements for the Leonid stream, but they were based on a poorly determined radiant so that, in this case, he was not able to identify the parent comet.

Meanwhile, on January 21, 1867, the director of the Paris observatory, Urbain Le Verrier, delivered a communication on the origin of meteors to the Academy of Sciences in Paris. Le Verrier set out to explain why the Leonids did not produce a good shower annually. He noted that, in time, planetary perturbations should spread the meteor stream particles uniformly around the orbit and a shower would be expected every year when the Earth passed through the stream. The fact that a strong Leonid meteor shower was not observed each year implied the stream particles were too young to have evolved entirely around the orbit. The relative youth of the Leonids, in their current orbit, could be explained if the stream had been severely perturbed by a planet, at some relatively recent time past.

197

Theodor Ritter von Oppolzer (1841–1886). (*Courtesy of the Mary Lea Shane Archives of the Lick Observatory.*)

Le Verrier assumed that the orbital period of the Leonid stream was 33.25 years and computed the remaining orbital elements from the 1866 meteor observations. He calculated the motion of the stream particles back to A.D. 126 when an arbitrary adjustment of the stream orbital node by 1.8 degrees and the longitude of perihelion by 4 degrees would bring the stream very close to Uranus. These arbitrary adjustments of the angular elements were felt justified as being within the errors of the analysis. Le Verrier further speculated that the original stream may have been moving in a direct orbit and he cited comet Lexell as an example of how great the perturbations can be during a single passage past a major planet. Le Verrier's calculations were made before the Leonid stream was identified with a specific comet and, given the uncertain nature of his initial stream orbit, his conclusions and analysis concerning its orbital history must be rejected. However, his orbit was an improvement over Schiaparelli's and an identification of the Leonid stream with comet Tempel-Tuttle, or 1866 I, was soon made.

As soon as Oppolzer published his orbit for comet 1866 I on January 7, 1867, it was obvious to at least three astronomers that its orbit was similar to

Le Verrier's orbit for the Leonid stream. Because his father was editor of the *Astronomische Nachrichten* and since he sent his new orbit there for publication, Carl F.W. Peters (1844–1894) was the first to identify Oppolzer's comet orbit with Le Verrier's Leonid stream orbit. Peters's comparison is shown in the table. Peters's work was dated January 29, 1867. In a letter dated February 2, 1867, Schiaparelli published an improved second set of orbital elements for the Leonid stream and identified it with Oppolzer's orbit for the comet 1866 I. Finally, in a letter dated February 6, 1867, Oppolzer himself made the identification between his orbit for the comet of 1866 I and the Leonid stream orbit of Le Verrier.

The Viennese astronomer Edmund Weiss (1837–1917) took an obvious step in 1867 when he calculated, for several comets that could approach the Earth's orbit, the time of the Earth's closest approach to the cometary orbits and their solar longitude and separation distance. Weiss also recorded the time of the year when meteor showers were observed and pointed out several possible connections between comets that approach the Earth's orbit and observed meteor showers. In particular, Weiss found that each year the Earth passed within 0.002 AU of the orbit of comet 1861 I on April 20 and within 0.018 AU of comet Biela's orbit on November 28. These two meteor shower events are now known as the April *Lyrids* and the *Bielids*, or *Andromedids*. The Bielids are no longer active, having nearly died out after 1899. In 1867, Johann Galle and Heinrich D'Arrest also noted the correct cometary associations for the April Lyrids and the Bielids respectively.

In 1868, Weiss predicted a Bielid meteor shower for 1872 or 1879 on or about November 28. An impressive Bielid display occurred on November 27, 1872, with an observer in Greenwich, England noting over 10,000 meteors from 5:30 to 11:50 in the evening. As the shower was ending, Wilhelm Klinkerfues, in Göttingen, Germany, reasoned that if the Earth had just passed through the debris of comet Biela, then the comet itself ought to be located in a celestial position exactly opposite the meteor shower radiant

Peters's Orbit Comparison

	Leonid Stream (Le Verrier)	Comet Tempel-Tuttle (1866 I) (Oppolzer)
Period	33.25 Years	33.18 Years
Semimajor axis	10.34017 AU	10.32479 AU
Eccentricity	0.904354	0.905420
Perihelion distance	0.98900 AU	0.97652 AU
Inclination (retrograde)	165.32 degrees	162.70 degrees
Longitude of descending node	51.30 degrees	51.43 degrees

point as it went away from the Earth. Since this point was in the northern hemisphere, the comet should be observable in the southern hemisphere. Finding the radiant point from 80 meteors, Klinkerfues sent a telegram to Norman Robert Pogson (1829–1891) in Madras, India. The telegram from Göttingen to Madras via Russia took only 1 hour 35 minutes. On November 30, 1872, Pogson received the telegram, which read, "Biela touched Earth on 27th: search near Theta Centauri." After waiting two days for a break in the clouds, Pogson recorded what he considered to be comet Biela on December 2, noting that it moved 2.5 arc seconds in 4 minutes and had a circular coma with a 45-arc-second diameter. It had a central nucleus region but no evidence of a tail. Pogson continued his observations the following day, recording the coma diameter as 75 arc seconds with a bright nucleus and a faint but distinct tail some 8 arc minutes in length. He was unable to continue observing it beyond December 3 because of poor weather and advancing twilight.

Karl Christian Bruhns (1830–1881), director of the Leipzig observatory, made it clear in 1875 that this comet was neither component of the double comet Biela, last seen in 1852. Acting on the telegram sent by Klinkerfues, Pogson had accidentally discovered a new comet.

The successful prediction of the Bielid meteor shower of November 27, 1872, showed dramatically that the debris resulting from the observed disintegration of comet Biela in 1846 and 1852 was responsible for the 1872 meteor shower. By the end of the nineteenth century, the origin of meteor streams was well understood and the prediction of periodic shower events like the Perseids and Leonids was thought to be fairly routine. The reports of the great Leonid meteor storms of 1799 and 1833, as well as the successful prediction of the 1866 Leonid shower, created a good deal of public excitement and anticipation for the expected 1899 Leonid display.

With an eye toward providing a shower prediction for 1899, G. Johnstone Stoney and A.M.W. Downing began with John Couch Adams' orbit for the Leonid stream and continued the integration of the meteor stream motion forward from 1866 to January 1900. The perturbative effects of Mars, Jupiter, Saturn, and Uranus were computed, while those of Venus and the Earth were found negligible. Compared with the orbits of those stream particles seen in the 1866 shower event, Stoney and Downing found that the period had increased by four months and the perihelion distance had decreased by 0.01 AU. Hence, particles from the same group responsible for the 1866 event could not be expected to provide a similar shower in 1899. However, if particles differing in orbital position from those in 1866 were moving on similar orbits and underwent similar planetary perturbations between 1866 and 1899, then a shower could be expected at 6 A.M. on November 15, 1899.

The public anxiously awaited this event. However, as the date of the predicted shower event drew closer, Stoney became more uncertain of his prediction and on November 10, 1899, he addressed the Royal Astronomical Society of London and stated that a shower could be expected only if the particle stream radius extended at least 0.014 AU from the central orbit path. As it turned out, Stoney's concern was well founded; there was no substantial shower event in 1899. In 1925, Charles Olivier recalled that in the face of great public anticipation and substantial press coverage, "the failure of the Leonids to return in 1899 was the worst blow ever suffered by astronomy in the eyes of the public."

Summary

As was true in the seventeenth and eighteenth centuries, nineteenth century astronomers concentrated their cometary research on the dynamics, rather than the physics, of comets. After the 1759 return of comet Halley, the next successfully predicted return was comet Encke in 1822, followed by comet Biela in 1832. Almost immediately, comet Encke's motion was found to deviate from purely Newtonian motion; its orbital size and period appeared to be decreasing with time. In order for a comet's period to decrease, there must be a force acting against its orbital motion. Johann Encke suggested that this opposing force was caused by the friction of a comet moving through a resisting medium surrounding the Sun. This resisting medium would cause a comet to spiral closer and closer to the Sun due to frictional drag effects.

Friedrich Bessel believed that reactive forces, acting on the comet itself, could cause its period to decrease with time. Bessel observed sunward jets, or emanations, arising from the head of comet Halley in 1835. He reasoned that these rocketlike emanations, whatever their cause, must affect the motion of the comet itself, much like the exhaust of a rocket drives the projectile forward. In the nineteenth century, Encke's resisting medium hypothesis was considered more likely than Bessel's rocket effect. Today, the reverse is true.

To contemporary astronomers, who still lionized Isaac Newton, the departure of comet Encke from pure Newtonian motion was disconcerting. To make matters worse, comet Biela was to demonstrate rather convincingly the error in Newton's perception of a comet as a permanent, solid body. Comet Biela was observed to split into two pieces in 1846 and after the double comet reappeared in 1852, it was never found again. However, when the Earth passed very close to comet Biela's orbit on November 27, 1872, a meteor shower was observed and this event clearly emphasized the connection between cometary debris and meteor showers.

The Bielid meteor shower of November 1872 brought the relationship between comets and meteors clearly into focus, and the evidence for this relationship had been accumulating ever since the great Leonid meteor shower of November 1833. It was primarily the November Leonid and the August Perseid meteor showers that prompted efforts to understand the relationship between comets and meteors. In a nearly parallel fashion, nineteenth century astronomers first recognized that the Leonid and Perseid showers were celestial rather than terrestrial phenomena and the two showers were recognized as periodic. Finally, by 1867, each meteor shower was identified with a parent comet.

NOTES

1. Once the orbit of a minor planet is accurately computed, it is given a number. Minor planet (78) Diana was the 78th minor planet to receive a number.

2. A century later, Brian G. Marsden and Zdenek Sekanina found that during the nineteenth and early twentieth centuries comet Faye's mean motion had a very small decrease with time.

3. It seems likely that Bessel was correct in assuming that the primary nongravitational effect is due to the rocketlike, sunward thrusting of a comet preferentially before or after perihelion.

4. Later in 1846, Challis would use John Couch Adams' (1819–1892) prediction of a new planet beyond Uranus and actually observe the planet as early as August 4, 1846. However he did not completely check his observations against the known star positions in the area until he received word that Johann Gottfried Galle and Heinrich Louis d'Arrest (1822–1875) had actually discovered Neptune on September 23, 1846, using the predictions of Urbain Le Verrier.

5. Periodic meteor showers are usually named after the constellation from which they appear to radiate. The November Leonid showers radiate from the constellation Leo and the August Perseid showers radiate from Perseus. Occasionally a meteor shower like Bielid is named after the parent comet, Biela.

6. A body in a direct orbit about the Sun travels eastward in the same direction as the planets. A body in a retrograde orbit travels about the Sun in the opposite direction, westward.

9

The Physics of Comets

Studies of cometary spectra begin to define the physical nature of comets. Comets shine from sunlight reflecting off dust particles and from the fluorescence of their atmospheric gases. Dust tails result from the pressure of sunlight on cometary dust particles. Interactions of cometary gases with the solar wind create ion tails. Relatively tiny, dirty iceballs are the source of the enormous cometary atmospheres.

THE WIDE DIVERSITY OF OBSERVED cometary phenomena requires an exacting physical model. It must be capable of explaining how the enormous, rarefied cometary atmosphere is generated from a source too small to be seen through a telescope. The tails are so diverse that the same comet can have both a long, thin, straight tail, as well as one that is short, wide, and curved. Its central region, termed the *nucleus*, can exhibit brightness flares or split into two or more pieces for no obvious reason. As they have for over two millennia, they seem to defy attempts to develop consistent mechanisms to explain their behavior.

It was not until the nineteenth century that scientific tools became available to untangle the diverse cometary phenomena. The *polariscope* established that at least some of the comet's light was polarized, and hence was reflected sunlight. The same conclusion was reached as soon as the first cometary spectra were observed, since the continuous spectrum of solar light was evident early on. It is now known that the presence of a continuous spectrum of all wavelengths, or continuum, indicates that dust in the comet's atmosphere is reflecting sunlight.

As well as the continuum emission observed in the spectra of many comets, it soon became apparent that some light originated with the comet itself. Spectral emissions were evident in most early cometary spectra, with bright bands often laid over the continuum emission. These features are due to atoms, molecules, and so-called *ions*—molecules that have lost at least one electron—in the comet's coma that have been excited by solar radiation. Ions are positively charged because one or more positively charged protons in the molecules are without a corresponding negatively charged electron. An ion has an entirely different spectral signature than the same molecule before it becomes ionized. For example, the carbon monoxide molecule, CO, in its neutral state has only been seen in the ultraviolet wavelength region for a few comets. However, the carbon monoxide ions, CO^+, emit a characteristic blue light in the visible wavelength region and are the most obvious molecules seen in the gas tail.

The spectra of ionized and neutral molecules are complex. Molecules emit photons with energies corresponding to discrete electronic transitions that are due to the deexcitation of electrons. This process is complicated by the rotations and vibrations of the particular molecule. Vibrational and rotational energies are generally smaller than those due to deexciting electrons but they can either add to or subtract from the larger energies of the electronic transitions. Hence, instead of the separate spectral line, or lines, expected for an excited atom, excited molecules produce spectral lines known as *molecular bands*. Generally, each band terminates sharply in a band head on either the long or short wavelength side.

The spectral bands characteristic of the molecules of two carbon atoms (C_2) are often referred to as *Swan bands,* after the English pioneer spectroscopist William Swan (1828–1914). Early cometary spectra observed in the visible wavelength region would often show bands due to one or more of the three prominent Swan bands whose heads had approximate wavelengths of 4737, 5165, and 5635 *Angstroms* (Å). The Angstrom unit, named after the Swedish physicist Anders Jonas Ångström (1814–1874), is defined as 10^{-10} meters. Because early photographs were more sensitive than the eye to the shorter blue wavelengths, photographic spectra, or *spectrograms*, would often show the band at 4050 Å due to the C_3 molecule.

An understanding of the complex structure and identification of various molecular bands did not progress very far until the development of quantum mechanics in the late 1920s. Hence the early interpretations of cometary spectroscopic observations were often confused. The C_2 bands in early spectra were thought due to carbon, or perhaps carbon in some combination with hydrogen. They were often referred to as *hydrocarbon bands.* Nevertheless, the mere presence of molecular bands in a cometary spectrum meant that gases were present in an excited state. The additional presence of continuum radiation in a comet's spectrum implied that its atmosphere contained solid reflecting surfaces as well.

The physical understanding of cometary phenomena was markedly increased subsequent to the intensive ground-based and space observations of comets Halley and Giacobini-Zinner in 1985 and 1986 (discussed in Chapter 10). For the most part, this chapter is confined to a discussion of the physical understanding of comets prior to this time.

The Introduction of Photography

Throughout the nineteenth century, the primary astronomical research tool was still the long-focus refracting telescope that allowed large, linear image sizes for relatively bright objects. The successful application of photography to astronomy required not large image sizes but rather the light gathering power of a *fast* telescope. Since this power is proportional to the square of the lens' aperture divided by the square of the objective's focal length, a fast photographic telescope requires a short focal length. The ratio of a photographic instrument's aperture to its focal length is termed its *f ratio*—the smaller the f ratio, the faster the instrument is for a given exposure time. To photograph a relatively faint comet, an especially fast instrument is desirable.

The first photograph of a comet was taken on September 27, 1858, not by an astronomer, but by an English portrait artist named Usherwood. From Walton Common, England Usherwood made a seven-second exposure of comet 1858 VI Donati with a f/2.4 focal ratio portrait lens. George P. Bond (1825–1865) made the second successful cometary photograph the next night. Using the long-focus f/15 refractor of the Harvard Observatory, Bond managed to get only an image of the comet's relatively bright inner coma with his six-minute exposure. Usherwood's exposure was more than 50 times shorter, but because of his short focal length system—and possibly a more sensitive collodion plate—his photograph showed part of the comet's faint tail as well as the brighter nucleus region. The details of Usherwood's and Bond's photographs are given in a letter dated June 11, 1859, wherein Bond thanks the English astronomer Richard C. Carrington for sending him a copy of Usherwood's photograph.

The next bright comet to be photographed was 1881 III Tebbutt. Using a fast f/3 camera with an 11-inch aperture, the director of the French observatory at Meudon, Pierre Jules César Janssen (1824–1907) used a 30-minute exposure on June 30, 1881, to obtain an image in which the comet's tail was evident up to 2.5 degrees from the nucleus. The same comet showed a 10-degree tail in a 2-hour, 42-minute exposure taken by Henry Draper (1837–1882) in New York City. Although a practicing physician, Draper is

Pierre Jules César Janssen, the first director of the French observatory at Meudon and a pioneer in astronomical photography.

remembered primarily for his contributions to astronomy. Like his father, John William Draper, Henry Draper was a pioneer in astronomical photography. In August 1872 he observed the bright star Vega and managed to take the first photographic impression of a stellar spectrum using his homemade 26-inch aperture telescope. On the night of June 24, 1881, Draper took a 17-minute exposure of the head of comet 1881 III Tebbutt through breaks in the clouds. In the next few days, he took successful photographs of its spectrum using a two-prism slit spectrograph that he had been using for stellar spectroscopy. Draper's spectrograms of 180, 196, and 228 minutes showed continuous spectra with three weak bands.

However the credit for the first cometary spectrogram must go to another amateur astronomer, William Huggins (1824–1910). Working at his private observatory at Upper Tulse Hill in London, Huggins had been using

Henry Draper, a practicing physician and pioneer in astronomical photography.

a slit spectrograph to analyze stellar spectra. On June 24, 1881, he positioned his spectrograph on the comet's head and took a one-hour exposure. For comparison, he also took a 15-minute exposure of the bright star Arcturus. Huggins' spectrum shows continuous emissions and three bands, which he attributed to carbon in some form, possibly in combination with hydrogen. In fact, Huggins' drawing of his original 1881 spectrogram shows cometary emission bands due to the molecules of cyanogen, CN, and carbon, C_3, and several dark absorption lines of the solar spectrum. Both sets of lines were superimposed on the solar continuum. Although he did not earn a living as an astronomer, Huggins took over his father's silk textile business in 1842 and was successful enough that by 1853 he could devote his full time to science. He was also an able violinist and owned a fine Stradivarius.

The Great Comet of 1882 II was even brighter than Tebbutt's comet, 1881 III, and it, too, was photographed. Working at the Cape of Good Hope, her majesty's astronomer, David Gill (1843–1914), borrowed a 2.5-inch aperture portrait camera, f/4.5, from a local photographer and mounted it on

William Huggins helped introduce the chemical analysis of celestial objects, including comets, by studying their spectra.

the existing equatorial telescope to allow the camera to follow the moving comet. He obtained several excellent photographs using exposures between 20 minutes and 2 hours. Gill was impressed with the cometary images he obtained, but he also noted the many stars that appeared in the background. He soon laid plans for the first major photographic star catalog, the *Cape Photographic Durchmusterung,* which gives the position and brightness estimates for nearly a half million stars observable from the southern hemisphere.

Working at the Harvard College observing station, which he had set up in 1891 at Arequipa, Peru, William H. Pickering (1858–1938) took nine objective prism spectrograms of comet 1892 I Swift. The best one was taken April 2, 1892, and shows bands that can now be identified as due to CN, C_3, and possibly C_2.

Until the twentieth century, the discovery of new comets was done with the naked eye or with visual telescopic searches. Many new comets are still discovered visually, but the majority are found on photographic plates or

This engraving shows William Huggins' telescope equipped for spectroscopic observations.

Edward Emerson Barnard and the Automatic Comet Seeker Hoax

Edward Emerson Barnard.

Edward Emerson Barnard (1857–1923) never found much humor in the hoax played on him by the Lick Observatory astronomers. Barnard was a serious, self-made man whose fame rested on his observations and discoveries of double stars and comets. Born in Tennessee, Barnard had only two months of formal early education, and just before turning nine years old he began work in a portrait studio to help support his impoverished family. In his spare time, he developed a strong interest in astronomy and gained a reputation as a comet hunter with extraordinary eyesight and zeal. During the 17 years he worked at the studio, Barnard also gained a knowledge of photography that was to aid his subsequent astronomical career. In 1876 Barnard purchased a telescope of five inches aperture for $380, a sum that represented about two-thirds his annual salary. From 1883 to 1887 he was associated with Vanderbilt University as a student and instructor. By 1887 he had made first discoveries of nine comets—1881 VI, 1882 III, 1884 II, 1885 II, 1886 II, 1886 VIII, 1886 IX, 1887 III, and 1887 IV—using either his own telescope or a six-inch aperture telescope at the university.

In 1887, Barnard was invited to join the Lick Observatory astronomers at Mount Hamilton, near San Jose, California. His observing programs continued at the observatory until 1895, when he became associated with the Yerkes Observatory in Williams Bay, Wisconsin. At Lick, Barnard added seven additional comet discoveries—1888 V, 1889 I, 1889 II, 1889 III, 1891 I, 1891 IV, and 1892 V—bringing his lifetime total to 16. His achievements in discovering comets were obtained largely as a result of his acute eyesight, his extensive knowledge of the night sky, and the diligence with which he spent night after night peering through the eyepieces of various telescopes.

On March 8, 1891, Barnard must have been astonished to read in the *San Francisco Examiner* an article describing his invention of an automatic comet seeker that

210

Discovers Comets All By Itself
The Meteor Gets in Range, Electricity Does the Rest.

The lengthy article that followed had just enough plausibility to convince most readers. After outlining the long hours and uncomfortable conditions that astronomers endure in their quest to discover comets, the author related how Barnard, assisted by Lick Observatory astronomers John M. Schaeberle and James E. Keeler, had taken advantage of the three bright spectral lines that characterize cometary spectra to fashion an automatic comet detector. A special motorized spectroscope had been constructed that could scan the night sky from east to west, then move slightly northward before scanning back from west to east. This scanning motion of the spectroscope would then repeat again and again so that the instrument would automatically cover much of the night sky without the need for human intervention.

The light of any object in the instrument's view was dispersed by a prism, passed through the spectroscope's objective lens, and fell upon a metallic diaphragm with slits cut exactly in the positions of the three cometary spectral lines. Behind this diaphragm was a strip of selenium, a substance electrically affected by the action of light. When light from any celestial object fell on the spectroscope, the prism dispersed it into a spectrum. Because this spectrum differed from that of a comet, nearly all of the object's light was blocked upon encountering the special diaphragm. However, when a comet came into view, the light from its spectrum passed easily through the diaphragm and illuminated the selenium. The selenium then generated a small electric current that closed a circuit and an alarm bell rang in Barnard's sleeping chambers. The astronomer then rushed to the site of the automatic comet seeker and noted the celestial position of the new discovery. The comet was trapped automatically.

After reading the article, Barnard wrote letters of denial to the *Examiner* and other San Francisco newspapers. However, newspapers' editors had already received warnings from perpetrators of the hoax and none would publish Barnard's frantic disclaimers. For the next two years Barnard received numerous requests for more details on his automatic comet seeker, including one from the veteran comet hunter Lewis Swift (1820–1913). It was not until February 5, 1893, that the *San Francisco Examiner* apologized to Barnard and acknowledged the nonexistence of the automatic comet seeker. The newspaper apology mentioned the astronomer Charles B. Hill as the hoax perpetrator but Barnard thought that Keeler might have had something to do with it as well.

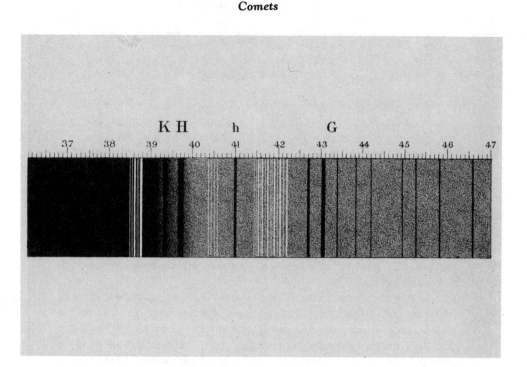

K H   h   G

37   38   39   40   41   42   43   44   45   46   47

The first photograph of a comet's spectrum, a spectrogram, was taken on June 24, 1881, by William Huggins. In this engraving of Huggins' spectrogram of comet 1881 III Tebbutt, the wavelength scale, in hundreds of Angstroms (Å), is given along the top edge. Reflected solar emission, at all wavelengths, a continuum, is present because the comet's dust particles reflect the Sun's light. This continuum emission can be seen as the grayish-white background that ceases for wavelengths shorter than about 3910 Å. The dark lines are absorption lines in the Sun's spectrum; they are referred to as *Fraunhofer lines* after the German physicist Joseph Fraunhofer who investigated them in the early 1800s. A solar absorption line is caused when radiation from excited atoms of a particular gas in the solar atmosphere passes through a layer of relatively cooler gas. Unexcited atoms in the cooler region absorb the radiation emitted from identical atoms in the hotter regions below. Thus absorption lines represent wavelength gaps in the solar radiation field. The Fraunhofer absorption lines due to solar calcium are denoted by K and H, while G represents absorption due to the CH radical. Emission from the comet's excited gases are present as bright molecular emission bands. Cometary spectral bands due to cyanogen, CN, are present with heads near 3883 and 4216 Å while the weaker emission band due to the carbon molecule, C_3, has its head near 4050 Å.

electronic detectors, either by chance or design. Edward Emerson Barnard (1857–1923) made a photographic discovery of a comet—by chance—when he was making long exposure photographs of the Milky Way with the six-inch Willard lens at Lick Observatory in California. On October 12, 1892, after a 4-hour, 20-minute exposure, Barnard found a short nebulous streak to the northwest of the star Altair. The detection was visually confirmed the following night with the 12-inch equatorial telescope at Lick. This first photographic discovery of a comet was the third periodic comet that Barnard found; periodic comet Barnard 3 was also the last of 16 comets the sharp-eyed Barnard would discover.

By the end of the nineteenth century, it was becoming increasingly obvious that photography would assume a major role in astronomical research. The long-focus refractors that served so well for visual observations of relatively bright objects began to be replaced by large-aperture reflecting telescopes of shorter focal length. One of the first to build a short focal length photographic camera for spectroscopic work was the French Count de la Baume Pluvinel who was assisted in this work by Henri Chrétien and A. Senouque. In the first few years of the twentieth century, Pluvinel built a short-focus, f/4, objective prismatic camera for recording the spectra of faint comets. This spectrographic camera was to serve as a model for future spectroscopic work on faint comets. His camera was first used on comet 1902 III Perrine, when a one-hour exposure on October 24, 1902, revealed bands that can now be attributed to C_2, methylidyne or CH, C_3, and CN.

Composition of the Cometary Atmosphere and Tail

Polarization Results

The direct observational study of cometary composition began when Dominique François Jean Arago (1786–1853) directed his recently developed polarimeter toward comet 1819 II Tralles. On July 3, 1819, Arago observed the comet's tail region with a doubly refracting prism attached to a small telescope. The two tail images, representing two states of polarization, were of slightly different intensities, indicating that at least some of the light coming from the tail was polarized—and hence reflected—sunlight. To rule out the possibility that the slight polarization might be due to the Earth's atmosphere, Arago observed the nearby star Capella and noted that its two images were of exactly the same intensity. In October 1835 Arago also detected polarized light coming from comet Halley.

In 1862, George P. Bond reported that polarized light coming from comet 1858 VI Donati was observed by George Airy at Greenwich, Karl Bruhns at Berlin, Jean Chacornac at Paris, Emmanuel Liais at Rio de Janeiro, Joseph Lovering at Harvard, and A. Poey at Havana, Cuba. Both Liais and Poey reported that the plane of polarization passed through the comet and Sun, indicating that some of its reflected light originated with the Sun. From his observations of comet 1861 II Tebbutt, Angelo Secchi noted that on July 1, 1861, the light from the comet's tail and nucleus jets was

Dominque François Jean Arago

strongly polarized, while on and after July 3 light from the region of the nucleus was also polarized. Secchi's polarization measurements of comet 1862 III Swift-Tuttle, from July 26 to August 28, showed the outer coma strongly polarized, but the region about the nucleus remained unpolarized until the end of the observation period.

Nineteenth Century Spectroscopic Results

The Florentine astronomer Giovanni Battista Donati (1826–1873) gets credit for being the first to obtain measurements on the chemical composition of a comet's atmosphere. Donati employed a spectroscope with a large 41-centimeter lens that had been presented to the Granduke Cosimo de'Medici in 1690 and had, at one time, been used by the English chemist Humphry Davy (1778–1829) to burn diamonds in an effort to investigate their nature. The slit of Donati's spectroscope was illuminated using a small cylindrical lens giving a line image of the observed object. The line spectra were then produced by a 60-degree prism. During the early morning hours of August 5 and 6, 1864, Donati used his spectroscope to observe and draw the spectrum of comet 1864 II Tempel. Donati's drawing of the first cometary spectrum shows the three molecular Swan bands of C_2 that often dominate a comet's visual spectrum. Donati noted the bands had heads at approxi-

214

George Bond, director of the Harvard College Observatory, drew this image of comet 1858 VI Donati as seen on October 4, 1858. Two straight gas, or ion, tails are present as is the more curved dust tail that sweeps the handle of the big dipper in Ursa Major.

mately 4737, 5165, and 5635 Å and that they resembled the spectra produced by metals. He did not try to identify them further.

The next comet to be observed spectroscopically was periodic comet Tempel-Tuttle (1866 I). William Huggins observed this comet on January 9, 1866 and noted that its coma had a broad, continuous spectrum indicating that most of its light was reflected sunlight. However, he also noticed that the region of the nucleus showed a bright point, implying that it was shining by its own light. This bright point showed no extent perpendicular to the dispersion of the spectrum—that is, no extent above or below the spread out spectrum. Huggins concluded that the bright nucleus region was too small to be shown in his telescope. He observed only one obvious band and made no attempt to identify it. On January 8, 1866, Angelo Secchi also made spectroscopic observations of comet Tempel-Tuttle and reported the presence of the three bands that Donati had seen in comet 1864 II. Their spectroscopic observations of comet Tempel-Tuttle in 1866 clearly showed it was shining by both reflected light and glowing gas.

Huggins continued his spectroscopic observations on periodic comet Tempel 1 (1867 II) and again noted a continuous spectrum with a hint of

Giovanni Battista Donati. (*Courtesy of the Mary Lea Shane Archives of the Lick Observatory.*)

two or three bands. Both Huggins and Secchi observed periodic comet Brorsen (1868 I) spectroscopically. Because no solar Fraunhofer lines were present, Secchi concluded that at least some of its light was intrinsic, not reflected sunlight. Huggins attempted without success to simultaneously compare its spectrum with those of magnesium, sodium, hydrogen, nitrogen, and the spectrum of an interstellar nebula. The three spectral bands he observed were laid over a faint continuum and resembled those first observed by Donati in 1864.

Huggins' spectroscopic observations of comet 1868 II Winnecke were made simultaneously with a carbon spectrum, and the side-by-side comet and carbon spectra matched very well. The carbon spectra were generated by the decomposition of olive oil with the spark from an induction coil. On June 23, 1868, and again two nights later, Huggins successfully identified the three cometary spectral bands as due to carbon, not only from the coincidence of the band positions but also from their relative brightness and general appearance. In 1965 the Belgian astronomer Polydore, or Pol, F. Swings (1906–1983) pointed out that Huggins was lucky because the profile of the

216

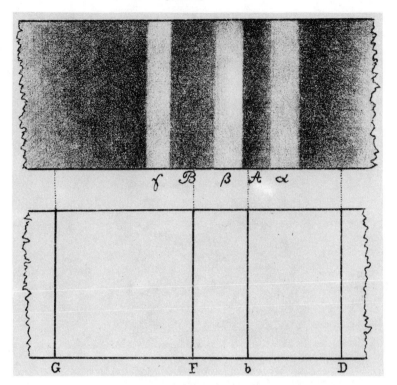

Giovanni Donati made the first visual spectroscopic observations of a comet, 1864 II Tempel, in the early morning hours of August 5 and 6, 1864. The three bands, labeled by Donati as γ, β, and α, are now known to be due to the carbon molecule, C_2, in the comet's atmosphere. Their band heads are located at 4737, 5165, and 5635 Å respectively. A solar comparison spectrum is drawn below that of the comet showing absorption lines due to CH, the Fraunhofer line G, hydrogen, F, magnesium b, and sodium, D.

cometary emission bands compared to bands experimentally produced in the laboratory is similar only for the C_2 molecule. In any case, Huggins identified the first chemical constituent of the cometary atmosphere.

Both Huggins and Secchi obtained photographic spectra, or spectrograms, for comet 1881 III Tebbutt. While he didn't obtain a spectrogram of this comet, Charles A. Young's visual spectroscopic observations made at Princeton not only showed the three C_2 bands on a continuum background, but he also managed to isolate a nucleus jetlike structure for spectroscopic scrutiny. Young's observations of June 29, 1881, showed the nucleus jet to have a continuous spectrum.

In 1882, two comets passed very close to the Sun and the intense solar heating allowed the identification of atomic sodium (Na) in both. Comet 1882 I Wells passed within 0.061 AU of the Sun on June 11 and the Great

Angelo Secchi. (*Courtesy of the Mary Lea Shane Archives of the Lick Observatory.*)

September Comet, 1882 II, came nearly eight times closer, at 0.0078 AU, on September 17, passing very close to the Sun's surface. Working at Ireland's Dun Echt observatory, Ralph Copeland and J.G. Lohse first identified sodium lines in the spectrum of comet 1882 I on May 28. A comparison with the spectrum of a sodium spirit lamp proved successful. On June 6, just five days before perihelion, the characteristic double lines of sodium were noted before the comet entered the solar glare and became unobservable. Around 1817 the German physicist Joseph Fraunhofer labeled the Sun's sodium emission lines with the letter D, and they are often referred to as *sodium D lines*. Copeland and Lohse also observed the sodium lines and lines they attributed to iron in the Great September Comet 1882 II. One day after perihelion, the two sodium D lines were observed, while on September 29, 12 days after perihelion, a single line was observed. Louis Thollon and A. Gouy, working at the observatory in Nice, France also observed the sodium D lines one day after perihelion and correctly concluded from the lines' shift in wavelength that the comet was receding from the Earth at 61 to 76 kilometers per second. This so-called *Doppler shift* results when the light source and the observer are moving relative to each other. Both the astronomers at Dun Echt and at Nice observed the sodium lines one day after perihelion in daylight.

In 1927, Sergey V. Orlov reanalyzed the carefully measured spectral features reported by Copeland and Lohse and confirmed their reported detection of sodium and iron. In addition, Orlov identified features due to nickel and carbon and found evidence for sodium, iron, nickel, and carbon in the spectral observations reported by Thollon and Gouy.[1]

Hermann C. Vogel (1841–1907), working at the Potsdam, Germany observatory, observed the spectra of periodic comet Pons-Brooks, 1884 I, from November 29, 1883, to January 15, 1884. On January 1, 1884, Vogel reported that its continuous spectrum was greatly enhanced during an observed brightness flare. He suggested the effect was just as if the comet threw off a great cloud of dust, thereby simultaneously enhancing both its visual brightness and its continuous spectrum. Vogel's interpretation has been confirmed in several subsequent cometary observations.

Twentieth Century Spectroscopic Results

By the early twentieth century, spectroscopes and short focal length spectrographs were capable of producing visual spectra and spectrograms that showed not only the three molecular bands associated with C_2, but also bands that would later be identified with CH, C_3, and CN. Sodium D lines were also identified when the comets of 1882 I and 1882 II passed very close to the Sun. Until 1907 all spectra and spectrograms were of the brighter coma region and not the much fainter tail regions.

The first reliable tail spectra were obtained by the French astronomers Henri Deslandres (1853–1948) and A. Bernard using a f/4.4 prismatic spectrograph while observing comet 1907 IV Daniel. By using 30- to 60-minute exposures on August 9, 15, and 20, 1907, they were able to identify weak bands in the wavelength regions 4000–4040, 4230–4290, and 4520–4580 Å. These weaker bands, now known to be due to singly ionized carbon monoxide (CO^+), were observed some 45 arc minutes down the tail from the comet's nucleus. Deslandres and Bernard suggested that these bands were due to a special vibratory mode of carbon gas or hydrogen alone. John Evershed (1864–1956) at Kodaikanal observatory, India observed the same bands in the tail region of comet 1907 IV Daniel. Assisted by his wife and using a short focal length objective prism spectrograph, Evershed was able to obtain spectrograms, in between monsoon rains, on August 28, September 3, and September 15, 1907. Using Pluvinel's short-focus objective prism camera, Fernand Baldet also obtained spectrograms showing bands due to CO^+ in the tail region of comet 1907 IV Daniel.

The almost pure ion tail of comet 1908 III Morehouse was observed by Deslandres and Bernard. Spectral bands due to CO^+ were identified as far as eight degrees from the nucleus, well down the long ion tail. They also found a

Henri Deslandres. (*Photograph courtesy of the Paris Observatory.*)

spectral band at 3914 Å, which was later identified as being due to the singly ionized nitrogen molecule (N_2^+). The laboratory work of Alfred Fowler (1868–1940), reported in 1910, finally allowed him to identify CO^+ with the tail bands observed in the spectra of comets 1907 IV Daniel and 1908 III Morehouse. He also identified the N_2^+ band at 3914 Å that had been observed in the latter comet.

Although cometary spectrograms were becoming more and more plentiful in the early twentieth century, a viable theory for how cometary spectral lines and bands were formed was not forthcoming until 1911 when Karl Schwarzschild (1873–1916) and Erich Kron suggested a fluorescence mechanism to explain the distribution of brightness in comet Halley's tail. They concluded that tail molecules are excited by solar radiation of a particular wavelength and subsequently emit radiation of the same wavelength to return to their deexcited state. In line with this explanation, Herman Zanstra proposed in 1929 that the bright line and band spectra observed in the coma of comets were produced by the absorption of solar radiation by coma molecules and the subsequent reemission of the same wavelength radiation, a process called *resonance fluorescence*. Both Schwarzschild and Kron, in 1911, and Zanstra, in 1929, suggested that excited cometary molecules could also

220

Photograph of Comet 1908 III Morehouse taken by E.E. Barnard on November 15, 1908. The tail of this comet is almost entirely a gas, or ion, tail with little evidence of dust. (*Photo courtesy of Yerkes Observatory.*)

emit radiation at longer wavelengths than the incident solar radiation, or fluorescence. However, it is now known that the strongest cometary emissions are of the resonance fluorescence type.

In his masterful compilation of data and results concerning the 1909 to 1911 apparition of comet Halley, Nicholas T. Bobrovnikoff (1896–1988) suggested that the nucleus jets of this comet were mostly cyanogen, or CN. However, he was careful to stress that this was not true for all comets. In various spectrograms of comet Halley, Bobrovnikoff also pointed out the presence of CO^+ and N_2^+ ions in the tail region, and in the coma region he reported the presence of C_2 Swan bands, sodium D lines, CN bands, the CH band near 4300 Å, and bands near 4000–4100 Å that he attributed to $C+H$. These $C+H$ bands were later shown to result from the C_3 molecule.

In 1938, Marcel Nicolet successfully compared spectra of CH produced in his laboratory with the corresponding, but previously unidentified, bands in the spectra of several comets, including Halley. Before Nicolet's work, only

221

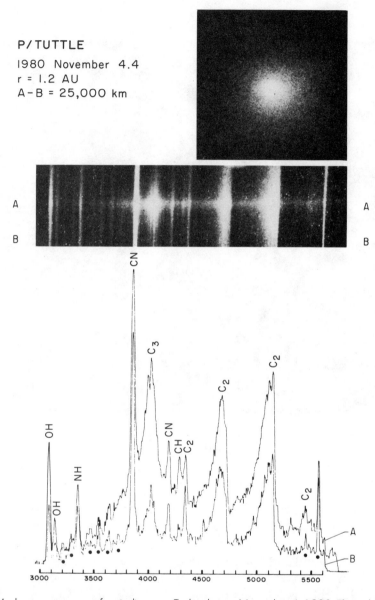

P/TUTTLE

1980 November 4.4
r = 1.2 AU
A–B = 25,000 km

Modern spectrogram of periodic comet Tuttle taken on November 4, 1980. The visible image of the comet is at the top and its spectrum below. At the bottom of the page is a traced spectrum showing the relative intensities of the various spectrum emission bands. The slit of the spectrograph has been placed on regions of the nucleus and the outer coma, and the resulting spectra are spread out, or dispersed, in a direction perpendicular to the slit length. Tracing A represents the dispersed spectrum of the comet's central nucleus region while the trace B represents the spectrum of the outer coma, some 25,000 kilometers from the central region. The cyanogen, CN, band is the most intense emission feature in both the nucleus and outer coma regions. Note that there is a hint of continuum radiation in the central region, but not in the outer coma, suggesting that the comet's dust is concentrated in the nucleus region. (*Courtesy of Stephen Larson, University of Arizona.*)

the single 4300 Å band of the CH molecule had been identified. The spectrograms of comet 1941 I Cunningham's nucleus region, taken by Pol Swings and his colleagues, revealed for the first time the ultraviolet bands due to the hydroxyl radical OH, at 3078–3100 Å, and NH, near 3360 Å. Their quartz optical system and the aluminized 82-inch mirror at the McDonald observatory allowed them to work at the high spectral resolution and short wavelengths necessary to identify these bands. They suggested that the OH probably arose from the dissociation of water vapor as a result of sunlight. The NH_2 band at 6300 Å was also identified by Swings' group in their spectra of comet Cunningham.

The identification of cometary spectral features with spectra produced in the laboratory is a difficult procedure. In general, the intensity distribution within a cometary molecular band does not match that produced in laboratory spectra. In fact, this distribution differs from comet to comet and even with a comet's position in its orbit.

The various separate lines within a spectral band had to be separated from one another, or resolved, and detailed comparisons made between the line structures in the unknown cometary spectral band and those corresponding to known molecular spectra produced in the laboratory. Identifying chemical species in cometary atmospheres by using their characteristic spectral bands is analogous to identifying an unknown person by comparing their

Polydore Swings. (*Photograph courtesy of Jean-Pierre Swings.*)

223

fingerprints with those identified with specific individuals. To make the analogy more nearly correct, however, we must imagine that people's fingerprints can change with time!

Pol Swings cleared things up considerably in 1941 when he concluded that cometary spectra depend on the Doppler shifted, exciting solar radiation field that is seen by the moving comet. Exciting solar radiation, and the Fraunhofer solar absorption lines that represent gaps in the exciting radiation, are Doppler shifted in wavelength due to the relative velocity of the comet with respect to the Sun. If the comet were not moving with respect to the Sun, atomic and molecular emissions from the Sun would excite their counterparts in the comet. However, the motion of the comet in its orbital path causes the wavelengths of exciting solar radiation, as seen by the moving comet, to be Doppler shifted away from their normal *rest wavelengths*. Hence, particular lines in a cometary spectral band can be either strong or faint depending on their positions relative to strong solar emission or absorption lines. In his honor, this phenomenon has been termed the *Swings effect*.

Pol Swings and Thornton Page took high-resolution spectrograms of comet 1948 I Bester in March and April 1948 and made the first identification of ionized carbon dioxide, CO_2^+, in the tail region; this ion was then added to the already identified tail ions of CO^+ and N_2^+. They were also able to identify spectral features corresponding to the molecules of OH, CH, NH, and a band at 4050 Å that was thought due to the CH_2 molecule.

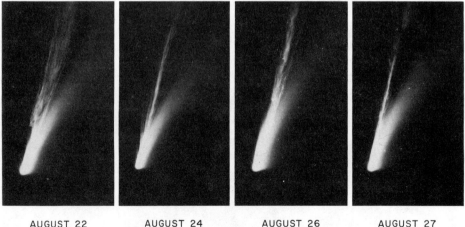

AUGUST 22   AUGUST 24   AUGUST 26   AUGUST 27

1957

Comet 1957 V Mrkos photographed in late August 1957 with the Palomar Mountain 48-inch aperture Schmidt telescope. Note the presence of both the ion tail to the left and the dust tail to the right. (*Courtesy The Observatories of the Carnegie Institute of Washington, Pasadena, California.*)

Gerhard Herzberg, who won the 1971 Nobel prize for chemistry, reported in 1942 that he had produced the 4050 Å spectral band in the laboratory. He tentatively attributed this band to the molecule CH_2 and for nearly a decade thereafter, the 4050 Å band was reported as due to CH_2. However, following a suggestion by Herzberg himself, Alexander E. Douglas demonstrated in 1951 that C_3 molecules were responsible for the 4050 Å spectral band.

High-resolution spectrograms of comet 1957 V Mrkos taken at Palomar Mountain observatory by Pol Swings and Jesse L. Greenstein allowed them to identify the *forbidden lines* of neutral oxygen near the wavelength 6300 Å. Forbidden lines are so called because under normal, terrestrial conditions these molecular emissions cannot occur. The time necessary for them to spontaneously deexcite and produce a spectral emission line is far longer than the time between collisions of the particular molecules, a process which causes them to lose their excess energy. Forbidden lines are only possible in extremely rarefied atmospheres where molecular collisions are infrequent. The forbidden oxygen lines are some of the very few cometary emissions that are not compatible with the resonance fluorescence mechanism.

Greenstein studied his high-resolution spectra of comet 1957 V Mrkos and noted the intensity ratios between individual rotational lines within the CN band were different on either side of the nucleus. This so-called *Greenstein effect* was attributed to coma gas motions of approximately 3 kilometers per second within 4000 kilometers of the nucleus itself. Radial motions of the coma gases with respect to the Sun apparently displace, or Doppler shift, the wavelengths of the exciting solar radiation, as seen by the coma gases, such that the fluorescence effect is slightly different from one side of the nucleus to the other. These radial motions in the inner coma region are due to expanding clouds of gas that give rise to components of motion toward and away from the Sun. This phenomenon is thus a special case of the Swings effect.

Using spectroscopic observations of comets 1957 III Arend-Roland and 1957 V Mrkos, William C. Liller observed that the sunlight scattered by particles in the cometary atmospheres was redder than the incident sunlight itself. He noted that his observations were consistent with the scattering of sunlight from particles in the size range from 0.25 to 5 microns. This was an early indication that cometary dust particles are approximately one micron (10^{-6}m) in size.

The sungrazing comet 1965 VIII Ikeya-Seki passed within 0.008 AU of the Sun, less than two solar radii, on October 21, 1965. Its naked-eye brightness near the Sun allowed infrared observations to be made by Eric E. Becklin and James A. Westphal. They concluded that their observations, made at wavelengths up to 10 microns, were compatible with the infrared

225

emission from solar heated iron particles. Several groups observed the comet very close to perihelion, and the intense heating of the cometary dust created several species of metallic atoms when the particles dissociated. The spectra of neutral iron, nickel, potassium, chromium, manganese, vanadium, cobalt, copper, and both neutral and ionized calcium (Ca and Ca^+) were reported as fairly definite; additional identifications were possible for neutral magnesium, aluminum, and ionized iron and strontium.

In 1964, Ludwig Biermann (1907–1986) and Eleanore E. Trefftz emphasized the importance of cometary ultraviolet observations. They reasoned that if Fred Whipple's icy conglomerate model for the cometary nucleus was correct, then the observed oxygen forbidden lines could be explained if water vapor was being dissociated by solar radiation, or *photodissociated*, to form the necessary atomic oxygen. They concluded that an atomic hydrogen coma must also be present and, for an active comet, one would expect this coma to supply approximately 10^{33} photons per second in the ultraviolet region of the spectrum (1216 Å). Since this short wavelength radiation cannot easily penetrate the Earth's atmosphere, its detection would have to rely on observations made largely above the Earth's atmosphere.

On January 14, 1970, Arthur D. Code and his colleagues directed the spectrograph aboard the second Orbiting Astronomical Observatory, OAO-2, toward comet 1969 IX Tago-Sato-Kosaka. They reported the first detection of the predicted neutral hydrogen coma that presumably surrounds all active comets. Edward B. Jenkins and David W. Wingert confirmed the OAO-2 observations with a rocketborne ultraviolet spectrogram taken on January 25, 1970. Their nearly circularly symmetric, diffuse image corresponds to a hydrogen coma nearly one million kilometers in diameter.

Jean L. Bertaux and Jacques E. Blamont followed up on the OAO-2 success with observations of the hydrogen coma surrounding comet 1970 II Bennett during April 1970 using a spectrometer aboard the fifth Orbiting Geophysical Observatory, OGO-5. On December 12, 1970, they also briefly observed a cloud of hydrogen surrounding periodic comet Encke. Code and his colleagues used a spectrometer aboard the OAO-2 satellite to observe the hydrogen coma surrounding comet 1970 II in April and May 1970, which was some three million kilometers in diameter. Along with emission from CN and NH molecules, they observed OH emissions, confirming their previous observations of OH in comet 1969 IX Tago-Sato-Kosaka. The presence of a hydrogen, H, and hydroxyl, OH, coma surrounding comets 1969 IX and 1970 II strongly suggested that the parent molecule for both was water, H_2O.

Spectral information on cometary dust most often comes from long wavelength, or *infrared*, radiation since, once warmed by the Sun, the tiny dust grains reradiate most efficiently at wavelengths approximately equal to their physical dimensions. From a combination of infrared and optical wavelength photometry of comets 1965 VIII Ikeya-Seki, 1969 IX Tago-Sato-Kosaka and 1970 II Bennett, Charles R. O'Dell suggested that the observed surface brightness of these comets implies the cometary dust had a particle radius of approximately 0.1 micron and an albedo of roughly 30 percent.[2]

From infrared observations of comet 1970 II Bennett, R. W. Maas and his colleagues noticed a sharp emission peak at 10 microns, which they attributed to silicate dust surrounding the comet. The chemical composition of this cometary dust would be composed of silicon, magnesium, and oxygen. Maas and his coworkers suggested that these dust particles were originally formed as condensation products in some circumstellar regions and later were incorporated into cometary bodies. Tragically, Maas died of a heart attack less than 24 hours after making his important observations. As early as the 1930s, Peter M. Millman used the spectra of meteors to identify silicon, magnesium, iron, nickel, calcium, aluminum, manganese, and chromium as constituents of particles that were largely cometary dust.

With the launch of the Infrared Astronomy Satellite, IRAS, on January 26, 1983, information was forthcoming on the nature of relatively large cometary dust particles that cannot be observed using ground-based techniques. After comet 1983 VII IRAS-Araki-Alcock was first detected by the IRAS on April 26, 1983, Russell G. Walker and colleagues observed it in the infrared wavelength regions from 12 to 100 microns. A dust tail, directed in the antisolar direction and made up of particles having radii from 5 to 30 microns, was evident to beyond 400,000 kilometers from the comet's nucleus.

Mark Sykes and his coworkers detected narrow trails of dust particles in the IRAS data that they identified with periodic comets Tempel 2, Encke, and Gunn. The brightest trail was attributed to the dust emitted by comet Tempel 2 near perihelion. These particles were submillimeter in size and spread out in the comet's orbit plane both ahead of and behind the position of the nucleus to a distance corresponding to 0.5 AU. For comet Tempel 2, the particle density in the dust trail was estimated as being 10^{-11} per cubic centimeter, or one particle in a box that is 50 meters on each side. As well as the dust trails definitely attributed to periodic comets Tempel 2, Encke and Gunn, there were tentative dust trail identifications for comets Tempel 1, Kopff, and Shoemaker 2, with another 100 or more trails for which there were no known comets.

Comet 1973 XII Kohoutek was discovered nearly 10 months prior to its perihelion passage on December 28, 1973, and its orbit was first determined a month after discovery. Hence enough time was available to make extensive observing plans before the comet became bright. Although the drop in this comet's intrinsic brightness after perihelion was disappointing, the extensive advance preparations allowed many important science results to be obtained.

In the ultraviolet spectral region, comet Kohoutek was observed by spectrometers aboard two sounding rocket flights, the Earth orbiting Skylab, the OAO-3 satellite, and even the Mariner 10 spacecraft on its voyage to the planet Mercury. Using the Finson and Probstein technique described in the next section, Zdenek Sekanina studied comet 1973 XII Kohoutek's dust tail and concluded that the particles in the short, sunward antitail were larger, 0.1 to 1 millimeter, than those particles in the conventional tail.

From the results of infrared photometry of four comets, Edward P. Ney pointed out that the 10-micron emission peak first seen in the spectrum of 1970 II Bennett was a common characteristic of dusty comets. He considered that the dust particles in a comet's coma and tail were less than two microns in diameter, with an albedo between 0.1 and 0.2. However, from the observed absence of the 10-micron silicate emission peak in his infrared observations of comet Kohoutek's antitail, Ney concluded that these dust particles were larger than 10 microns.

Observations of comets at radio wavelengths began in 1957 with attempts to observe comet 1957 III Arend-Roland. H.G. Müller and colleagues made observations at 21 centimeters, Raymond A. Coutrez and colleagues tried 50-centimeter observations and 11-meter observations were made by John D. Kraus. While all these observers reported successful detections, the first definitive observations of a comet at radio wavelengths came during the 1973 to 1974 return of comet Kohoutek.

The first reported radio detection of a molecule in the atmosphere of comet Kohoutek was methyl cyanide (CH_3CN) in early December 1973. Bobby L. Ulich and Edward K. Conklin made these observations using the 140-foot antenna of the National Radio Astronomy Observatory (NRAO) in Green Bank, West Virginia. Walter F. Huebner, Lewis E. Snyder, and David Buhl, also working at NRAO, reported observations of hydrogen cyanide (HCN) in mid-December, 1973, and in early January 1974 John H. Black and colleagues reported observing the CH molecule. The most certain radio detection for comet Kohoutek was that of the OH radical. François Biraud and colleagues at Nancy, France observed OH in December 1973 and one month later, when the comet had passed perihelion. Barry E. Turner, using the NRAO antenna, also reported successful OH observations in early December 1973. William M. Jackson and colleagues reported a marginal radio observation of the neutral water molecule in comet 1974 III Bradfield while Wilhelm J. Altenhoff and colleagues reported radio emissions from the molecules of both water and ammonia during the close Earth approach of comet IRAS-Araki-Alcock in May 1983.

Didier Despois and colleagues observed the 1665- and 1667-megahertz, MHz, emissions of the hydroxyl radical, OH, strongly in comets 1976 VI West and 1977 XIV Kohler and less strongly in comets 1975 IX Kobayashi-Berger-

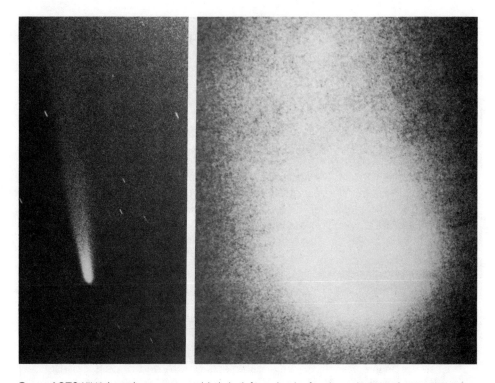

Comet 1973 XII Kohoutek as seen in visible light (left) and in the far ultraviolet light of atomic hydrogen (right). The visible light photograph was taken by a Johns Hopkins University camera onboard an Aerobee rocket flown on January 4, 1974. The far ultraviolet image was taken by a Naval Research Laboratory camera during a similar flight three days later. The ultraviolet image, to the same scale as the visible image, reveals that the hydrogen halo of comet Kohoutek had a diameter of more than 5 million kilometers, nearly four times the diameter of the Sun. (*This photograph is courtesy of Chet Opal and George R. Carruthers, Naval Research Laboratory.*)

Milon and 1978 VII Bradfield.[3] David Meisel and Richard A. Berg made infrared observations of the OH radical in comet 1973 XII Kohoutek, while Jacques Blamont, Michel Festou, as well as Charles Fehrenbach, and Yvette Andrillat, observed it in the ultraviolet wavelength region of the comet's spectrum. In the visible wavelength region, Peter Wehinger, Susan Wychoff, Gerhard Herzberg, George Herbig, and Hin Lew identified the bands of the singly ionized water molecule (H_2O^+) in comets 1973 XII Kohoutek and 1974 III Bradfield. With the optical observations of oxygen, O, and the ionized water molecule, the ultraviolet observations of hydrogen, H, and the observations of the hydroxyl radical at ultraviolet, infrared, and radio wavelengths, the source of these species seemed most likely to be water, H_2O. Upon photodissociation and ionization, the parent water molecules could form all of these daughter products. These observations strongly supported Fred Whipple's icy-conglomerate model of the cometary nucleus.

229

Forgotten Cometary Infrared Observations

Although Eric Becklin and James Westphal's observations of comet 1965 VIII Ikeya-Seki have generally been considered to be the first successful attempt to observe a comet in the infrared wavelength region, Carl O. Lampland deserves this honor for observations that were made as early as 1927. On December 15, 1927 an employee of a timber company in Flagstaff, Arizona noted the presence of a comet near the setting Sun and notified astronomers at nearby Lowell Observatory, who were unaware of it. After this minior embarrassment, the Lowell astronomers began studying the comet the following day. It was about 5 degrees north of the Sun and described as being several times brighter than Venus.

Lampland had been studying infrared radiation from the planet Jupiter with a thermocouple when the new comet offered a unique opportunity to observe it in the infrared. Using the 42-inch aperture reflector, Lampland measured comet 1927 IX Skjellerup-Maristany in four wavelength regions between 0.3 and 14.5 microns from December 16 to 19. Since the comet reached perihelion on December 18, Lampland's observation interval included measurements before and after its closest approach to the Sun. Although he was unsuccessful in attempts to observe periodic comet 1927 VII Pons-Winnecke the previous June, Lampland succeeded this time, in spite of what he described as strong cold winds, blowing snow, and miserable seeing. The comet was so close to the Sun that parts of the telescope, including the primary mirror, had to be shielded from direct sunlight.

After reviewing Lampland's work in 1985, Arthur Hoag concluded that thermal radiation from the region of the nucleus was indeed observed. At the beginning of the observation interval, approximately 60 percent of the measured radiation was shorter than 1.2 microns, while at the end of the interval about 80 percent was longer than 1.2 microns. With hindsight, it is clear that the observations before perihelion indicated pure reflected sunlight, while after perihelion much of the measured radiation was reradiated by the comet's nucleus and surrounding dust.

The Formation of Cometary Tails

Explanations for the cometary tail phenomenon were confused well into the twentieth century. The confusion arose because there are two types

Fedor Aleksandrovich Bredikhin. (*Photograph courtesy of N.A. Belyaev.*)

of tails, gas, or ion, and dust tails. Each type is formed by an entirely different process. Dust tails of comets are composed primarily of submicron-sized particles that are driven back into the tail region by solar radiation pressure. This mechanism seems well understood. Less well understood are gas tails, which seem to be controlled by interactions with the so-called *solar wind,* a high-velocity cloud of protons and electrons spiraling away from the Sun.

Dust Tails

As mentioned in Chapter 3, Johannes Kepler may have had something like the pressure of sunlight in mind to explain the antisolar nature of cometary tails. In 1812, Wilhelm Olbers suggested that comet tails were formed of minute particles driven away in an antisolar direction by a solar repulsive force, which he supposed to be electric in nature. The interplay of these forces with gravitational attractions between the Sun, comet, and dust particles was responsible for the various tail shapes. Friedrich Bessel further developed Olbers' work on the mechanical theory of cometary tails in 1836.

The mechanical theory of tail formation was extensively analyzed by the Russian astronomer Fedor Aleksandrovich Bredikhin (1831–1904). Bredikhin classified comet tails into three groups according to the ratio of the antisolar repulsive force to the Sun's gravitational attraction. This ratio is usually denoted as $1-\mu$. Type I comet tails, with the solar repulsive force exceeding the solar gravitational attraction by about 11–18 ($1-\mu = 11$–18),

231

were long, straight tails that Bredikhin attributed to hydrogen gas. Type II tails ($1-\mu = 0.7$–2.2) were wider, more curved, and supposed to consist of hydrocarbon gas and light metals. The short, stubby tails of Type III ($1-\mu = 0.1$–0.3) were assumed due to heavier material than the first two types, perhaps vapor of iron. A fourth type of tail is occasionally seen directed straight toward the Sun. These latter sun-pointing *antitails* were correctly attributed by Bredikhin to projection effects seen when comets cross the Earth's orbital plane.

While his analysis could explain the nature of dust tails fairly well, Bredikhin considered that he was dealing with gas tails. He assumed a repulsive interaction between the Sun's electric charge and the like charge of the cometary gas molecules. The lighter molecules, like hydrogen, were repulsed most strongly to form Type I tails, while the heavier molecules formed the more curved Type II tails. Although Bredikhin's suppositions on the chemical makeup of each type of tail were incorrect, Type I is a term still used to describe long, narrow ion tails and Type II is used to describe broad, curved dust tails.

In 1873, the Scottish physicist James Clerk Maxwell (1831–1879) published his classic *Treatise on Electricity and Magnetism*. One important conclusion of his theoretical electromagnetic theory was that light exerts a radiation pressure. In 1900, the Swedish chemist and physicist Svante August Arrhenius (1859–1927) suggested the repulsive force acting on cometary tail

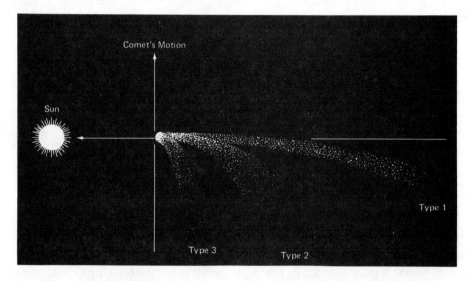

This schematic diagram illustrates the three types of comet tails proposed by Fedor Bredikhin in the late nineteenth century. A fourth type of antitail is illustrated later in the chapter.

Photograph of comet 1957 III Arend-Roland taken on April 24, 1957 by Freeman Miller with the Curtis-Schmidt telescope at the University of Michigan. The sunward spike, or antitail, appears to point directly toward the Sun, while the regular tail points nearly in the opposite direction. (*Photograph courtesy of the University of Michigan.*)

particles was light pressure from the Sun. A few months later, the German astronomer Karl Schwarzschild investigated Arrhenius' suggestion and showed that the magnitude of the light pressure was related to particle size and density. Schwarzschild also demonstrated that for particle diameters between 0.07 and 1.5 microns, light pressure could explain a repulsive force of as much as 20 times the Sun's gravitational attraction $(1-\mu = 20)$ but no more than that. This was an important result, because in many comets the tail particles suffered a repulsive force far greater than 20 times the solar attraction. Clearly some other mechanism, besides radiation pressure, was affecting some cometary tails.

In 1901, the Russian physicist Pëtr Nikolaevitch Lebedev (1866–1912) made the first experimental detection and measurement of light pressure on solid bodies in the laboratory. Observations of the same experimental effect were also claimed by Ernest Fox Nichols (1869–1924) and Gordon Ferrie Hull the same year.

The 1968 analysis of Michael L. Finson and Ronald F. Probstein on cometary dust particle dynamics was a very successful application of fluid dynamics concepts. Dust particles having a wide distribution of particle sizes were assumed to be continuously released from the comet nucleus, then accelerated rapidly away from the nucleus by the drag effects of the expanding

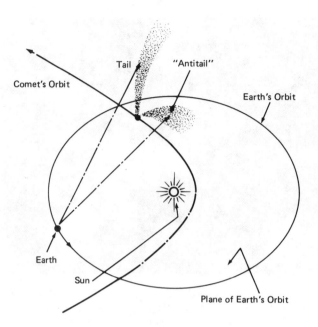

This diagram shows how a cloud of antitail particles, somewhat larger than those in the ordinary tail of comet 1957 III, is less affected by the Sun's radiation pressure and lags behind the comet as it exits the solar neighborhood. These larger particles can then appear, in projection, as a sunward spike. The effect was only noticeable for comet 1957 III Arend-Roland when the thin sheet of antitail particles was viewed edge-on as the comet passed through the Earth's orbit plane in late April 1957.

gas coma. Once in the tail region, the only forces acting on the dust particles were assumed to be solar gravity and radiation pressure. Finson and Probstein derived theoretical expressions for the dust tail surface densities. By comparing the calculated distribution of surface density with the measured distribution of light intensity from photographic plates, it was possible to determine dust and gas production rates as a function of time, dust particle size distribution, and emission velocity from the inner coma region as a function of particle size and time. Finson and Probstein applied their technique only to comet 1957 III Arend-Roland, but others have used it more extensively. Zdenek Sekanina and Freeman D. Miller applied the technique to comet 1970 II Bennett and concluded that just prior to perihelion, this comet's gas production rate was twice that for dust and the gas drag forces were due to water vapor from the vaporization of water snow on the comet's surface.

Ion Tails

Although the Bessel-Bredikhin theory for the development of cometary tails could explain the mechanics of what we now know to be dust tails, the theory could not be applied to the motion of gas tails. Neither Bessel nor Bredikhin made a distinction between the two types of tails, but by the end of the nineteenth century evidence began to mount for two separate mechanisms to explain cometary tail phenomena.

In a little known work by George Fitzgerald in 1883, James Maxwell's treatment for a theoretical light pressure was used to show that this effect on small gas molecules was slight. Fitzgerald computed that a hydrogen molecule would have to absorb 96 percent of the solar radiation that fell on it for the light pressure to equal the solar gravitational attraction on the molecule at 1 AU from the Sun. In a 1943 review article on the nature of comets, Karl Wurm demonstrated that the repulsive radial force due to radiation pressure on the molecules CN, C_2, and CO^+ could only be 0.69, 1.66, and 46.9 times the solar gravitational attraction. These values were far too small to explain observed values of 151 for comet 1908 III Morehouse and 70 to 90 for comet 1910 II Halley. Light pressure was not a viable mechanism for forming gas tails. An alternative theory was required.

In 1893, John M. Schaeberle (1853–1924) suggested that repeated impulses of invisible solar coronal particles on the comet's atmospheric matter were responsible for the cometary tail phenomena. Upon encountering the cometary atmosphere, the coronal particles were slowed down, became more dense, and were then evident, along with the cometary material itself streaming back in the antisolar direction. Schaeberle considered antitails to be observable when the comet receded from the Sun with a velocity greater than the coronal stream. The main tail, in the antisolar direction, was due to a coronal stream moving faster than the comet. Schaeberle's hypothesis was purely mechanical, involving no electromagnetic interactions. However, his suggestion of a high-velocity stream of particles from the solar corona was an interesting anticipation of the currently accepted theory. In a similar vein, it is interesting to note that Benjamin Peirce (1809–1880), in 1859, and William Pickering, in 1895, suggested that the same effect that gives rise to auroral displays would have some influence on the observed development and changes in comet tails.

A major breakthrough in explaining cometary ion tails came in 1951 when the German astronomer Ludwig Biermann suggested that the structural form and ion motions of Type I tails could be explained if one postulated a continuous outflow of ionized particles from the solar corona. The momentum transferred from this high-velocity solar wind to the cometary tail ions

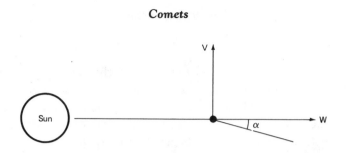

The orientation of a comet's ion tail is governed by the solar wind velocity, W, at the distance of the comet and the comet's orbital velocity, V, perpendicular to the Sun-comet line. The angular deviation of the ion tail from the extended Sun-comet line is described as an aberration angle (α).

could explain the observed repulsive accelerations of 100 or more times the solar gravitational attraction. The angle between the Type I ion tail and the extended Sun-comet line is termed the *aberration angle,* α. This angle depends directly on the comet's orbital velocity component, V, perpendicular to the Sun-comet line and inversely as the radial velocity of the solar wind, W, at the comet's heliocentric distance. The approximate relationship is:

$$\tan{(\alpha)} = \frac{V}{W}$$

Since the quantities α and V can be determined for comets exhibiting ion tails, the solar wind velocity could be deduced at various heliocentric distances. Thus Biermann, and later John C. Brandt and his colleagues, investigated solar wind velocities by observing cometary ion tails. At the Earth's distance from the Sun, the radial bulk velocity of the solar wind protons and electrons is a rather extraordinary 400 kilometers per second.

As pointed out by Hannes Alfvén in 1957, ionized gases moving radially outward from the cometary nucleus and then interacting with the solar wind would tend to form Type I ion tails with parabolic envelopes. However, observed Type I tails have nearly rectilinear envelopes. Alfvén postulated that solar wind particles drag solar magnetic field lines with them and, on encountering a comet, wrap around the comet's ionized gas coma and fold behind it in a parallel, rectilinear fashion—rather like the spokes of an opened umbrella being gradually closed until they are parallel to one another. Propelled by momentum transfers from the solar wind particles, the cometary ions are guided down the rectilinear ion tail by the folded magnetic field lines.

The German astronomer Klaus Jockers and colleagues correlated observations of ion tail features for two comets with measurements of solar wind parameters. Using ground-based observations of comets 1969 IX Tago-Sato-Kosaka and 1970 II Bennett and simultaneous satellite observations of the solar wind, Jockers found correlations between peculiarities in the cometary ion tails and solar wind events. These ion tail peculiarities included knots and kinks that appeared to be moving away from the comet's head. From a series of photographs, Ludwig Biermann, Rhea Lüst, and Klaus Jockers measured the motions of these features and found their velocities to be approximately 20 to 250 kilometers per second, which probably represents the bulk motion of ion tail material down the tail axis.

By monitoring the flow of some irregularities down the ion tail of comet 1973 XII Kohoutek, Charles L. Hyder and his colleagues inferred a velocity of 250 kilometers per second for ion tail material. They concluded that if the

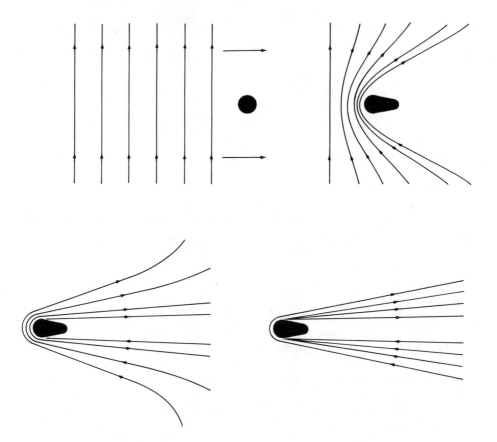

The solar magnetic field, represented here as magnetic field lines, is embedded in the ionized plasma of the solar wind particles. When these field lines encounter the ionized cometary plasma, they wrap around the comet and form nearly parallel field lines directed in the antisolar direction. Guided by these folded magnetic field lines, the ionized cometary particles form the comet's ion tail as they are pushed away from the Sun by the much faster solar wind particles.

Comet Hunter, Civil War Hero, and Embezzler

At the Oakwood Cemetery in Falls Church, Virginia, the grave of one of the most celebrated nineteenth century comet hunters lies unmarked. Like his grave, the interesting life of Horace P. Tuttle (1837–1923) was not recognized until the sleuthing of Richard Schmidt of the U.S. Naval Observatory in Washington, D.C.

Horace Tuttle and his older brother, Charles W. Tuttle (1829–1881), began their careers as assistant astronomers at the Harvard College Observatory. Because of failing eyesight, Charles resigned his position in 1854 to take up law. Only seven years later, he was admitted to the bar of the U.S. Supreme Court. Horace followed his brother to the Harvard College Observatory, became an assistant astronomer in 1857,

Horace Tuttle, a member of Company D of the 44th Massachusetts volunteer infantry in 1862 and 1863. (*Photograph courtesy of Richard Schmidt, U.S. Naval Observatory.*)

and immediately began sweeping the skies using the observatory's Merz comet seeker. By April 1857, Horace had discovered his first comet, periodic comet Brorsen, 1857 II, a comet originally discovered 11 years earlier by Theodor Brorsen. Since Karl Bruhns, in Berlin, had reported this comet earlier, on March 18, 1857, Horace Tuttle is credited with a later independent codiscovery.

On January 4, 1858, Tuttle made a first discovery of comet 1858 I. Orbit computations by his brother Charles and a new assistant at Harvard, Asaph Hall, showed the comet to be identical to comet 1790 II, which was discovered by Pierre Méchain. This comet is now named periodic comet Tuttle. Nineteen years later, Asaph Hall would discover the two moons of Mars at the U.S. Naval Observatory. For his part, Tuttle would be credited with four comet discoveries—periodic comets 1858 I Tuttle and 1858 III Tuttle-Giacobini-Kresák and long-period comets 1858 VII and 1861 III. He has been credited with independent codiscoveries of comets 1857 II, 1857 V, 1857 VI, 1858 VI, 1859, 1860 III, and 1888 V and periodic comets 1862 III Swift-Tuttle and 1866 I Tempel-Tuttle. These latter two are the parent comets of the Perseid and Leonid meteors. Tuttle also discovered minor planet 66 Maja on April 9, 1861, and 73 Klytia on April 7, 1862. In March

1859, Tuttle was awarded the Lalande Prize in astronomy by the Academy of Sciences in Paris for his discovery of three comets in 1858.

With the American Civil War a year old, Tuttle left his position at Harvard in 1862 and served nine months in the 44th Massachusetts infantry. The following year, he secured an appointment as an acting paymaster in the Union Navy and requested a transfer to one of the iron-clad ships engaged in the blockade of Charleston Harbor, South Carolina. August 1864 found Tuttle making observations of comet 1864 II Tempel on the deck of the U.S.S. *Catskill,* an iron-clad ship with a single turret shielded with 11 inches of iron. According to Tuttle's own account, he was instrumental in the capture of the English blockade runner *Deer* when on February 19, 1865, he went ashore and put up a red light that the *Deer* took as a safe signal to proceed into harbor. Tuttle and a group of men then boarded the vessel and claimed it as a prize for the U.S.S. *Catskill.* While Tuttle's account of the *Deer's* capture may have exaggerated his own role in the incident, this was perhaps the high watermark of Tuttle's career.

Tuttle's success in observing comets and capturing enemy blockade runners did not extend to his abilities as a Navy paymaster. In 1869, one year after receiving an honorary Master of Arts degree from Harvard, Tuttle's account books were found deficient by $8800— nearly four times his annual salary. In 1873 while trying to catch up with the ship to which he had recently been assigned, Tuttle illegally cashed a Naval bill of exchange for £150, then alleged that £120 of it had been stolen from him. Navy auditors caught up with him and he was charged with embezzling nearly $6000 and scandalous conduct tending to the destruction of good morals. Tuttle was found guilty, his sentence was approved by President Ulysses S. Grant, and on March 23, 1875, he was dismissed from the Navy.

Though disgraced, Tuttle's astronomical career didn't come to an end. In fact, three weeks after being dismissed from the Navy, he was appointed astronomer to the Rocky Mountain Region of the U.S. Geographical and Geological Survey and in the summer of 1877 helped run the boundary line between Wyoming and the Dakota Territory. The Navy apparently forgave his checkered career as a Navy paymaster because around 1884, Tuttle returned to Washington, D.C. and began to carry out a number of observing programs at the U.S. Naval Observatory. In his later years, he worked as a contractor for the U.S. mail service and supplemented his income by writing popular articles on astronomy. In August 1923, Tuttle died with an estate of only $70 and was buried in an unmarked grave. An appropriate epitaph might have been—Horace P. Tuttle (1837–1923), Comet Hunter, Civil War Hero, and Embezzler.

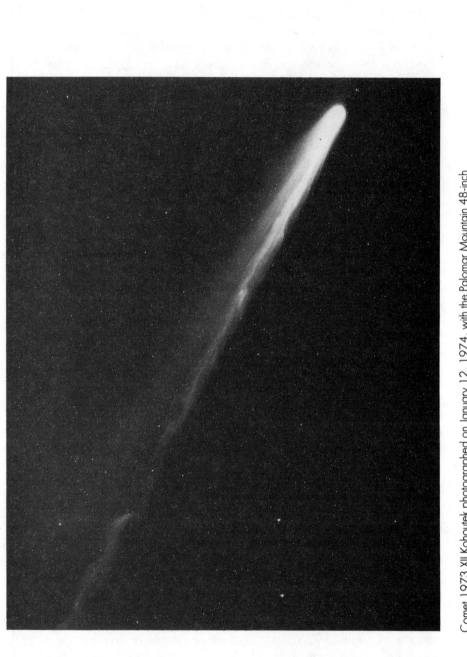

Comet 1973 XII Kohoutek photographed on January 12, 1974, with the Palomar Mountain 48-inch aperture Schmidt telescope. Note the wavelike features in this comet's ion tail. (*Photograph courtesy of The Observatories of the Carnegie Institute of Washington, Pasadena, California.*)

motion of these ions was due to instabilities resulting from electric currents flowing along the tail's axis, then the magnetic field in the tail would be 100 *nanoTeslas*, nT, or more. For comparison, the strength of the Earth's magnetic field at the poles is approximately 60,000 nT. From their physical model published in 1976, Wing Ip and Devamitta Mendis also concluded that the folding of interplanetary magnetic field lines around comets could lead to magnetic fields of up to 100 nT. However, the physical models developed by Alexander I. Ershkovich predicted magnetic fields in ion tails of 10 nT or less, very close to the ambient interplanetary magnetic field of about 8 nT. As noted in Chapter 10, the magnetic field strengths measured both by the International Cometary Explorer spacecraft in comet P/Giacobini-Zinner and by the Giotto spacecraft in comet P/Halley reached maxima of approximately 60 nT.

The Cometary Nucleus

Before discussing the nature of the cometary nucleus, it should be noted that very few details are known with any certainty. In his review of cometary nuclei in 1988, Michael F. A'Hearn wrote that there are probably more theories describing details of cometary nuclei than there are well-determined properties.

Structure and Composition

Prior to the late nineteenth century, those who considered comets to be celestial phenomena generally asumed its head, or nucleus, was a single coherent body. In 1687, Isaac Newton treated the cometary nucleus as a single body affected by the same dynamical laws that govern planetary motions.

Pierre-Simon de Laplace recorded his views on the cometary nucleus in the third and fourth editions of his *Exposition du système du monde*, published in 1808 and 1813. Laplace stated that cometary material frozen near aphelion vaporizes when the comet passes near perihelion. The formation of a cometary atmosphere in the Sun's proximity was a likely consequence of the vaporization of nucleus surface fluids, and the resultant cooling must moderate the excessive heat of the Sun. He explained the cometary phenomena using packed snow or ice as an example of how the Sun's heat could first reduce the ice to liquid, then vaporize it into a gas. During each of these two phase changes, the temperature would remain constant. For example, despite the amount of heat available, the temperature of the water-ice mixture could get no warmer until it was all reduced to water. Apparently, Laplace did not realize that in the near vacuum of space, an ice would sublimate directly from a solid to a gas without going first through a liquid phase. Still, he had suggested an essentially correct view of the cometary nucleus.

In his publications of 1836, Friedrich Bessel considered the cometary nucleus as different from solid bodies, such as the Earth, Moon, and planets. Bessel thought that the nucleus must be a volatile substance, able to vaporize easily under the influence of heat or some other property of repulsion. He pointed out that the volatility must first show itself on the sun-facing side. Bessel noted Laplace's suggestion that the active cometary mass was protected from destruction, because once the vaporization temperature has been reached, all the Sun's heat was used in the vaporization process and none went to further increase the temperature of the nucleus itself. Although both Laplace and Bessel had come very close to the correct explanation for the cometary nucleus, it would be well over a century before these ideas would again be considered in Fred Whipple's icy-conglomerate model.

During the mid and late nineteenth century, the established connection between meteor streams and comets led to a belief that the cometary nucleus was nothing more than a swarm of small, solid particles in the same orbit. The *flying sandbank model*, as it was dubbed, was championed by Richard A. Proctor (1837–1888) and later by Raymond Arthur Lyttleton. Adsorbed gases within these particles were thought to be liberated when the swarm approached the Sun, thus producing the observed coma and tail. This model was seriously considered by many astronomers throughout the first half of the twentieth century, though there were problems. The small particles would lose all of their gases after only a few revolutions around the Sun, and interplanetary space was not a sufficient source of gas to resupply them between returns to the Sun. The fact that several sungrazing comets survived passages to within a few solar radii implied that they were not made up of individual particles. If they were, tidal forces would have scattered them or they would have been completely vaporized. During the spacecraft flybys of comet Halley in March 1986, images of its solid nucleus dramatically validated the icy-conglomerate model as the correct one (see Chapter 10).

In the mid-1930s, Karl Wurm showed that many observed cometary radicals—such as C_2, C_3, CH, and NH—were very active chemically and could not be expected to last the comet's lifetime before recombining into more stable molecules. This led Wurm to suggest that many of the observed species are byproducts, or so-called *daughter molecules*, that are dissociated from more stable *parent molecules*. Under the influence of solar radiation, the parent molecules quickly dissociate into the daughter products and it is these latter species that are observed spectroscopically.

In his 1948 study of periodic comet Encke, Pol Swings observed spectral features due to molecules of C_2, CN, CH, OH, NH, NH_2 and the bands

242

at 4050Å that he attributed to CH_2. They were later identified as due to the C_3 molecule. Swings suggested that water (H_2O), ammonia (NH_3), methane (CH_4), nitrogen (N_2), carbon monoxide (CO), and carbon dioxide (CO_2) were the likely parent molecules that, under the influence of sunlight, dissociated into the daughter molecules observed spectroscopically. Significantly, Swings noted that when the comet was distant from the Sun, the parent molecules would be in the form of ice.

In two papers published in 1950 and 1951, Fred Whipple proposed what has become known as the *icy-conglomerate model* for a cometary nucleus. Whipple considered the nucleus to be a single coherent body, a few kilometers in extent, and made up of various ices with embedded meteoric dust. When the icy-conglomerate nucleus approaches the Sun, the surface ices begin to sublimate. The embedded dust particles are entrained in the vaporizing gases and escape the nucleus' tiny gravitational field along with the gases themselves. Once ionized by solar radiation, the gases are swept back by the solar wind to form the comet's ion tail and the liberated dust particles are propelled in an antisolar direction by radiation pressure—forming the dust tail. One of the principal reasons for Whipple's introduction of this model was to explain the obvious nongravitational forces affecting the motion of comet Encke. The vaporization of the icy nucleus causes a rocketlike effect on the nucleus (see Chapter 8). The various ices that Whipple suggested for his icy-conglomerate model included H_2O, CH_4, NH_3, CO_2, and HCN. These parent molecules were not directly observable at the time, but they were thought necessary to explain the species that had been observed: for example, OH, CH, NH, CO^+, CN, C_3, C_2.

While CH_4 and other possible cometary ices were considered necessary to explain the sources of some daughter molecules, they themselves are highly volatile. Hence, a parent molecule like CH_4 wouldn't be expected to last more than a few revolutions about the Sun before the supply was completely exhausted. In 1952, Armand H. Delsemme and Pol Swings suggested that highly volatile CH_4 molecules and other volatile species might be embedded in the crystalline structure of water ice, as so-called *clathrate hydrates*. The crystalline structure of water ice has empty cavities that make it less dense than liquid water, which explains why it floats on water. Volatile gases embedded in these cavities last as long as the less volatile water ice. Since the crystal lattice structure is different, the ice density is even less when a clathrate forms.

Whipple originally suggested that water ice would predominate in the icy nucleus, and many subsequent observations of radio, visual, and ultraviolet spectra have borne out this conclusion. However, Paul D. Feldman, Michael A'Hearn and John J. Cowan pointed out that the outgassing observed during the apparition of comet 1976 VI West implied the vaporization of a more volatile ice than water. Carbon monoxide or carbon dioxide were considered strong possibilities; the former has been observed in a neutral state in comets 1976 VI West, 1979 X Bradfield, and periodic comet 1986 III Halley. Although it is thought to be only a trace parent molecule, the sulfur molecule, S_2, was detected in the inner region of comet 1983 VII IRAS-Araki-Alcock by Michael A'Hearn, Paul Feldman and David Schleicher. The observations were made with an ultraviolet spectrometer aboard the Earth-orbiting International Ultraviolet Explorer (IUE) spacecraft. As it leaves the nucleus, the S_2 molecule has a very short lifetime—less than eight minutes—before being dissociated by solar radiation, hence it is observable only in the innermost region of the comet. The observations of S_2 in comet IRAS-Araki-Alcock were successful because they were made during its close approach to the Earth, to within 0.04 AU on May 11 and 12, 1983. The presence of sulfur as a common trace constituent of comets has been established by the detection of carbon monosulfide, CS, in nearly all the dozens of comets observed by the IUE spacecraft.

The observation that long-period comets are consistently more active at large distances from the Sun than the more evolved short-period comets was explained by G. David Brin and Devamitta Mendis in 1979. They suggested that comets slowly develop a crust, or mantle, as they continually pass near the Sun. As a comet's volatiles sublimate in the solar neighborhood, the dust particles that do not get blown off its nucleus by the vaporizing gases remain to form an insulating, dark crust. This mechanism could also explain the fairly frequent phenomenon of a long-period comet being intrinsically brighter approaching perihelion and relatively fainter on its way out. In this regard, it should be pointed out that as early as 1950 Fred Whipple had suggested that an evolved nucleus would be covered by a dust mantle.

In a work published in 1976, Donald E. Brownlee and coworkers described micrometeorite particles that were captured with a device flown on a high-altitude aircraft. Interplanetary particles were found that were low-density, fluffy aggregates, approximately 10 microns across. Their mineral composition, which was often similar to certain types of stoney meteorites, included mainly oxides of such elements as silicon, iron, and magnesium. Because of their fragile nature and composition, they are good candidates for cometary dust. In a rather prophetic last line of a 1982 review paper, P. Fraundorf, Donald Brownlee and Robert M. Walker pointed out that these particles appear similar in size and absorptivity to soot, so when they mix with cometary material they give a cometary nucleus a rather dark appearance in spite of the presence of ice. When spacecraft flew past in March 1986, the nucleus of comet Halley was observed to be blacker than coal (see Chapter 10).

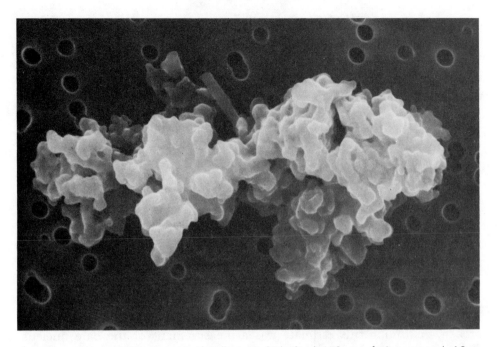

Interplanetary dust particle captured with collector on high-altitude U-2 aircraft. Approximately 10 microns across, this refractory particle's image has been enlarged 8200 times with an electron microscope. It is a very black, low density, friable object and may represent a cometary dust particle. (*Photograph courtesy of Donald Brownlee, University of Washington.*)

Size and Albedo

There has been no successful attempt to resolve a comet's nucleus with ground-based telescopes. However, assuming that cometary nuclei are solid, attempts have been made to place upper limits on their extent by establishing a limit on the angular size of what appears to be the nucleus and converting it to a linear size using the known distance between the comet and Earth. For example, a comet nucleus 1 AU from the Earth that subtended an angle of 1 arc second would have a linear dimension of 725 kilometers or 450 miles.

Harvard astronomer George Bond observed Donati's comet in 1858 and noted that the solid part of the nucleus subtended an angle less than 1 arc second at a time when the comet was 1 AU from the Earth. He correctly stated that the diameter of the nucleus was less than 500 miles. On May 18, 1910 comet Halley passed before the Sun at a distance of only 0.18 AU from the Earth. Apparently assuming that telescopes of the time could not resolve an image less than 0.4 arc second, Nicholas Bobrovnikoff concluded in his 1931 monograph that Halley's nucleus could not contain a piece as large as 50 kilometers since the comet was not observed during the solar transit.

On June 26, 1927, periodic comet Pons-Winnecke passed within 0.039 AU of the Earth and Vesto M. Slipher observed it just before its closest approach with the 0.63-meter refracting telescope at Lowell Observatory in Arizona. He compared the comet with neighboring stars that were similar in brightness, then compared the four Galilean satellites of Jupiter and a different set of stars that were near Jupiter yet similar in brightness to the comet's comparison stars. Thus, he indirectly compared the comet's brightness with the Galilean satellites. Since the sizes of the latter objects were known, he was able to conclude that the nucleus of Pons-Winnecke could not be larger than 3 to 5 kilometers. Fernand Baldet observed the comet with the 0.83-meter refracting telescope at Meudon, France. He determined that a disk nucleus would have been seen if it subtended an angle of only 0.16 seconds. Since he could not resolve the nucleus, Baldet concluded that the limiting diameter of the comet's nucleus was approximately 5 kilometers. He reached a similar conclusion using an entirely different approach, noting that the comet's nuclear brightness was magnitude 13 during the Earth close approach. Assuming this estimate was a measurement of the bare nucleus alone, it could only be a function of its illuminated surface area and reflectivity. Of course the comet's magnitude also depended on the Sun-comet and Earth-comet distances and phase angle, Sun-comet-Earth angle, but these quantities were known from its ephemeris computations. By comparing the comet's nuclear magnitude to the Sun's absolute brightness, Baldet derived an expression that allowed the determination of its diameter if its reflectivity, or albedo, was assumed. He used the values of 0.05 and 0.10 for the albedo of the nucleus and calculated a diameter of only 620 and 420 meters respectively. Had Baldet taken the phase angle into account, the diameters would have been somewhat larger.

Applying a similar analysis to 29 comets in 1966, Elizabeth Roemer assumed two extreme values for the albedo, 0.02 and 0.7, and found that most periodic comets have diameters less than 10 kilometers. Baldet and Roemer assumed the observed nucleus magnitudes used in their analyses included no contributed light from gas or dust surrounding the comet. If this assumption was not valid, their deduced nucleus diameters would have to be revised toward yet smaller values. By celestial standards, the nuclei of comets seem very small indeed.

In 1973, Armand Delsemme and David A. Rud made estimates of the diameter and albedo for several comets. They used brightness measurements of comets at large distances from the Sun to estimate the product of the cross-sectional area of the nucleus, S, and its albedo, A. Assuming that the comet's outgassing near the Sun was dominated by the vaporization of water

Armand H. Delsemme

ice, they used ultraviolet observations of hydrogen, H, and the hydroxyl radical, OH, to estimate its total vaporization rate, which they considered to be directly proportional to the product of the cross section, S, and the energy absorbed, $1-A$. Thus the two determined quantities AS and $S(1-A)$ could be used to determine S and A separately. For the nuclei of comets 1969 IX Tago-Sato-Kosaka and 1970 II Bennett, they determined albedos of 0.63 and 0.66 with corresponding radii of 2.2 and 3.8 kilometers respectively. For periodic comet Encke, the method resulted in inconsistencies, which they resolved by assuming vaporization resulted from only a small portion of the surface, a result supported by all subsequent work on this comet. Michael A'Hearn confirmed in 1988 that the method is inappropriate for short periodic comets because the effective areas for vaporization and reflected light are likely to be different and at least some of the incident energy is reradiated rather than used in its entirety during vaporization.

Martha S. Hanner and colleagues apparently observed the bare nucleus of comet 1983 VII IRAS-Araki-Alcock using infrared measurements on May 11 and 12, 1983. From their observations, they concluded that the nucleus was approximately 5 kilometers in radius and warmer than 300 degrees K, or 81 degrees F. Although the nucleus was large and warm, it was relatively inactive, so their observations were best explained if they assumed that only a small fraction of its surface was exposed ice.

Simultaneous measurements of a comet's brightness variations in the infrared and visual wavelength regions have been used to study the time history of their thermal emission and reflected optical brightness. If the observed variations in visual and infrared brightness vary in phase with each other, one can assume that the measured variations are due to differing surface areas being observed as the nucleus rotates, rather than the comet's varying outgassing activity. Using this data, Humberto Campins, Robert L. Millis, Lucy-Anne McFadden, David Schleicher, and Michael A'Hearn derived radii, geometric albedos, and axial ratios for comets P/Neujmin 1, P/Arend-Rigaux, and P/Tempel 2. They found the effective radii were respectively 10.6, 5.2, and 5.9 kilometers with corresponding geometric visual albedos of 0.02–0.03, 0.028 and 0.022. Since these objects were far from spherical in shape, only their effective radii could be determined; the square of the effective radius is equal to the product of their long and short axes. Lower limits for the ratios of their longest to shortest axes are respectively 1.45, 1.6, and 1.9. Estimated periods of rotation for these objects are given in the following section.

Radar signals have been successfully bounced off the nucleus regions of comets P/Encke in November 1980, P/Grigg-Skjellerup in May and June 1982, 1983 VII IRAS-Araki-Alcock in May 1983, 1983 V Sugano-Saigusa-Fujikawa in June 1983, and Halley in November 1985. For comet Encke, Paul D. Kamoun and his colleagues were able to infer a nucleus radius of 0.5–3.8 kilometers. Interpreting their radar data for comets IRAS-Araki-Alcock and Halley, John K. Harmon, Donald B. Campbell, and colleagues concluded that both comets were surrounded by large, centimeter-sized particles. Richard M. Goldstein and coworkers also detected large particles surrounding the nucleus of 1983 VII IRAS-Araki-Alcock and estimated a radius of 3–4 kilometers for the nucleus, whereas Harmon and his coworkers estimated a 2.5–8 kilometer radius, depending upon the nature of the comet's surface.

Rotation

If the nuclei of comets are discrete entities, there is no reason why they should not rotate about a spin axis like the Sun, Moon, and planets. In 1836, Friedrich Bessel suggested that the nucleus of comet Halley oscillates back and forth about a spin axis normal to its orbital plane and with a period of 4.6 days. Bessel thought this type of oscillatory motion, perhaps due to a magnetic Sun-comet interaction, would explain the motion of the jetlike phenomenon he observed near the comet's nucleus in late 1835.

American astronomer Benjamin Peirce noted in 1859 that the envelopes observed issuing from the nucleus of comet 1858 VI Donati might be

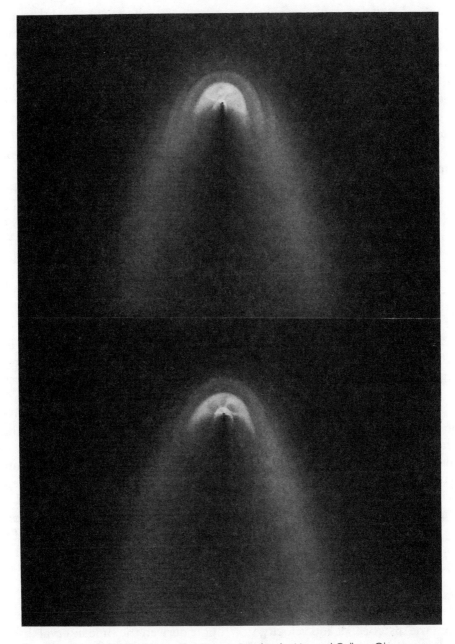

Drawing of comet 1858 VI Donati by George Bond at the Harvard College Observatory. These images were drawn on October 9 (top) and October 10, 1858 (bottom) and show the shells of material expanding away from the region of the nucleus.

explained if the comet were rotating about an axis directed toward the Sun with each envelope arising from a different point on its surface. George Bond used his beautifully drawn illustrations of Donati's comet to derive a period between successive envelopes that varied irregularly between 4 days 16 hours and 7 days 8 hours. Bond's 1862 monograph on comet Donati won the gold medal of the Royal Astronomical Society of London, the first ever awarded to an American. In 1930 Boris Aleksandrovich Vorontsov-Vel'aminov concluded that the tail movements and forms of comet 1893 IV Brooks could be explained by assuming the tail was a beam continuously emitted by a nucleus rotating about its axis every 3.793 days.

Stephen M. Larson and R.B. Minton took some very impressive photographs of the inner regions of comet 1970 II Bennett. They revealed a spiral structure that clearly suggested the comet's nucleus was in direct rotation. From the curvature of the spiral features, and the assumption of an outward expansion velocity of 0.6 kilometer per second on March 28, 1970, they derived a rotation period of 1.4 to 1.5 days. Beginning with the 1979 work of Zdenek Sekanina, it became apparent that active, localized sources on the surfaces of nuclei were responsible for fan- and spiral-shaped structures in the inner comae of some comets. These structures could be explained when certain spin axis orientations and active source locations were assumed. In fact, Sekanina used some observed inner coma structures to infer their rotational properties, along with the number and location of the active sources on their surfaces.

In a 1982 publication, Fred Whipple outlined an interesting technique for determining cometary rotation periods. In an extension of work published by Johann Schmidt in 1863, Whipple assumed that envelopes, or halos, issued from a cometary nucleus whenever an active source on the nucleus rotated to face the Sun. By measuring the angular separation between successive envelopes, the linear distance, d, between envelopes could be determined because the Earth-comet distance at the time of the observation was known. By using a relationship developed from Nicholas Bobrovnikoff's data, the envelope expansion velocity, v, was related to the Sun-comet distance, r, by the following expression:

$$v \ (km/s) = 0.535 \ r^{-0.6} \ (r \ in \ AU)$$

From the distance, d, and velocity, v, Whipple derived the time interval dt = d/v between envelope formations. The so-called *zero date* of the envelope initiation was then the observation time less the interval dt. Each zero date should be spaced by an integer number of the comet's rotation period. The appropriate integer and rotation period were found from the best fit to the data. Whipple's derived periods, for 47 comets, ranged from 4.1 hours for

250

1955 V Honda to 120 hours for periodic comet Schwassmann-Wachmann 1. For comets 1970 II Bennett, 1858 VI Donati, and periodic comets Halley, d'Arrest, and Encke, the derived rotation periods were 28.0, 4.62, 10.3, 5.2, and 6.5 hours respectively. The derived periods for comet Halley and at least some others are likely in error. The technique was not definitive as it assumed that only one surface source on the nucleus was causing the issuance of the envelopes.

Using brightness, or photometric, observations of periodic comet d'Arrest made in early August 1976, Theodore D. Fay and Wieslaw Wisniewski derived a periodic light variation history, or light curve, for this comet that had an amplitude of 0.15 magnitude in the visual wavelengths. Assuming this variation was due to different cross-sectional areas presented to the observer as the object rotated about its axis, they deduced a rotation period of 5.17 hours. A problem with this analysis is that the derived rotation period rests on the assumption that the observed light variation is due only to light reflected from the rotating nucleus, with little or no contribution from the comet's outgassing. For a relatively inactive comet, with a large amount of data spread over a lengthy time interval, this assumption should be valid. For example, the rotation period of periodic comet Tempel 2 seems reliably determined at about 9 hours. Using photometric data taken during April 9 to 15, 1988, David Jewitt and Jane Luu determined a period of 8.95 hours, while Wisniewski determined a period of 8.97 hours from his data taken during May 20 to 22, 1988.

Luu and Jewitt observed periodic variations in the brightness of comet Encke when it was near aphelion in 1988, and hence inactive. From these light variations, which had a range of 0.6 magnitude, they deduced a rotation period of 15.1 hours.

As mentioned in the previous section, simultaneous measurements of a comet's brightness variations in the infrared and visual wavelengths have been used to study the rotational properties of a few relatively inactive comets. This technique, which is perhaps the most reliable of those available to Earth-based observers, has been used to determine a rotation period of 12.67 hours for comet P/Neujmin 1, 13.47 hours for comet P/Arend-Rigaux, and 8.9 hours for comet P/Tempel 2. This latter value agrees with the estimates determined from the comet's variability in the visual region only.

Nongravitational Forces

A pivotal conclusion in the theory of cometary physics was the realization that some comets are affected by perturbative forces other than those due to the gravitational attractions of the Sun and planets. In Chapter 8, efforts of Johann Encke and Friedrich Bessel to explain these nongravitational forces were outlined. By the end of the nineteenth century, it was becoming clear that Encke's resisting medium hypothesis was untenable. In addition to the evidence against the resisting medium based on the dynamics of solar system objects, Gustave A. Hirn showed it was incompatible with the behavior of comet tails. With Encke's resisting medium unable to explain the observed nongravitational forces, it was inevitable that Bessel's original ideas would be reintroduced. On the basis of his observations of comet Halley in 1835, Bessel noted that the explosive phenomena observed emanating from the nucleus could cause a rocketlike thrust on the nucleus itself. The comet's motion would then deviate from predictions computed without allowing for these nongravitational forces.

In 1948 Soviet astronomer Alexander D. Dubiago suggested that the expulsion of dustlike material from the nucleus, in a direction differing from the radial Sun-comet line, could either add or subtract energy from a comet's orbital motion. For example, an impulsive thrust in the same direction the comet was traveling could add energy to its orbital motion and cause its period to increase slightly from one return to the next. Although he developed his model for the comet's outgassing much more fully, Fred Whipple used the same mechanism in his icy-conglomerate model to explain the nongravitational forces affecting the motion of comet Encke.

In 1968 Brian G. Marsden studied the orbits of 18 short-period comets observed at three or more apparitions and found that 15 of them were definitely affected by nongravitational forces. The following year he modeled these forces in the comet's equations of motion using an empirical expression. Together with Zdenek Sekanina and Donald K. Yeomans in 1973, Marsden published results on several short- and long-period comets in which the nongravitational force model simulated the water snow vaporization rate as a function of heliocentric distance. The vaporization rate expression was based upon the 1971 work of Armand Delsemme and David C. Miller. The success of this model was one of the first indications that cometary outgassing was due primarily to the vaporization of water snow. Attempts by Marsden and colleagues to model these nongravitational forces with frozen gases more or less volatile than water snow were largely unsuccessful.

Whipple and Sekanina, in 1979, used the time variation of comet Encke's nongravitational forces to model the comet's nucleus as a precessing oblate spheroid. They inferred the history of this comet's spin axis location over its entire observed interval and concluded that the spin axis has been nearly in the orbit plane over its entire observed history. In 1988 Sekanina reinvestigated comet Encke's motion, applying different assumptions on the nucleus outgassing rates as a function of its orbital position. He then concluded that outgassing from two source regions causes the comet's spin pole to move, or precess, 1 degree per orbit period as an average, for a total of 34

Zdenek Sekanina, Fred Whipple, and Brian Marsden surround Whipple's automobile with its distinctive license plate. (*Photograph courtesy of Fred L. Whipple.*)

degrees over the period between 1868 and 1984. The pole direction has evolved from 21 degrees from the orbit plane in 1868 to 14 degrees in 1984. Over its observed lifetime, it seems that comet Encke has been laying nearly on its side within its orbit plane, issuing jets of gas and dust from two active source areas whenever these icy vents were exposed to sunlight.

Summary

Prior to the extraordinary international efforts to mount observing campaigns for comet Halley in 1985 and 1986, the generally agreed upon model of a comet nucleus was that of an iceball a few kilometers in extent with embedded meteoric dust particles. The cometary ices were thought to be predominantly composed of water with substantial amounts of carbon dioxide or carbon monoxide ices. As the comet nucleus approached the Sun's neighborhood, the ices began to vaporize and escape the tiny gravitational field of the nucleus. Micron-sized silicate dust particles were thought to escape the nucleus through their entrainment in the vaporizing gases and to then form the dust tail when blown antisunward by solar radiation pressure. After being ionized, the gases were swept antisunward by the effects of high-

253

speed solar wind particles. Guided down solar magnetic field lines that wrapped around the ionized plasma of the comet's coma, these ionized molecules formed the ion tail.

While there were many surprises as a result of the extensive comet Halley observing programs, the cometary model that emerged just prior to comet Halley's 1986 return was, in large part, consistent with subsequent Halley observations.

NOTES

1. Presumably, these metallic chemical species arose as a result of the complete vaporization of cometary dust particles during the very close solar approach.

2. The albedo, or more properly the *Bond albedo,* is defined as the ratio of energy reflected and refracted by a particle in all directions to the energy incident on the geometric cross section. The geometric albedo is defined as the ratio of energy scattered by the particle at 180 degrees phase, or backscattering, to that scattered by a white disk of the same geometric cross section. The geometric albedo is slightly larger than the Bond Albedo.

3. Astronomers observe in the infrared, visible, and ultraviolet wavelength regions generally use Angstrom units to specify wavelengths. However, radio astronomers often prefer to use the frequency of the radio radiation. The frequency of a particular radiation in cycles per second, or *Hertz,* is equal to the speed of light, 3×10^{10} centimeters per second, divided by the wavelength. The unit megahertz, MHz, is one million Hertz. Thus the radio frequency corresponding to a wavelength of 18 centimeters would be 1667 MHz.

10
Comet Halley

Efforts are made to predict comet Halley's returns in 1835, 1910, and 1986. The ancient returns of comet Halley are traced using observations back to 240 B.C. Astronomers search for comet Halley in 1982. The International Cometary Explorer spacecraft flies past comet Giacobini-Zinner in September 1985, and six spacecraft fly by comet Halley in February 1986. Ground-based and spacecraft observations confirm Fred Whipple's model of a cometary nucleus as a dirty iceball. However, the nucleus is both larger and blacker than expected, and surprisingly small motes of dust are discovered.

THE INTERNATIONAL CAMPAIGN TO observe comet Halley during its 1986 return to perihelion was extraordinary, both for the number of scientific investigations that were undertaken and for the increased knowledge of cometary processes that resulted. When the comet passed through the Earth's orbital plane in March 1986, some 50 scientific instruments on six separate spacecraft were there to greet it. Measurements were also made from Earth orbit, during rocket flights above most of the Earth's atmosphere, and even from a spacecraft orbiting Venus. From the ground, thousands of professional and amateur astronomers from more than 50 countries conducted coordinated observing programs.

After a discussion of the attempts to determine the comet's predicted times of perihelion prior to the 1835, 1910, and 1986 returns, and the comet's long-term motion, the scientific results of the 1982 to 1990 observing campaigns will be discussed in this chapter. However, since the number of observations and results were so extensive, no attempt will be made to give a detailed summary. Note will be made of only those results that added significant new understanding to the cometary phenomena.

The Returns of Comet Halley

From Clairaut's 1758 work through 1910, all efforts to compute the perturbed motion of comet Halley were based on the variation of orbital elements technique first developed by Joseph Lagrange in 1785 (see Chapter 7). The effects of planetary perturbations on the comet's orbit were computed at various times and changes in orbital elements were added to those of the comet's reference ellipse. The comet's motion was then represented by the new reference ellipse, until it was once again rectified by another set of changes, or perturbations, later on in its orbit. Various works on comet Halley's motion differ only in how many perturbing planets were included in the calculations, how many orbital elements were allowed to change, and how many times per revolution the reference ellipse was rectified by adding the perturbations in elements. Until after the 1909 to 1911 apparitions, no attempt was made to link observations of two or more of them into one orbital solution.

Predicting the 1835 Return of Comet Halley

Fifteen years prior to the expected 1835 return of comet Halley, the French astronomer Marie Charles Théodore Damoiseau (1768–1846) computed the perturbative effects of Jupiter, Saturn, and Uranus on the comet over the interval 1682 to 1835 with an eye toward predicting the comet's perihelion time. Since the actual time of perihelion passage in 1835 was November 16.4, Damoiseau's initial prediction of November 17.15 was remarkable. However in 1829, Damoiseau added the perturbations due to the Earth and revised his prediction to November 4.81. Working over the period 1830 through 1835, Philippe Gustave Le Doulcet comte de Pontécoulant (1795–1874) considered the perturbative effects of Jupiter, Saturn, and Uranus on comet Halley during the interval 1682 to 1835 and the Earth's perturbative effects near the comet's 1759 perihelion passage. His predictions for the 1835 perihelion passage times were, successively, November 7.5, November 13.1, November 10.8, and finally November 12.9.

The most complete work leading to the 1835 return was undertaken by a former student of Friedrich Bessel, Otto A. Rosenberger (1800–1890). After a complete reduction of available observations, Rosenberger recomputed orbits for comet Halley during its 1759 and 1682 apparitions, then computed the effect on all the orbital elements from the perturbations of the seven known planets from 1682 to 1835. Assuming the comet's motion was unaffected by any resisting medium surrounding the Sun, Rosenberger's prediction for the 1835 perihelion passage time was November 12.0. Another

Philippe Gustave Le Doulcet comte de Pontécoulant. (*Courtesy of Mary Lea Shane Archives of the Lick Observatory, University of California at Santa Cruz.*)

German astronomer, Jacob W.H. Lehmann, investigated the motion of comet Halley over the 1607 to 1835 interval, taking into account the perturbative effects of Jupiter, Saturn, and Uranus. However, his perihelion passage prediction was late by more than 10 days.

Comet Halley was first recovered in the constellation of Taurus just before morning dawn on August 6, 1835 by Father Étienne Dumouchel at the Collegio Romano observatory in Rome, Italy. Except for a period from mid-November to mid-December 1835 when it was in the same part of the sky as the Sun, the comet was regularly observed through early 1836; the last observation was made on May 19, 1836, by Palm Heinrich Ludwig von Boguslavsky at Breslau, now Wroclaw, Poland. Among the extensive sets of observations were those made in Estonia by Friedrich G.W. Struve at Dorpat, now Tartu, in Germany by Bessel at Königsburg and Johann Encke at Berlin, and in South Africa by Thomas Maclear at the Cape of Good Hope and John Herschel at Feldhausen.

Work Leading up to the 1909 Recovery

To anticipate the next apparition of comet Halley, Pontécoulant, in a work published in 1864, took into account the perturbative effects of Jupiter, Saturn, and Uranus before predicting May 24.36, 1910 as the next time of perihelion passage for comet Halley. The actual time turned out to be April 20.18.

In 1907 and 1908 two English astronomers at the Greenwich observatory, Philip H. Cowell (1870–1949) and Andrew C.D. Crommelin (1865–1939), published their preliminary calculations, which were undertaken to see if Pontécoulant's prediction was correct. Their computations included perturbations by all the planets from Venus to Neptune—except Mars—and they predicted a return to perihelion on April 8.5. In their subsequent work published in 1910, Cowell and Crommelin abandoned the variation of elements technique and studied the comet's motion using direct numerical integration, whereby perturbed heliocentric positions were obtained directly at each time step. Thus the effects of planetary perturbations on the comet's motion were computed far more often. In this improved work, they computed the planetary perturbations and used an integration time step that varied from 2 to 256 days. They predicted a 1910 perihelion passage time of April 17.11.

The comet's recovery was made on September 11, 1909 by Maximilian Franz Joseph Cornelius Wolf (1863–1932) at Heidelberg, Germany. Subsequently, a careful reexamination of photographic plates taken by Harold Knox-Shaw on August 24 at the Khedivial observatory in Helwan, Egypt also revealed the comet's image. Curiously, the guiding telescope of the photographic reflecting telescope at Helwan was one of the telescopes used by John Herschel to observe the comet from the Cape of Good Hope in 1836.

From the 1909 observations of comet Halley, it was obvious that Cowell and Crommelin's prediction had to be corrected by three days. Recovery observations in hand, Cowell and Crommelin improved the accuracy of their work by reducing the time steps to one-half, carrying an additional decimal place, and correcting errors in the previous work. Their post-recovery prediction was revised to April 17.51. They concluded that at least two days of the remaining discordance was due to causes other than errors in the calculations or the planetary positions and masses. It should be noted that the best predictions for the 1835 perihelion passage time, by Rosenberger and Pontécoulant, as well as the 1910 prediction by Cowell and Crommelin, were too early by 4.4, 3.5, and 2.7 days respectively. With hindsight, one would expect these predictions to be early by a few days since none of them included the effects of the so-called *nongravitational* forces, the rocketlike thrusting of an outgassing cometary nucleus.

A 1909 British cartoon by William Heath Robinson showing astronomers at Greenwich observatory looking for comet Halley. A number of improbable methods to augment the telescope's power are being employed. (*Courtesy of Janet Dudley, Royal Greenwich Observatory.*)

In 1910, the arrival of comet Halley inspired this advertisement for Moët & Chandon champagne illustrated by the Austrian artist Marquis de Bayros.

The 1982 Recovery of Comet Halley

Looking back on the work of Friedrich Bessel during the 1835 to 1836 apparition of comet Halley and the consistently early perihelion passage time predictions for 1835 and 1910, it is clear that its motion is influenced by more than solar and planetary gravitational perturbations. In 1968, Herman F. Michielsen pointed out that perihelion passage predictions based on strictly gravitational perturbation calculations required a correction of

The tail of comet Halley was predicted to have swept the Earth on May 18 and 19, 1910 and the fear of possible poisonous gases in the tail caused more than a little concern. This contemporary French postcard poked fun at the fears by showing several ways of leaving Earth before the fateful day. This card's reverse proclaims this item as an "official souvenir of the end of the world."

+4.4 days over the past several revolutions. In 1972, Tao Kiang determined a mean correction of +4.1 days.

To account for this 4-day discrepancy between the actual period of comet Halley and that computed using perturbations from known planets, some unorthodox solutions have been proposed. Joseph L. Brady, in 1972, suggested the influence of a massive trans-Plutonian planet, and Hans Q. Rasmusen, in 1967, adjusted the ratio of the Sun-to-Jupiter mass ratio from the accepted value of 1047 to 1051. Both suggested solutions must be rejected because they would have produced effects on the motion of the known planets that were not supported by observation.

In their work published in 1967, Joseph Brady and Edna Carpenter suggested a 1986 perihelion passage time of February 5.37 based on a trial-and-error fit to the observations during the 1835 and 1910 returns. Four years later they modified their prediction to February 9.39 after introducing an empirically determined term in the comet's equations of motion to account for nongravitational effects. This empirical term had the unrealistic effect of decreasing the solar gravity with time. In 1979 Rasmusen derived a 1986 perihelion date of February 5.46 from a fit to the observations in 1835 and 1910, then added +3.96 days to yield a 1986 perihelion passage time prediction of February 9.42. It is now clear that the actual 1986 time (February 9.46) was accurately predicted by Rasmusen as well as Brady and Carpenter. However if the comet's orbit is to be accurately computed throughout a particular apparition or its motion traced back to ancient times, the mathematical model used to represent obvious nongravitational effects must be based on a realistic physical model, not empirical mathematical devices.

In introducing the icy-conglomerate model for a cometary nucleus in 1950, Fred Whipple recognized that comets may undergo substantial perturbations due to nongravitational forces or rocketlike effects acting on the nucleus itself (see Chapter 8). To accurately represent the motions of many short-period comets, Brian Marsden began to model these nongravitational forces with semiempirical terms in the comet's equations of motion. In 1973 Marsden and colleagues modified the nongravitational force terms to represent the vaporization flux of water ice as a function of heliocentric distance. To account for the rocketlike, nongravitational effects, the cometary nucleus was assumed to be an outgassing snowball made primarily of water ice.

Using the mathematical model devised by Marsden and his colleagues and observations from 1607 to 1911, Donald Yeomans computed an orbit for comet Halley and predicted a perihelion passage time of February 9.66, 1986, with an uncertainty of 0.25 days. This analysis also demonstrated that the observations were most easily represented when the model of the nucleus assumed it consisted of water ice and rotated in the same direction it moved about the Sun. Using Yeomans' ephemeris, the comet was recovered on October 16, 1982, by David Jewitt and Edward Danielson using the 200-inch telescope at Palomar Mountain, California.

The Identification of Early Comet Halley Apparitions

Until the twentieth century, all attempts to identify ancient apparitions of comet Halley were done either by determining orbits directly from observations or by stepping back in time at roughly 76-year intervals and noting whether observations of a particular comet could be represented by adjusting the perihelion passage of an approximate, predetermined orbit for comet Halley. In his 1783 to 1784 catalog of cometary apparitions, Alexandre Pingré confirmed Edmond Halley's suspicion that the comet of 1456 was an earlier apparition of comet Halley. Édouard C. Biot (1803–1850) pointed out in 1843 that a previous orbit by Johann K. Burckhardt (1773–1825) for the comet of 989 closely resembled that of comet Halley. By examining back records of approximate 76-year intervals, Biot identified observations of comet Halley in 1222, 1145, 1066, 989, 912, 837, 684, 451, and 12 B.C. Paul A.E. Laugier (1812–1872), in two works published in 1843 and 1846, correctly identified comets seen by the Chinese in 451, 760, and Autumn of

Comet Halley

Circular No. 3737

Central Bureau for Astronomical Telegrams
INTERNATIONAL ASTRONOMICAL UNION

Postal Address: Central Bureau for Astronomical Telegrams
Smithsonian Astrophysical Observatory, Cambridge, MA 02138, U.S.A.

TWX 710-320-6842 ASTROGRAM CAM   Telephone 617-864-5758

PERIODIC COMET HALLEY (1982i)

D. C. Jewitt, G. E. Danielson, J. E. Gunn, J. A. Westphal, D. P. Schneider, A. Dressler, M. Schmidt and B. A. Zimmerman report that this comet has been recovered using the Space Telescope Wide-Field Planetary Camera Investigation Definition Team charge-coupled device placed at the prime focus of the 5.1-m telescope at Palomar Observatory. Five exposures of 480-s effective duration each (in seeing measured to be $1{.}''0$ fwhm) were taken on Oct. 16 through a broad-band filter centered at 500 nm. Definite images near the expected position and having the expected motion of P/Halley were noted. No coma was detected, and the object had a Thuan-Gunn magnitude of [g] = 24.3 $\pm$ 0.2 (corresponding to V ~ 24.2; and presumably B ~ 25). Two exposures were also made in the [r] band. Preliminary representative positions, which have an estimated external error of $\pm$ $0{.}^{s}35$ in α and $\pm$ 5″ in δ but greater internal consistency, follow:

1982 UT	α_{1950}	δ_{1950}
Oct. 16.47569	$7^{h}11^{m}01{.}^{s}9$	$+ 9° 33' 03''$
16.49097	7 11 01.8	+ 9 33 02
16.52153	7 11 01.7	+ 9 33 00

The object is located some $0{.}^{s}6$ west of the position predicted by D. K. Yeomans (1981, <u>The Comet Halley Handbook</u>), suggesting that T = 1986 Feb. 9.3 UT. Confusion with a minor planet would be extremely unlikely. An attempt to confirm the recovery on Oct. 19 was successful in the sense that no objects were detected at the Oct. 16 locations and that the comet's image would then have been in the glare of a star; the dense stellar field has in fact thwarted other attempts to recover the comet during the past month. The recovery brightness indicates that the 1981 Dec. 18 attempt (cf. IAUC 3688) failed to record the comet by a very small margin and for an assumed geometric albedo of 0.5 leads to a radius of 1.4 $\pm$ 0.2 km. The comet's heliocentric and geocentric distances at recovery were 11.04 and 10.93 AU, respectively.

NOVA SAGITTARII 1982

<u>Corrigendum</u>. On IAUC 3736, line 17, the first astrometric position should be attributed to J. Hers, Sedgefield.

1982 October 21   Brian G. Marsden

International Astronomical Union Announcement Card 3737 announcing recovery of comet Halley on October 16, 1982. The 1982 attempt to recover comet Halley at Palomar observatory was led by David Jewitt and G. Edward Danielson. Jewitt, Alan Dressler, and Donald Schneider were at the 200-inch telescope on October 16, 1982, and Maarten Schmidt gave up some of his telescope time three nights later to allow a second look at the position where the comet was first seen. James Gunn and James Westphal had been responsible for the development of the CCD detector as well as earlier attempts to recover the comet at Palomar and Barbara Zimmerman provided the data acquisition system for the CCD camera.

263

1378 as comet Halley. Laugier also noted in 1842 that four of the five parabolic orbital elements for the comet seen in 1301 were close to Halley's.

Analyzing previous European and Chinese observations, John Russell
Hind (1823–1895) attempted to identify comet Halley apparitions from 11
B.C. to 1301. Approximate perihelion passage times were often determined
directly from observations, and an identification was suggested if Halley-like
orbital elements could satisfy existing observations. Although most of
Hind's identifications were correct, he seriously erred in his suggested times
in 1223, 912, 837, 608, 373, and 11 B.C.

Using a variation of orbital elements technique, Cowell and Crommelin's
1907 publication represented the first effort to actually integrate the comet's
equations of motion backward in time. They assumed that the orbital eccentricity and inclination were constant with time, but the argument of perihelion and longitude of the ascending node changed uniformly with time.
Their rates were deduced from the values computed over the 1531 to 1910
interval. Using Hind's times of perihelion passage, or computing new values
from the observations, Cowell and Crommelin deduced preliminary values
of the orbital semimajor axis for the perturbation calculations. The motion
of the comet was accurately carried back to 1301 by taking into account perturbations in the comet's period from the effects of Venus, Earth, Jupiter,
Saturn, Uranus, and Neptune. Using successively more approximate perturbation methods, Cowell and Crommelin then carried the comet's motion
back to 239 B.C. At this stage, their integration was in error by nearly 1.5
years in the perihelion passage time and they adopted a time of May 15, 240
B.C., not from their integration but from a consideration of the observations
themselves.

After a complete and careful analysis of the European and Chinese observations, Tao Kiang used the variation of elements technique to investigate
the motion of comet Halley over the interval from 240 B.C. to A.D. 1682. In
his 1972 publication, Kiang computed the times of perihelion passage directly from the Chinese observations and calculated other orbital elements
by considering the perturbations from all nine planets.

In their 1971 work, Joseph Brady and Edna Carpenter were the first to
apply direct numerical integration to the study of comet Halley's ancient
apparitions. Using an empirical expression to represent its nongravitational
effects, they initiated an integration with an orbit determined from observations from 1682 through 1911. The comet's motion was taken back to 240
B.C. in one continuous computer run. Because their integration was tied to
no observational data prior to 1682, the early perihelion dates they computed diverged from those Kiang had determined from Chinese observations. Using an orbit determined by Brady and Carpenter, Yü-che Chang

Babylonian clay tablet with Halley observations. The text describes the comet's first appearance in the region of the Pleiades and the constellation Taurus in 164 B.C.

integrated the comet's motion back to 1057 B.C. However, this integration was not based upon any observations prior to 1909, nor were nongravitational effects taken into account.

In results reported in 1981, Yeomans and Kiang studied comet Halley's past motion with an orbit based on the 1759, 1682, and 1607 observations, and numerically integrated its motion back to 1404 B.C. Planetary and nongravitational perturbations were taken into account at each step. In nine cases, perihelion passage times calculated by Kiang in 1972 from Chinese observations were redetermined and the unusually accurate observed perihelion times in A.D. 837, 374 and 141 were used to constrain the computed motion. The dynamic model, including nongravitational effects, successfully represented all existing Chinese observations of comet Halley. Subsequently, Richard Stephenson and his colleagues located Babylonian tablets in the British Museum that documented observations of the comet in 87 B.C. and 164 B.C. These observations also agreed with the perihelion passage times computed by Yeomans and Kiang, who also concluded that the comet's ability to outgas had not decreased significantly during the 2000 years it has been observed. Peter Broughton's 1979 work showed its intrinsic brightness remained nearly constant over roughly the same time. In 1986 an additional long-term integration of comet Halley's motion back to 466 B.C. was published by Werner Landgraf and another by Grzegorz Sitarski in 1988. Within the uncertainties of the Chinese observations, these results are consistent with the earlier work by Yeomans and Kiang.

In a 1989 study into the feasibility of predicting the past and future motion of comet Halley, B.V. Chirikov and V.V. Vecheslavov concluded that the comet's motion cannot be extrapolated more than a few perihelion passages beyond the period of observational data used to determine its orbit. Because of Jupiter's perturbative effects, two nearly identical numerical extrapolations of comet Halley's motion will rapidly diverge from one another after only a few revolutions. The motion of comet Halley is said to be dynamically chaotic. Table 10.1 presents the orbital elements of comet Halley from A.D. 2134 back to 466 B.C., a period over which there should be little problem with the extrapolation accuracy. As noted in this book's appendix, the observed returns of comet Halley extend only as far back as 240 B.C.

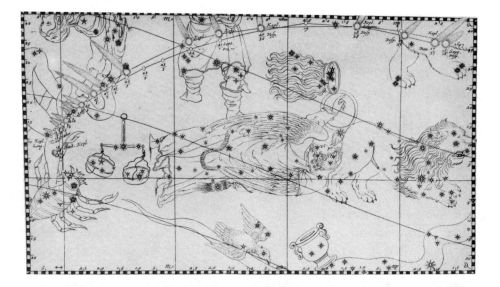

During its 1607 apparition, comet Halley was seen in Ursa Major during late September. It then traveled through Boötes in early October and before passing into solar conjunction in late October, it entered Ophiuchus. From Stanilaus Lubienietski's *Theatrum Cometicum* (Amsterdam, 1668).

Searching for Comet Halley

The last faint photographic image of comet Halley was taken in mid-June 1911 as it raced toward the outer regions of the solar system. Sixty-six years later, the search resumed. After making a slow U turn beyond Neptune in 1948, comet Halley was still near the orbit of Uranus when astronomers, using two of the world's largest telescopes, began the race to recover this most famous of all comets. First to attempt recovery were Michael J.S. Belton and Arthur Hoag, working with the 4-meter aperture telescope at Kitt Peak National Observatory in Flagstaff, Arizona. A 30-minute photographic exposure on November 13, 1977, showed no image in the comet's expected position. On November 16 and 17, 1977, James Westphal and Jerome Kristian attempted to recover the comet using the 5.1-meter (200-inch) telescope at Palomar Mountain, California. Like the Kitt Peak astronomers, they failed to detect an image. The Palomar telescope was not only larger than its counterpart, it had recently been fitted with a charged coupled device, CCD, that was significantly more light sensitive than the best photographic emulsions. The CCD device was a prototype detector developed for the Hubble Space Telescope, which was launched into Earth orbit in April 1990 after numerous delays.

Though no recovery attempts were reported in 1978, both Kitt Peak astronomers and those working at the 4-meter telescope in Cerro Tololo, Chile made unsuccessful attempts in late November, 1979. During 1981 and 1982, additional unsuccessful attempts were made at Palomar Mountain, Kitt Peak, the Canada-France-Hawaii 3.6-meter telescope on Hawaii, the Anglo-Australian 3.8-meter telescope in Siding Springs, Australia, and at the European Southern Observatory's 1.54-meter telescope in La Silla, Chile. Recovery attempts continued into February 1982, with astronomers at the University of Texas reporting unsuccessful results using the 2.1-meter telescope at the McDonald Observatory. After additional fruitless attempts in Siding Springs during April 1982, the comet slipped behind the Sun for several months. It would not be well placed for additional attempts before October 1982. By this time the giant Kitt Peak telescope had been fitted with a video camera that was very sensitive to light so the stage was set for a showdown with Palomar Mountain. Brian Marsden bet Fred Whipple $10 that the comet would not be recovered at all until 1983, but both groups of astronomers were hoping that Marsden was being too pessimistic.

Mike Belton and Harvey Butcher were awarded five half nights of observing time during the 1982 October-November-December time interval and they carefully planned their strategy by selecting the first observing date on October 18, when the comet's predicted position was well away from a rel-

Table 10.1 Orbital Elements for Comet Halley, 466 B.C. to A.D. 2134*

Year	T (E.T.)	q (AU)	e	ω	Ω	I	Epoch
2134	2134 Mar. 27.82625	0.5932158	0.9666429	113.98859	60.59196	161.74573	Mar. 15
2061	2061 Jul. 28.71195	0.5927821	0.9665774	112.03331	58.67599	161.96220	Aug. 4
1986 III	1986 Feb. 9.45895	0.5871036	0.9672769	111.84656	58.14339	162.23925	Feb. 19
1910 II	1910 Apr. 20.17829	0.5872094	0.9673038	111.71762	57.84546	162.21558	May 9
1835 III	1835 Nov. 16.43961	0.5865645	0.9673962	110.68465	56.80119	162.25569	Nov. 18
1759 I	1759 Mar. 13.06333	0.5844878	0.9675306	110.70817	56.54679	162.38297	Mar. 21
1682	1682 Sep. 15.29064	0.5826372	0.9677987	109.24246	54.86670	162.27537	Aug. 31
1607	1607 Oct. 27.53437	0.5836877	0.9673329	107.57718	53.06780	162.91222	Oct. 24
1531	1531 Aug. 26.23846	0.5811975	0.9677499	106.95724	52.34044	162.91385	Aug. 14
1456	1456 Jun. 9.63257	0.5797014	0.9679974	105.81647	51.15021	162.88607	Jun. 28
1378	1378 Nov. 10.68724	0.5762013	0.9683723	105.27668	50.30348	163.10897	Nov. 5
1301	1301 Oct. 25.58194	0.5727097	0.9689307	104.48199	49.43575	163.07179	Nov. 9
1222	1222 Sep. 28.82294	0.5742108	0.9688444	103.83087	48.58845	163.18782	Oct. 15
1145	1145 Apr. 18.56090	0.5747921	0.9687853	103.68573	48.33830	163.22004	Apr. 2
1066	1066 Mar. 20.93405	0.5744956	0.9688655	102.45543	46.90873	163.10814	Mar. 8
989	989 Sep. 5.68757	0.5819144	0.9678887	101.46581	45.84533	163.39474	Aug. 19
912	912 Jul. 18.67429	0.5801559	0.9680692	100.75913	44.93122	163.30679	Jul. 14
837	837 Feb. 28.27000	0.5823182	0.9678055	100.08403	44.21516	163.44258	Mar. 10

Year	T	e	q	ω	Ω	i	
760	760 May 20.67126	0.5818368	0.9678541	99.98016	43.97218	163.43860	Jun. 2
684	684 Oct. 2.76682	0.5795841	0.9681495	99.13197	43.08465	163.41338	Sep. 29
607	607 Mar. 15.47581	0.5808315	0.9680396	98.78209	42.54593	163.47190	Mar. 18
530	530 Sep. 27.12998	0.5755915	0.9687113	97.56504	41.26006	163.38977	Oct. 8
451	451 Jun. 28.24911	0.5737438	0.9689123	97.01122	40.49602	163.47468	Jun. 25
374	374 Feb. 16.34230	0.5771940	0.9685857	96.49409	39.86451	163.53760	Mar. 1
295	295 Apr. 20.39842	0.5759148	0.9687528	95.22565	38.39767	163.36268	Apr. 25
218	218 May 17.72347	0.5814660	0.9679755	94.13158	37.19436	163.56891	Apr. 29
141	141 Mar. 22.43405	0.5831377	0.9678439	93.67835	36.50620	163.43259	Mar. 24
66	66 Jan. 25.96014	0.5851046	0.9675458	92.63672	35.41600	163.57158	. Feb. 6
12 B.C.	-11 Oct. 10.84852	0.5871999	0.9673664	92.54339	35.19064	163.58392	Oct. 8
87 B.C.	-86 Aug. 6.46171	0.5856047	0.9676769	90.76383	33.30553	163.33505	Aug. 23
164 B.C.	-163 Nov. 12.56604	0.5845470	0.9676686	89.09882	31.35152	163.69946	Nov. 15
240 B.C.	-239 May 25.11796	0.5853647	0.9675871	88.09919	30.09811	163.46207	Jun. 7
315 B.C.	-314 Sep. 8.52367	0.5874295	0.9673085	86.86997	28.83174	163.59479	Sep. 29
391 B.C.	-390 Sep. 14.36897	0.5880489	0.9672597	86.80007	28.61248	163.59935	Sep. 28
466 B.C.	-465 Jul. 18.23879	0.5902897	0.9671425	85.23459	26.86861	163.25878	Jul. 4

*Following the year, the orbital elements are, respectively, the time of perihelion passage, T, in ephemeris time; the eccentricity, e; the perihelion distance in AU, q; the argument of perihelion, ω; the longitude of the ascending node, Ω; and the inclination, I. The last column gives the epoch of osculation where the decimal of a day is .0, ephemeris time. The angular elements are given in degrees and referred to the *ecliptic*, equinox 1950.0. Prior to 1582, the Julian calendar is used.

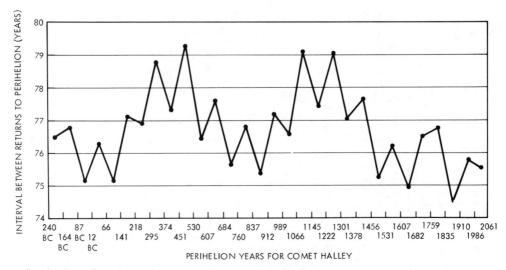

This plot shows the time intervals, in years, between Comet Halley's successive times of perihelion passage from 240 B.C. to A.D. 2061. The differences in the comet's period are largely due to the perturbations of Jupiter.

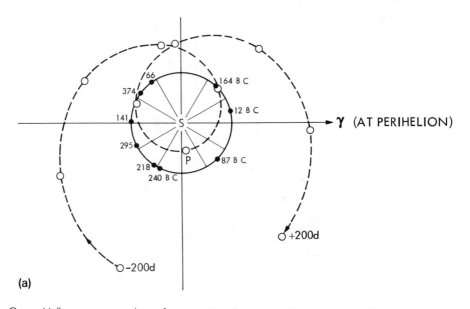

(a)

Comet Halley viewing conditions for apparitions between 240 B.C. and A.D. 374, a, between A.D. 451 and 1145, b, and between A.D. 1222 and 2061, c. These figures are drawn in a rotating reference system so that, for a given apparition, the Earth and Sun positions remain fixed, and only the comet's apparent motion is depicted. The open circles on the comet's apparent path represent the comet's position before (−) or after (+) perihelion, P, in 40-day increments. The position of the vernal

270

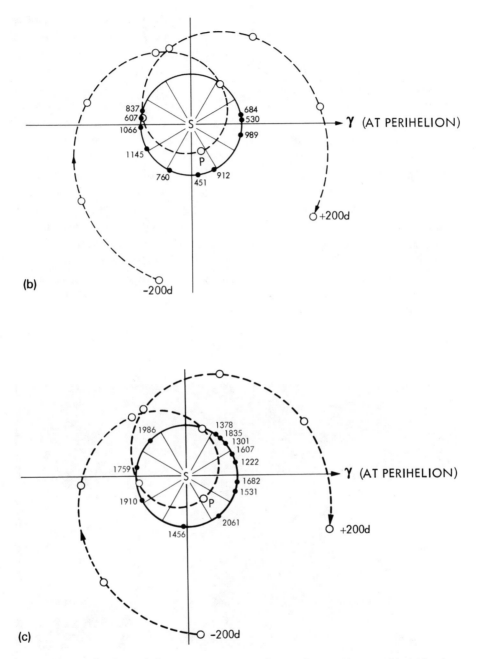

(b)

(c)

equinox (γ) is given for the perihelion passage time in each case. Because the comet's orbit has been projected onto the ecliptic plane, the viewing conditions can only be considered approximate. Using the A.D. 141 apparition as an example, note that the Earth remains fixed at the 9 o'clock position; the comet was a difficult object for viewing a few weeks before perihelion because it was behind the Sun and it passed close to the Earth about a month after perihelion.

atively bright star that could have easily obscured the faint comet. The meticulous planning paid off when the comet was successfully observed on both October 18 and 20. However, Belton and Butcher's recovery observations weren't the first. That honor goes to David Jewitt, then a graduate student at Caltech, the California Institute of Technology, and G. Edward Danielson, a staff astronomer at Caltech. Jewitt and Danielson had borrowed a few hours of observing time from Alan Dressler and Donald Schneider, who were observing extragalactic quasars and galaxy clusters on the evening of October 15 and the early morning hours of October 16. Just after 3 A.M. Pacific time on October 16, the CCD detector on the Palomar telescope was exposed to the comet's feeble light for 8 minutes at a time on six separate occasions, with each exposure separated by 10 to 15 minutes. To avoid the glare of a neighboring star—the same one that caused the Kitt Peak astronomers to postpone their observations for two days—Dressler, Schneider, and Jewitt masked a portion of the light reaching the CCD detector with two razor blades. After the digital CCD data were processed by computer, Jewitt and Danielson detected the comet at visual magnitude 24.2, only 8 arc seconds from its predicted position. Once the recovery was announced, the comet's image was also detected on photographs taken October 16 at the Canada-France-Hawaii telescope three hours after the initial recovery at Palomar.

G. Edward Danielson and David Jewitt express their pleasure at recovering comet Halley on October 16, 1982. This photograph, taken by Eleanor Helin, shows Jewitt pointing out the comet's image at the American Astronomical Society's Division of Planetary Sciences meeting on October 21, 1982.

After five years of competition, the race to recover comet Halley was won by the razor thin margin of three hours.

The Halley Spacecraft Armada, the International Halley Watch, and the International Cometary Explorer

The Halley Armada

Within a three-week period in March 1986, an armada of six spacecraft flew past the nucleus of comet Halley at distances that ranged from 28 million to less than 600 kilometers. Spacecraft Sakigake and Suisei were launched by Japan, VEGA 1 and VEGA 2 by the Soviet Union, and Giotto by the European Space Agency, a consortium of 11 countries. By the time it reached the neighborhood of comet Halley, a sixth spacecraft, the International Cometary Explorer, ICE, had already been used in a five-year study of the charged particle and magnetic field environments of the Earth's neighborhood and had passed through the tail of comet Giacobini-Zinner on September 11, 1985. Professional and amateur astronomers, organized into nine observing disciplines of the International Halley Watch, monitored comets Halley and Giacobini-Zinner using mostly ground-based instrumentation. The scope and size of the international scientific community organized to study comet Halley were unprecedented in the history of science.

It was ironic that though the U.S. scientific community had been recommending a space mission to a comet for at least 15 years, the United States would play a secondary role in the spacecraft observations of comet Halley. NASA study teams and advisory boards had recommended that a comet rendezvous mission was superior to a fast flyby mission because only the rendezvous mission mode would allow extensive scrutiny of the nucleus as the spacecraft flew alongside for an extended time. While it was recognized that comet Halley had enormous public appeal, the comet's motion through the inner solar system was retrograde, so any spacecraft launched from Earth on a direct intercept orbit in the ecliptic plane would fly past comet Halley at extremely high velocities, 65 to 80 kilometers per second.[1] Assuming the spacecraft would survive high-velocity collisions with cometary dust particles, such extraordinary velocities would allow only a quick peek at the nucleus, not an in-depth study. Thus in 1974, the Space Science Board of the U.S. National Academy of Sciences advised NASA that the primary scientific objective of an American space mission to a comet should be to study its nucleus. They advised against a fast flyby mission.

In 1976 and 1977, the Jet Propulsion Laboratory, JPL, identified two advanced spacecraft propulsion systems that could effect a rendezvous with comet Halley. One concept involved a *solar sail,* which used a large, thin sheet of reflective plastic to harness the radiation pressure of sunlight. The other used an *ion drive* engine that first ionized heavy mercury ions and then electromagnetically accelerated them to provide a weak, but continuous, rocket thrust. Both concepts were significantly different from the spacecraft that NASA had previously flown and it was feared that the new designs would be relatively expensive and risky. Having received authorization in fiscal year 1978 to proceed with a mission to Jupiter, the Galileo mission, as well as a large Earth orbiting telescope, the Hubble Space Telescope, NASA decided against requesting fiscal year 1979 authorization for the Halley rendezvous mission. With the long flight times required, NASA's decision effectively ruled out any rendezvous with comet Halley.

In 1978 and 1979, another mission option was identified. An ion drive spacecraft would first fly by Comet Halley in late 1985, drop off a separate probe into Halley's atmosphere, and continue on to rendezvous with periodic comet Tempel 2 beginning in July 1988. To decrease costs and increase the international scientific constituency, the European Space Agency was invited to provide the Halley atmospheric probe. Unfortunately, the Halley flyby/Tempel 2 rendezvous mission did not receive NASA's top billing for authorization in either the 1980 or 1981 budget years. Both the Gamma Ray Observatory and the Venus Orbiter Imaging Radar mission, later modified to the Magellan mission, were given higher priority despite the fact that only the comet mission had to be launched at a specific time. With the failure to get an authorization to begin work on the ion engines in fiscal year 1981, the Halley flyby/Tempel 2 rendezvous mission was effectively killed. The U.S. comet science community led by Cornell University astronomer Joseph Veverka, and JPL's director Bruce Murray, continued to lobby for a U.S. flyby mission to comet Halley, but it was a rearguard action. NASA's comet science advisory group had already said that fast comet flybys were of limited value and NASA continued to place a higher priority on its Venus Orbiter Imaging Radar mission.

In January 1980, European scientists considered modifying the Halley probe portion of the Halley flyby/Tempel 2 rendezvous mission for their own Giotto flyby mission to Halley. The term *Giotto* was named after the early fourteenth century Italian painter who included a Halley-like image as the star of Bethlehem in his fresco entitled "The Adoration of the Magi."[2] At first, ESA, the European Space Agency, sought use of an American Delta expendable launch vehicle, which was being phased out by NASA in favor of the Space Shuttle. By June 1980, after NASA initially refused to delay phasing out the Delta launch vehicles and ESA refused to redesign the Giotto spacecraft to be compatible with a Shuttle launch, the Europeans decided to use the French Ariane launch vehicle. This limited American involvement to participation on some of the 11 European experiments. Some Giotto space-craft tracking support was provided using the large radio antennas of NASA's Deep Space Network. Giotto was launched with an Ariane 1 rocket from Kourou, French Guiana on July 2, 1985, and approached to within 596 kilometers of Halley's nucleus on March 14.0, 1986, Greenwich mean time, at a relative velocity of 68 kilometers/second. Giotto, the first ESA inter-planetary mission, provided the most dramatic images of Halley's nucleus.

The Japanese launched their first and second interplanetary spacecraft to encounter comet Halley. The Sakigake and Suisei spacecraft were launched on January 8 and August 18, 1985, aboard Japanese rockets from the Kagoshima launch facility. *Sakigake* and *Suisei* mean pioneer and comet in Japanese. The Sakigake spacecraft made its closest approach to comet Halley on March 11, 1986, at a distance of nearly 7 million kilometers. It car-ried three instruments to measure the solar wind interaction with the comet's atmosphere. Suisei arrived within 151,000 kilometers of the comet on March 8 carrying a solar wind particle detector and an ultraviolet spec-trometer designed to image the comet's enormous, neutral hydrogen atmosphere.

The two Soviet spacecraft were launched from Tyuratam, Kazakhstan aboard Soviet Proton rockets on December 15 and 21, 1984, and flew past the planet Venus on June 11 and 15, 1985. Each one deployed an atmo-spheric balloon in the Venusian atmosphere and an instrument package de-signed to operate briefly on the planet's super-heated surface. After their pass by Venus, the spacecraft—called VEGA, a contraction of two Russian words, Venera, or Venus, and Gallei, or Halley—were targeted for March 6 and 9, 1986, encounters with Comet Halley. As with Giotto, a number of in-ternational scientists participated in VEGA's 14 scientific experiments. VEGA 1 and VEGA 2 passed by comet Halley at distances of 8890 kilome-ters and 8030 kilometers on March 6 and 9 respectively. The flyby velocities, relative to the comet, were 79 and 77 kilometers/second.

The International Halley Watch

Although the U.S. Congress did not appropriate funds for a Halley flyby mission in fiscal year 1982, they did appropriate funding for U.S. scien-tists to participate on some Giotto experiment teams and for the formation of the International Halley Watch, IHW. The International Halley Watch organization began in 1979 with an effort led by Louis Friedman, then at the Jet Propulsion Laboratory, JPL. In 1980 a NASA science working group was formed to establish the objectives of the organization and NASA Headquar-

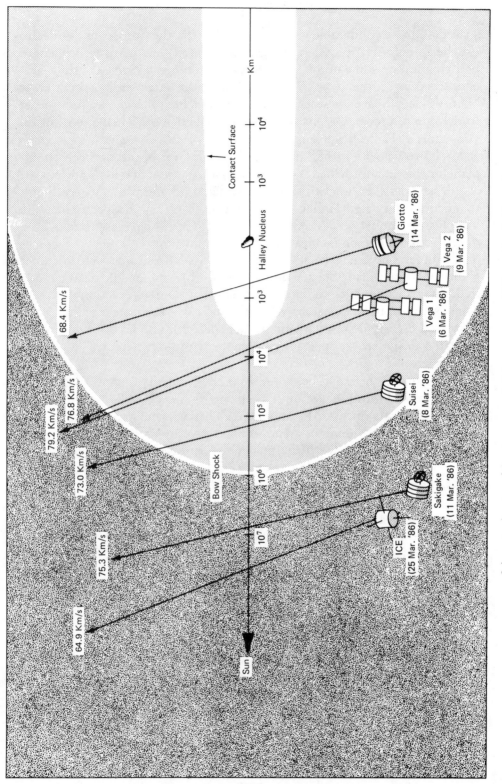

Schematic drawing of the six spacecraft flying past comet Halley in March 1986.

ters established two lead centers, one at JPL directed by Ray L. Newburn Jr. and the other in Bamberg, Germany directed by Jürgen Rahe. Ultimately, nine observing disciplines, or networks, were formed to encourage, coordinate, and archive the scientific data resulting from the observations of comet Halley during its apparition from 1982 to 1989.

Participation in the International Halley Watch was extraordinary: more than 1000 professional astronomers in 51 countries and almost 1200 amateur astronomers, some using professional equipment, in 54 countries provided comprehensive ground-based observations. IHW coordination and data-archiving activities included observations of both comets Halley and Giacobini-Zinner. In addition to providing ephemerides and recommending standardized equipment and techniques to ground-based observers, the IHW supported the various projects making Halley observations from Earth-orbiting spacecraft or rocket flights. From Earth orbit, the Solar Maximum Mission, SMM, satellite observed Halley in visual light. In the ultraviolet spectrum, Earth orbital observations were made using three satellites, the International Ultraviolet Explorer, or IUE, the Dynamic Explorer 1, or DE-1, and the Soviet ASTRON spacecraft. Ultraviolet observations were also recorded from four U.S. rocket flights into the Earth's upper atmosphere and from the Pioneer Venus Orbiter (PVO) orbiting the planet Venus.

The International Cometary Explorer

A joint ESA and NASA investigation, the Third International Sun-Earth Explorer, ISEE-3, spacecraft was launched on August 12, 1978, as part of a three-spacecraft mission to study the solar wind interactions with the Earth's charged particle and magnetic field environment. The spacecraft was placed in a small orbit about the so-called *Lagrange point* located 240 Earth radii from the Earth in the sunward direction. A spacecraft placed near this point requires occasional thruster firings to maintain its position, but due to the relative positions of the spacecraft, Earth and Sun, far fewer thruster firings are required near this point than at any other point along the Sun-Earth line. Before launch, the spacecraft was equipped with a supply of thruster fuel that turned out to be far more than was required.

In 1981, Robert W. Farquhar, a flight dynamicist at NASA's Goddard Space Flight Center, began to consider how excess fuel could be used in maneuvers that combined spacecraft thruster firings with Earth and lunar swingbys to first move the spacecraft to the other side of the Earth from the Sun, then fly by comet Giacobini-Zinner in September, 1985. A thruster firing aboard the ISEE-3 spacecraft on June 10, 1982, began five extraordinary lunar swingbys designed to reshape its orbit. After a close approach to within 120 kilometers of the Moon's surface on December 22, 1983, the space-

Periodic comet Giacobini-Zinner displays a narrow ion tail in this photograph taken by Elizabeth Roemer on October 26, 1959. (*Official U.S. Naval Observatory photograph.*)

craft's new orbit took it out of the Earth-Moon system and put it on an intercept trajectory with comet Giacobini-Zinner. After this last lunar swingby, the spacecraft's name was changed from ISEE-3 to International Cometary Explorer, ICE. Seven of the thirteen spacecraft experiments were suitable for investigating the comet's charged particle and magnetic field environment.

Since the ICE spacecraft carried no camera, it was targeted to pass through the comet's tail rather than the sunward side of its atmosphere, as the Halley spacecraft were to do. On September 11, 1985, the ICE spacecraft successfully passed through the plasma tail of the comet, at a speed of 21 kilometers/second, some 780 kilometers from the nucleus.

One of the principal results of the ICE mission was the dramatic confirmation of the plasma tail model put forward by Hannes Alfvén in 1957. Edward J. Smith and his colleagues used ICE magnetometer measurements to show that the magnetic field polarity reversed as the spacecraft traveled from one side of the plasma tail to the other—just as Alfvén's theory predicted. The solar magnetic field lines embedded in the charged particle stream of the solar wind were found wrapped around the comet's own ionized atmosphere so that two magnetic lobes of opposite polarity were detected separated by a current-carrying neutral sheet. The ICE's magnetometer recorded two magnetic field maxima of 60 nanoTeslas, nT, on either side of the central tail axis, where the field strength fell to about 5 nT.

As early as 1964, W. Ian Axford predicted the existence of a thin, stationary shock front on the sunward side of active comets. This boundary would develop when the solar wind ions, traveling at supersonic speeds of approximately 400 kilometers/second, encountered the much slower cometary ions and were slowed to subsonic velocities. Inside the shock front, Axford predicted the plasma and magnetic field of the solar wind would be disordered. The effect is analogous to the formation of a bow wave formed by a pier in a swiftly flowing river. Although no clear shock was detected in the comet's neighborhood, a broad region was detected some 70,000 to 120,000 kilometers from the comet's nucleus, where the smooth solar wind flow became more turbulent upon encountering the cometary ions.

Another important result from ICE came from the ion composition experiment designed by Keith Ogilvie and his colleagues. The principal ions were found to be H_2O^+ of the water group. The predominance of these ions suggested that water ice was the primary volatile constituent of the comet's nucleus. Ground-based and Earth orbital spectroscopic observations revealed the presence of neutral hydrogen, the hydroxyl radical, OH, and the ion H_2O^+. These species were also the likely byproducts of water vapor from the nucleus. After passing through the tail of comet Giacobini-Zinner, the ICE spacecraft flew on for a distant, 28 million kilometer encounter with comet Halley in late March 1986. Even at this enormous distance, the influence of charged particles from comet Halley were reported by Frederick Scarf (1925–1986) and colleagues using their plasma wave experiment and also by T.R. Sanderson and colleagues using their energetic proton experiment. However, whether the ICE spacecraft detected charged particles from comet Halley is still uncertain. Upon examination of ICE magnetometer measurements, Bruce Tsurutani and his colleagues could find no evidence of heavy ions from comet Halley.

More than a decade after its launch, the incredible ICE spacecraft was still operating and making scientific measurements in interplanetary space. Robert Farquhar has determined that the spacecraft will return to the Moon's neighborhood in August 2014. Only half in jest, he has commented that another series of orbital acrobatics might allow the spacecraft to be captured in Earth orbit, brought back to Earth with the Space Shuttle, then placed in the National Air and Space Museum.

Halley's Dust and Gas Atmosphere

The Plasma Environment and Ion Tails

Although often considered rather esoteric, the study of charged particles, or plasma, is an important area for understanding cometary phenomena. Well over 90 percent of all material in the universe is in the form of plasma and approximately half of the 50 scientific instruments flown past comet Halley in March 1986 were to measure some aspect of the plasma environment. Two principal instruments for measuring this environment were the magnetometer, which measures magnetic field strength and direction, and the ion mass spectrometer, IMS, which determines the mass of charged particles encountered. Strictly speaking, it measures the ratio of the ion's atomic mass to its total charge but since nearly all cometary ions have lost only one electron, they are singly charged. Thus, the measurements identify the atomic mass of each molecule that strikes it. A molecule or an ion's atomic mass, in atomic mass units, AMU, is nearly equal to the total number of protons and neutrons that comprise it. For example, the hydrogen atom has but one proton, so it would weigh one AMU, or 1.66×10^{-24} grams.

As far as 8 million kilometers from Halley's nucleus, plasma detectors aboard the Giotto and VEGA spacecraft began to detect ions distinct from those in the solar wind. Closer to the comet, the high-speed solar wind plasma, with its embedded magnetic field, encountered the much slower charged particles from the comet. Since these particles cannot cross magnetic field lines, they are dragged along with the solar wind in an antisolar direction. Hence the solar wind slows as it approaches the comet because it must pick up and drag along cometary ions, many of which are relatively heavy.

Just as a type of shock front had been detected during the ICE spacecraft encounter with comet Giacobini-Zinner, the plasma instruments aboard the Giotto, VEGA, and Suisei spacecraft encountered a broad shock front at approximately 1.1 million kilometers from the comet's nucleus. It

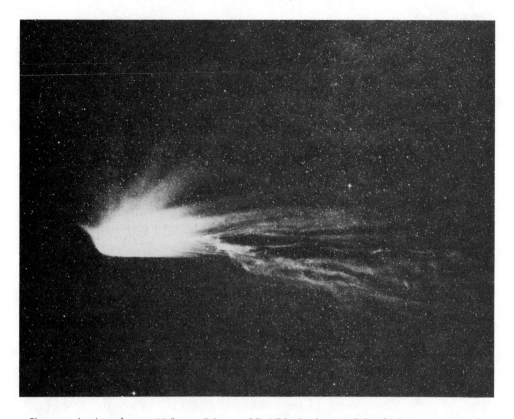

Photograph taken of comet Halley on February 22, 1986 by the U.K. Schmidt telescope in Australia. Note the presence of an antitail pointing toward 10 o'clock, several dust tail structures pointing from 11 o'clock through 3 o'clock and the wavy ion tail directed toward the general direction of 4 o'clock. (*Copyright © 1986, Royal Observatory, Edinburgh.*)

was not an abrupt boundary, but a wide region of some 40,000 kilometers where the smooth solar wind flow became more turbulent upon encountering the cometary ions.

The final barrier that keeps the solar wind from streaming closer to the nucleus is termed the *ionopause* or contact surface. At this boundary, the slow moving cometary molecules and ions prevent the faster but less dense stream of solar wind particles and the embedded solar magnetic field from progressing any farther toward the nucleus. Both the solar wind particles and magnetic field then "pile up" at the boundary. Hence the magnetic field increases up to the ionopause, then drops to nearly zero inside it. The magnetometer aboard the Giotto spacecraft measured a maximum field strength of 60 nT about 16,000 kilometers from the nucleus—7.5 times the ambient interplanetary value. At about 4700 kilometers, Giotto's magnetometer showed the magnetic field dropping to zero on the inward path. Closer to the nucleus than this ionopause boundary, the region contained only cometary particles and a negligible magnetic field. Beyond the shock front, 1.1 million kilometers from the nucleus, exists the undisturbed, supersonic solar wind flow with its embedded solar magnetic field. Between these boundaries is a mixture of subsonic solar wind protons and electrons with an embedded magnetic field and cometary ions.

Obvious changes were observed in the plasma tails of comet Halley in 1985 and 1986. Perhaps the most dramatic were the so-called *disconnection events* whereby the ion tail dropped off, or disconnected, and another tail formed to take its place. Ion tails are visible because fluorescing ions spiral around the solar wind magnetic field lines that have draped themselves about the comet's ionosphere. At least 20 prominent disconnection events were observed during comet Halley's active period in 1985 and 1986.

Malcolm B. Niedner and John C. Brandt suggested in 1978 that these events are caused when a comet passes from a region of one magnetic polarity in the solar wind to a neighboring region of the opposite polarity. Normally the solar magnetic field lines drape themselves around the comet's ionosphere like a folding umbrella and the ion tail remains firmly rooted to the comet. Frequently, however, the solar magnetic field reverses, causing a corresponding reversal in the field embedded in the solar wind. When this new, reversed magnetic field drapes itself around a comet, the two sets of magnetic field lines are of opposite polarity and hence unstable. The disconnection event occurs when the field lines reconnect to one another to form a

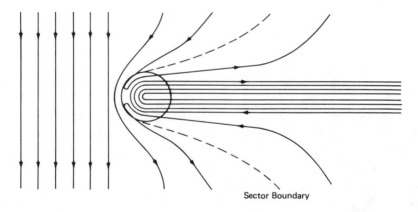

Sector Boundary

Schematic drawing showing disconnection event circumstances. As the solar magnetic field lines move toward the comet's ionized atmosphere, they drape around the comet to form a nearly linear ion tail. However, when the solar magnetic field reverses direction at a sector boundary, the magnetic field lines on either side of the boundary connect to one another near the head, and a portion of the ion tail disconnects from the comet.

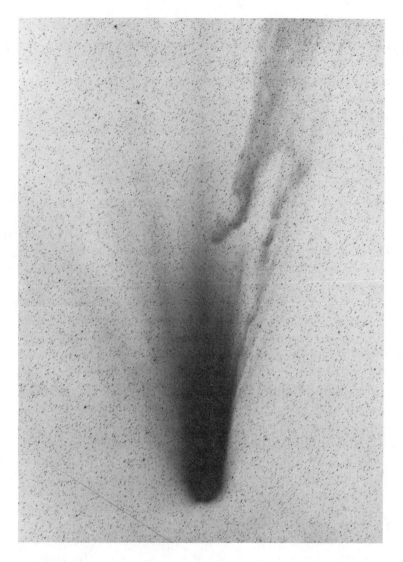

A photograph (negative print) of comet Halley on March 10, 1986 shows a spectacular disconnection event halfway up the ion tail. The distance from the comet's head to the disconnection feature is approximately 7.3 million kilometers. (*Photograph taken by the one-meter Schmidt telescope of the European Southern Observatory in La Silla, Chile.*)

more stable situation. One example occurred on March 8 to 10, and it was unambiguously connected to the passing of a solar wind magnetic sector boundary detected by magnetometers on VEGA 1 spacecraft on March 7, 1986. This was the first direct confirmation of the suspected cause of these events.

The Dust Atmosphere of Comet Halley

The dust grain population surrounding comet Halley's nucleus appears to be composed of a mixture of light substances consisting of carbon, hydrogen, oxygen, and nitrogen together with heavier, stony material consisting mainly of magnesium, silicon, iron, and oxygen. From grain to grain, there are wide differences in the ratio of light to stony material. Two broad groups of dust particles were identified. The so-called *silicate* grains are rich in silicon, Si; magnesium, Mg; and iron, Fe. The *CHON* particles are composed of carbon, C; hydrogen, H; oxygen, O; and nitrogen, N. Some of the heavier silicate grains may have a mantle of the lighter CHON material.

The composition of the Halley dust particles was analyzed by the dust impact mass spectrometers aboard the Giotto and VEGA spacecraft. A foil target of silver or platinum at the instrument's entrance broke dust particles down into clouds of ions and their atomic masses were measured to determine the chemical constituents. After analyzing data from their dust impact mass spectrometer aboard the VEGA and Giotto spacecraft, Jochen Kissel and his team reported what could be the existence of a mantle of carbon-based, or organic, CHON material covering the cores of many silicate grains. If this is the correct interpretation of the data, these observations would validate J. Mayo Greenberg's earlier laboratory model of a cometary grain that has a silicate core with a mantle of CHON-like material. Kissel's group also noted the existence of pure CHON material without silicate cores, which have masses of 10^{-16} grams or less—so tiny that they form a sort of transition between heavy molecules and small dust grains.

Using airborne and ground-based observations in December 1985 and April 1986, Jesse D. Bregman and colleagues examined the infrared spectra of comet Halley in the region between 5 and 13 microns and identified the spectral features as coming primarily from olivine, a mineral compound made up of magnesium, iron, silicon, and oxygen. This interpretation was verified by Michel Combes and his team using the data from the VEGA-1 infrared spectrometer, the so-called *IKS instrument*. As noted in Chapter 9, silicate grains had been discovered to be a component of the cometary dust population in 1970. The CHON particles were identified much later, and only as a result of the Halley spacecraft data. The evidence for these tiny, organic particles comes from the dust analyzers aboard the Giotto and VEGA spacecraft and possibly also from an infrared emission feature near 3.4 microns first observed by the IKS instrument on VEGA-1. This feature was subsequently observed in comet Halley and long-period comet 1987 I Wilson using ground-based infrared measurements. One possible interpretation as to its origin is thermal radiation from solar heated CHON particles.

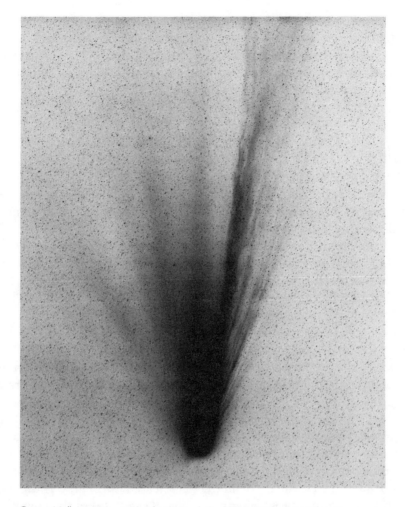

Comet Halley as it appeared on March 8, 1986. The several ion tail structures are directed toward one o'clock while several diffuse dust tail structures are seen from 10 o'clock through 12 o'clock. This image is a negative print of a photograph taken by the one-meter Schmidt telescope of the European Southern Observatory in La Silla, Chile.

Evidence for large dust particles, approximately 1 centimeter in size, in the neighborhood of Halley's nucleus was evident in the radar measurements reported by Donald Campbell and coworkers. A relatively large dust particle that hit the Giotto spacecraft was the likely cause of an intermittent loss of Giotto's radio signal 7.6 seconds before its closest approach and continuing for 32 minutes thereafter. According to the Giotto project scientist, Rudiger

Reinhard, the impact of a large particle on the outer edge of the spacecraft's dust shield could have induced a spacecraft wobble sufficient to cause Giotto's narrow radio beam to miss Earth-based antennas. A dust particle, with a mass of 0.17 grams or less, is believed to be the culprit. Such hits were also blamed for the extensive damage to both VEGA spacecraft near their close approaches to the comet's nucleus.

The Gas Atmosphere of Comet Halley

By far the largest gas cloud surrounding comet Halley's nucleus was that of neutral hydrogen, which was observed out to 10 million kilometers by spectrometers working in the ultraviolet wavelength region. Instruments with fields of view wide enough to capture most of Halley's enormous hydrogen cloud included those on the Japanese Suisei spacecraft; the Dynamics Explorer-1, DE-1, in Earth orbit; the Pioneer Venus Orbiter, PVO, orbiting Venus; and two National Research Laboratory, NRL, sounding rockets. Additional ultraviolet observations were carried out with two sounding rockets flown by Johns Hopkins University, JHU, scientists and instruments aboard the International Ultraviolet Explorer, IUE, and the Soviet ASTRON satellites.

One of the primary goals of the ultraviolet observations was to monitor the total water production rate of the comet as a function of time. From the ultraviolet observations of hydrogen and a relatively straightforward model of the radical outflow of water molecules sublimating from the cometary nucleus, it was possible to infer the total rate at which this sublimation occurred. Since water comprises the majority of all cometary gases, the total water production rates indicated the activity of the comet at any particular time and allowed an estimate of all the gas released during the entire apparition. This latter quantity, when combined with estimates of the comet's size, allowed an approximation of its remaining lifetime.

Estimates of the water production rates were inferred from measurements of the dissociation products of water. These included ultraviolet observations of the OH radical by IUE, ASTRON; neutral hydrogen by the DE-1, PVO, and NRL rockets; and neutral oxygen by the JHU rockets. The OH radical also emits at 18 centimeters in the radio wavelength region, and Eric Gérard and his colleagues extensively monitored it from July 1985 through July 1986 using the Nançay radio observatory in France. OH emission at 3090 Å and oxygen emission at 6300 Å were also monitored using ground-based observations. These observations were used to infer the comet's water production rates at various points in its path around the Sun. When its nucleus was at its maximum outgassing rate near perihelion and for two weeks thereafter, its production rate reached $1-1.7\times10^{30}$ water molecules per second, or approximately 33 to 56 tons of water per second.

The direct observation of water molecules in the atmosphere of comet Halley was noted as spectral emission at 2.7 microns by the VEGA-1 infrared IKS spectrometer and also at 1.38 microns by the VEGA-2 TKS infrared spectrometer. The neutral mass spectrometer aboard the Giotto spacecraft also detected a sharp mass peak at 18 atomic mass units, corresponding to water, and Dieter Krankowsky and his team concluded that water vapor made up some 80 percent of Halley's gases. Roger F. Knacke and colleagues also detected infrared spectral bands of water at 1.4 and 1.9 microns using ground-based infrared observations. However, the first definitive, direct detection of neutral water was made using an infrared spectrometer flown aboard a high-altitude aircraft on December 22, 1985. On this date, two days later, and again on March 22, 1986, Michael Mumma and colleagues obtained high-resolution infrared spectra of comet Halley's coma near the 2.7-micron spectral band of water.

After water, the next most abundant gas in comet Halley appears to be carbon monoxide, CO. Using data from the IUE satellite, Michel Festou and his colleagues concluded that the carbon monoxide production rate during the 1986 March 9 to 16 interval was 10 to 20 percent of the water production rate but there was little evidence for carbon dioxide, CO_2 emission. From the JHU rocket data, Thomas N. Woods and his coworkers found the CO production rate to be approximately 17 percent that of water. Infrared spectra taken with the VEGA 1 IKS spectrometer were interpreted by Michel Combes and his team as showing that the carbon dioxide production rate was approximately three percent of the corresponding water output during the spacecraft flythrough on March 6, 1986. This was the first direct detection of neutral carbon dioxide molecules in a comet.

On the basis of their data from Giotto's neutral gas mass spectrometer, Peter Eberhardt and colleagues concluded that the production rate of carbon monoxide at 1000 kilometers from the nucleus was less than or equal to 7 percent of the water rate, but oddly this ratio increased to 15 percent when the spacecraft made measurements at 20,000 kilometers from the nucleus. Apparently, there was a secondary source of carbon monoxide within the coma itself so that this molecule was actually more evident at greater distances. They proposed that the secondary source was the dissociation of the very tiny, heat resistant, or refractory, CHON dust particles. This was to become one of the more surprising results concerning the Halley gas coma; at least some of the parent molecules, of CO, apparently originate not in the ices of the nucleus but from the dissociation of dust particles in the coma itself.

Additional evidence in support of the CHON particles acting as secondary sources of gas was provided by the ground-based observations of Michael A'Hearn and his coworkers. Using filters to isolate the spectral light peculiar to the CN and C_2 molecules, they noted that the inner coma region exhibited persistent spiral gas jets during the months of March and April 1986. These gas jet structures were evident out to 50,000 kilometers from the nucleus and were not connected with the continuum emission that results from the usual micron-sized dust particles in cometary atmospheres. A'Hearn's group concluded that the observed CN and C_2 gas jets were best explained by assuming they originated as decay products of CHON particles that were already in spiral patterns far from the nucleus. Once emitted from vents on the surface of a rotating nucleus, these submicron-sized particles could remain in persistent spiral jet structures, whereas gas molecules from the vents would quickly diffuse in random directions. The dissociation of the CHON particles by sunlight apparently provided the secondary source of the CN and C_2 molecules that A'Hearn's group observed.

Although there were no convincing, direct observations of methane or ammonia in the atmosphere of comet Halley, Mark Allen and his coworkers inferred the presence of both volatile gases in the coma from the ion mass

Michael F. A'Hearn, an authority on comets at the University of Maryland.

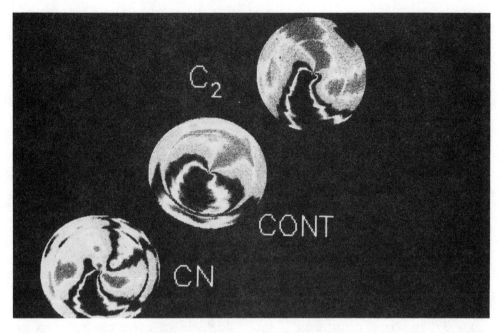

Computer enhanced images of comet Halley on April 23, 1986, in the light of the C_2 molecule, the continuum at 4845 Å, and the CN molecule. The pinwheel emission features seen in the light of C_2 and CN are thought to come from C_2 and CN molecules that formed from the vaporization of tiny CHON dust particles. (*Photograph courtesy of Michael F. A'Hearn.*)

spectrometer data aboard Giotto. Using their physical model, Allen's group could best simulate the data between the atomic mass units of 15 and 19 if they assumed the production rates of ammonia (NH_3) and methane (CH_4) were 1 to 2 percent and 2 percent of water, respectively.

Because of its importance as a likely parent molecule, four groups of radio astronomers led by F. Peter Schloerb, D. Brockelée-Morvan, D. Despois, and Anders Winnberg, observed the hydrocyanic acid (HCN) molecule in Halley's coma in late 1985, with two groups continuing their observations into 1986. This molecule was initially detected in comet Kohoutek in December 1973 and January 1974, and was thus confirmed as a result of Halley observations. Though its production rate is less than 0.1 percent that of water, this molecule is thought to be the parent molecule of CN, which was one of the first to be identified during the comet's 1985 and 1986 return. Susan Wyckoff and colleagues noted its presence in spectra taken on February 17, 1985, when the comet was still 4.6 AU from the Sun.

One of the surprises from the Giotto data was the apparent detection of identical formaldehyde molecules, H_2CO, attached to one another to form complex molecular chains, or *polymers*. Data from Giotto's Positive Ion

Cluster Composition Analyzer (PICCA) showed evidence for molecules of atomic weights equal to 45, 61, 75, 91, and 105 atomic mass units. Walter Huebner and a group led by David L. Mitchell suggested that these molecules are likely fragmentation products of a formaldehyde polymer called polyoxymethylene, or POM for short. A POM chain attached to a hydrogen atom has chemical bonds between the carbon, C, and oxygen, O, atoms that have an equal probability of being broken by sunlight. Thus the fragments have atomic masses separated by the mass of oxygen, which is 16, or of carbon, 14. The free, or unsaturated, chemical bonds at the end of each chain can easily attach not only to neutral hydrogen and the OH radicals on dust grain surfaces, they can also attach themselves to hydrogen and OH ions in the gas coma. Long POM chains will twist and form bundles, while shorter chains have an affinity for attaching themselves to graphite or silicate dust grains, thus forming whiskerlike structures.

If these observations of formaldehyde chains turn out to be correct, it would be the first detection of a polymer in space. Additional support for the detection of formaldehyde was evident in a weak spectral line at 3.6 microns in the VEGA infrared spectrometer data. Michel Combes and his team noted that if their identification of formaldehyde was correct, it was likely to be a parent molecule coming directly from the nucleus source region rather than originating as fragments of the polymer chains. The detection of formaldehyde molecules in the atmospheres of comets Halley and 1988 XV

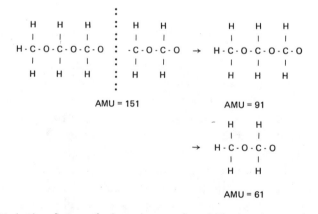

Under the influence of solar radiation, a formaldehyde polymer molecule, POM, can break down into smaller fragments. In this schematic drawing, a POM chain with 151 atomic mass units, AMU, breaks down into fragments having 91 AMU and 61 AMU. The second fragment is seen with the addition of a new hydrogen atom.

Machholz was also reported by Lewis Snyder and colleagues using 6-centimeter radio wavelength data from the Very Large Array, VLA, radio telescope facility in New Mexico.

The parent molecules reported as originating in the nucleus ices of comet Halley are presumably the source for nearly all observed atoms, molecules, and ions that have been identified in comet Halley's spectra. The identification of methane and ammonia as parent molecules has been inferred from sophisticated physical models of Halley's coma chemistry, but spectral features of these molecules have not been directly observed. Hence their relative abundance in comet Halley is quite uncertain. The total production rate of all gasses is approximately 1.3×10^{30} molecules per second. If one assumes the relative gas production rates seen near the time of the spacecraft encounters (Table 10.2) were typical for other times as well, then 78 percent of all parent molecules in the nucleus are water. The corresponding percentages for carbon monoxide, formaldehyde, carbon dioxide, methane, and ammonia are respectively 13, 4, 2, 2, and 1 percent. The ices of hydrocyanic acid, sulphur-bearing molecules, and all other gases make up less than one percent of the total.

From an analysis of the measurements made by the Giotto dust detectors, J. Anthony M. McDonnell and his team deduced that the amount of all dust particles released into the comet's atmosphere was equal to, or greater than, the total amount of gas released. Hyron Spinrad estimated that the

Table 10.2 Comparison of Parent Molecule Production Rates Near the Times of Spacecraft Encounters in March 1986

Parent molecule	Production rate (molecules/sec)	Ratio to water production	Reference
Water (H_2O)	1.0×10^{30}	1.00	Combes et al. (1988)
Carbon monoxide (CO)	$1-2 \times 10^{29}$	0.17	Woods et al. (1986)
Formaldehyde (H_2CO)	5.0×10^{28}	0.05	Mumma and Reuter (1989)
Carbon dioxide (CO_2)	2.7×10^{28}	0.03	Combes et al. (1988)
Methane (CH_4)	2.0×10^{28}	0.02	Allen et al. (1987)
Ammonia (NH_3)	$1-2 \times 10^{28}$	0.01	Allen et al. (1987)
Hydrocyanic acid (HCN)	9.0×10^{26}	0.001	Schloerb et al. (1987)

total amount of gas and dust that comet Halley lost during three months on either side of perihelion would be $1.5 - 2.0 \times 10^{14}$ grams. Assuming that Halley's total gas and dust output was approximately 2×10^{14} grams during each revolution, and taking 6×10^{17} grams as a crude estimate of its total mass, Halley's remaining lifetime is some 3000 revolutions or about 225,000 years. This estimate could easily be in error but it does suggest that Halley will be delighting sky watchers for many generations to come.

Comet Halley's Nucleus

When visible comets enter the neighborhood of the Sun and Earth, they actively throw off gas and dust, effectively hiding the central nucleus that is the source of all cometary phenomena. As comets recede from the solar neighborhood, their nuclei become inactive—hence exposed—but they are too far away to be directly resolved by Earth-based telescopes. Thus, one of the primary goals of the VEGA and Giotto missions was to directly image the heart of comet Halley, the nucleus.

The Size and Albedo of the Nucleus

Prior to the arrival of the spacecraft, the nucleus' dimensions and reflectivity were estimated by Neil Divine and his colleagues and by Dale P. Cruikshank and coworkers. Using ground-based infrared observations in February 1985, Cruikshank's group noted that the color of the nucleus was similar to that of a group of asteroids residing in the outer region of the asteroid belt between Mars and Jupiter. Assuming that all the comet's observed light was reflected sunlight and that its albedo was the same as the asteroid group, 0.04, they estimated the comet's nucleus radius as 10 kilometers. The other estimate, by Neil Divine's group, coming from what became known as the "Divine model" of comet Halley, suggested the comet's albedo was 0.06 with a radius of 3 kilometers. These latter figures were a compromise between several independent results, including an estimated radius of 3 to 5 kilometers by David Jewitt and G. Edward Danielson based on the comet's magnitude at the time of the October 1982 recovery.

The two VEGA spacecraft obtained the first television images of Halley's nucleus, but the actual solid surface was difficult to distinguish from the bright clouds of dust and gas issuing from it. This was less of a problem with the images obtained by the Giotto spacecraft. However, both the VEGA and Giotto images had to be processed by computers to enhance the contrast and resolution that eventually allowed the nucleus' shape to be characterized as peanut- or potatolike with dimensions estimated as $15 \times 8 \times 8$ kilometers. Using VEGA 1 infrared spectrometer measurements, Michel

292

Combes and his team estimated the nucleus temperature as warmer than 300 degrees K, or 81 degrees F. Because this temperature was at least 100 degrees warmer than one would expect from a sublimating, icy nucleus at the same distance from the Sun, it became clear that the nucleus did not consist entirely of exposed surface ices. The albedo of the nucleus was estimated as 0.04. Only 4 percent of the incident light is reflected, making the nucleus of comet Halley among the blackest objects in the solar system. Perhaps the popular description of the nucleus as a potato should be changed to a burned potato.

Characteristics of the Nucleus

One of the best representations of comet Halley's nucleus resulted from a composite of six separate Giotto spacecraft images. To produce this photograph, the Giotto imaging team, headed by H. Uwe Keller, carefully combined images taken from distances between 14,430 kilometers and 2,730 kilometers. Because the Giotto camera was instructed to follow the brightest

An image of comet Halley's nucleus taken by Giotto's Multicolor Camera at a distance of 25,600 kilometers. (*Copyright (1986) Max-Planck-Institut für Aeronomie, Lindau/Harz, Germany. Photograph courtesy of Horst Uwe Keller.*)

293

A composite of six separate images of Halley's nucleus taken by Giotto's Multicolor Camera at distances from 14,430 kilometers to 2730 kilometers. (*Copyright [1986] Max-Planck-Institut für Aeronomie, Lindau/Harz, Germany. Photograph courtesy of Horst Uwe Keller.*)

The Nucleus of Comet Halley

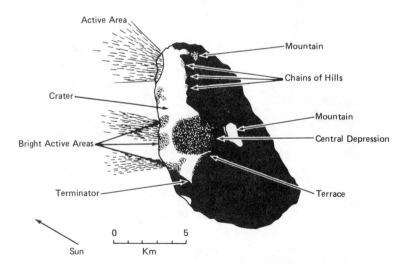

Nucleus schematic showing surface features of Halley's nucleus visible on the previous two illustrations.

feature in its field of view, the uppermost portion of the nucleus was imaged when Giotto was closest; hence, this portion has the most detail. The linear resolution, or the smallest detail discernable, varies from about 400 meters in the lower right to less than 100 meters near the top left end. In this image, north is up and the Sun is to the left.

The border, or *terminator*, between the sunlit and dark portions of the nucleus appears irregular, indicating that the surface is rough. At least one large, shallow crater is evident on the sunlit side and an elevated area, or hill, rises above the surface to catch the twilight Sun. Sublimating gases carry dust particles sunward, and these bright dust jets are the most obvious features in the image. However, they emanate only from selected portions of the nucleus where, presumably, the dark surface crust is absent, revealing the interior ices to the sunlight. Only about 10 percent of the nucleus surface area was active during the period when the VEGA and Giotto spacecraft flew past.

None of the scientific data taken during comet Halley's return to perihelion probed the subsurface structure of the nucleus. One indication of its interior would be its bulk density. Is this density more or less than that of

Paul D. Feldman, an expert on the ultraviolet observations of comets, and Horst Uwe Keller, head of the Giotto Multicolor Camera Team.

295

water ice, or one gram per cubic centimeter? If the comet's mass and volume could be estimated, its bulk density could be determined by dividing the mass by the volume. Since the rough dimensions of the nucleus were known, the volume was estimated at 500 to 550 cubic kilometers. Hans Rickman then used a clever technique to estimate the comet's mass. Rickman knew that, at any given time, the total mass of the nucleus multiplied by acceleration due to the thrust of escaping gas and dust was equal to the rate at which mass was lost by outgassing multiplied by the velocity of the escaping material. The so-called *nongravitational acceleration* could be estimated from the comet's motion, the mass loss rate could be deduced from observed water production rates, and the velocity of escaping material could be estimated. Rickman concluded the comet's bulk density was a surprisingly low 0.1 to 0.3 grams per cubic centimeter—roughly the density of a loaf of bread. R.Z. Sagdeev and colleagues used a similar technique and estimated that the density was in the range between 0.2 and 1.5 grams per cubic centimeter. Later, Stanton J. Peale noted that the estimates of both mass and volume were so poorly known that tight constraints could not be placed on the nucleus density, apart from noting that it was roughly 1 gram per cubic centimeter.

The Rotation of the Nucleus

A few months prior to the spacecraft encounters, Zdenek Sekanina and Stephen M. Larson finished their image processing on 1910 photographs of comet Halley in an effort to enhance the spiral dust features in its inner coma. By assuming these features were due to dust emission from discrete areas on the nucleus surface that activated when the comet's rotation brought them into sunlight and turned off during cometary night time, they were able to estimate a rotation period of approximately 2.2 days.[3]

The 2.2-day rotation period was seemingly verified using spacecraft data from Suisei by E. Kaneda. The brightness of the hydrogen cloud surrounding the comet varied in intensity with a period of 2.2 days. Wolfhard Schlosser and his coworkers identified the same periodicity in the expansion and contraction of the coma as seen in the light of the cyanogen molecule. Gas would presumably spew into the coma region every 2.2 days when the comet's rotation brought an active area into sunlight. Roald Z. Sagdeev and his team noted that this period was consistent with nucleus orientations seen by VEGA 1 on March 6, 1986, and VEGA 2 three days later. K. Wilhelm and his colleagues drew the same conclusion by comparing the orientation of the nucleus seen by VEGA with those seen by the Giotto spacecraft on March 14, 1986.

For a few months after the spacecraft encounters, the 2.2-day rotation period seemed verified. However, another period of 7.4 days was then suggested by Robert L. Millis and David G. Schleicher. Before and after perihelion, they monitored the coma's brightness intensity using filters to isolate the continuum emission from the dust particles and fluorescent emissions of the CN and carbon, C_2, molecules. They concluded that all their observations of the observed brightness variations were consistent with a rotation period of 7.4 days. A. Ian F. Stewart drew the same conclusion after analyzing intensity variations of the comet's hydrogen cloud using data from the Pioneer Venus Orbiter. Although the data spans did not overlap, the same observations from the Suisei spacecraft had been used earlier to verify the 2.2-day period.

In 1980 Schleicher and Schelte J. Bus found the 7.4-day period was also evident in the 1910 brightness estimates of comet Halley. Apparently the same active area on the nucleus must face the Sun every 7.4 days. M. J. S. Belton and William H. Julian noted that one possibility for explaining various observations of the comet's changing appearance would have the nucleus spinning about its longest axis so that the same point on the nucleus faced the Sun every 7.4 days while the longest axis itself revolved in space with a period of half that amount. There have been several additional estimates for the comet's rotation period as well. Unfortunately, the observations made during Halley's brief visit to the inner solar system do not allow a definite conclusion to be drawn concerning the true nature of its rotation.

The general picture of comet Halley's nucleus that emerged from the various ground-based and spacecraft studies was that of a potato-shaped object whose dimensions were approximately $15 \times 8 \times 8$ kilometers. A jet-black surface crust of organic material was broken in a few small areas, revealing a subsurface layer that consisted mostly of water ices embedded with black dust particles. As the comet rotated on its axis every few days, the ices were exposed to the warming sunlight. They then sublimated into gases and escaped the nucleus, dragging the entrained dust particles with them. From this relatively tiny, jet-black, dust-encrusted nucleus issued all the gases and dust that formed the comet's immense atmosphere and tails.

Chemical Clues for the Origin of Comet Halley

Comet Halley appears to have formed from the same chemical reservoir from which the rest of the solar system formed. According to the analysis of Armand Delsemme, the chemical abundance of various elements in comet

Table 10.3 Chronology of Major Events during Comet Halley's 1982–1989 Apparition. The date is followed by the Sun-comet distance (r) and the Earth-comet distance (d).

Date	r(AU)	d(AU)	Event
1982 Oct. 16	11.1	10.9	Recovery at Palomar Mountain.
1983 Jan. 10	10.5	9.6	One-magnitude light variation noted.
Oct. 10	8.8	8.7	First spectrum recorded.
1984 Feb. 4	8.0	7.2	Spectra shows continuum emission due to reflected sunlight.
1984 Sep. 25	6.1	6.2	Coma becomes visible.
1985 Feb. 17	4.8	4.5	Spectra shows cyanogen, CN, and oxygen, [OI], from outgassing.
Jul. 12	3.3	4.2	First photographs showing ion tail.
Nov. 8	1.8	0.9	First naked-eye sighting by Charles Morris and Stephen Edberg.
Nov. 27	1.5	0.6	Preperihelion close approach to Earth.
Dec. 9	1.4	0.7	Two 3- 4-degree ion tails noted.
1986 Jan. 9–11	0.9	1.3	Major ion tail disconnection event.
Feb. 9	0.6	1.6	Perihelion.
Feb. 15	0.6	1.5	First photograph taken after perihelion by Richard M. West.
Feb. 22	0.7	1.4	Multiple dust tails and antitail noted.
Mar. 6	0.8	1.2	VEGA 1 flyby at distance of 8890 kilometers.
Mar. 8	0.8	1.1	Suisei flyby at 151,000 kilometers.
Mar. 9	0.8	1.1	VEGA 2 flyby at distance of 8030 kilometers.
Mar. 9–11	0.8	1.1	Major ion tail disconnection event.
Mar. 11	0.9	1.0	Sakigake flyby at 7 million kilometers.
Mar. 14	0.9	1.0	Giotto flyby at 596 kilometers.
Mar. 20–22	1.0	0.8	Major ion tail disconnection event.
Mar. 25	1.1	0.7	ICE closest approach at 28 million kilometers.
Apr. 11	1.3	0.4	Closest approach to Earth, 0.42 AU.
Apr. 11–12	1.3	0.4	Major ion tail disconnection event.
May			Antitail noted pointing in the rough direction of the Sun.
Jun. 14	2.3	2.2	Last photo showing ion tail.
1987 Feb. 2	4.8	4.1	Last spectrum showing gas emission, cyanogen, CN, and carbon, C_3.
Apr. 1	5.4	4.6	Last CCD image showing faint tail.
1988 May	8.6	8.6	Dust cloud apparent surrounding nucleus region of 23rd magnitude.

Table 10.3 *(continued)*

Date	r(AU)	d(AU)	Event
1989 Jan. 1–9	10.1	9.5	R.M. West notes an asymmetric coma whose diameter is 80 arc seconds, surrounding a nucleus region whose magnitude is 23.5.
1990 Feb.	12.5	11.6	R.M. West records image of Halley at magnitude 24.4. The coma is no longer visible.

Halley are very similar to their solar abundance. An exception is the most volatile element, hydrogen. During the cometary formation process when the precometary particles agglomerated, hydrogen was presumably lost as a result of processing by galactic cosmic rays or stellar ultraviolet radiation.

The presence of the parent molecules carbon monoxide, CO, ammonia, NH_3, and methane, CH_4, in the coma suggests the nucleus has an abundance of these very volatile ices, together with the less easily vaporized water ice. Thus, it appears unlikely that the nucleus could have solidified in the relatively warmer regions inside Neptune's orbit. Compared with the chemical mixtures found on Earth and in meteorites, comet Halley has a larger percentage of the more volatile elements, a result consistent with its formation in the cold regions of the outer solar system.

One possible clue to the origin of comet Halley was offered by Susan Wyckoff and coworkers using their high-resolution spectral analysis of the CN molecule in the coma of comet Halley. They deduced that the ratio of carbon, atomic mass = 12 atomic mass units, AMU, to its isotope, atomic mass = 13 AMU, is the value 65. Since the corresponding value for the solar system is closer to 89, they suggested that Halley formed farther from the sun than 100 AU.

The initial formation of cometary dust may well occur when heavy chemical elements flow out from stars, then cool and condense into solid grains. These elements are created in thermonuclear processes within the stars themselves. Bertram Donn and then Paul Feldman and Michael A'Hearn have pointed out that one of the few compositional differences between individual comets is their relative proportions of gas and dust. Even so, the differences in the gas-to-dust ratio between comets do not offer any clues to age. This ratio is not significantly different for comets that have made numerous, or few, passages by the Sun, suggesting that the gas and dust are uniformly mixed throughout a comet's nucleus.

While the dust and gas that make up comets and all bodies of our solar system originally came from the interstellar medium, chemical evidence from comet Halley's gas and dust suggests that the agglomeration process that formed its solid nucleus probably took place within the outer regions of the early solar system. The evidence for the origin of comets, from considerations of their motions, will be examined in the next chapter.

NOTES

1. The muzzle velocity of a high-speed bullet is approximately one kilometer/second.

2. Enrico Scrovegni, a wealthy prince of extravagant tastes, chose Giotto to decorate the chapel of his palace to help redeem the activities of his family, who had grown rich from usury. Scrovegni achieved a certain immortality by being referred to as the usurer in Dante's *Divine Comedy*.

3. The rotation periods estimated from Earth-based observations are apparent, or *synodic,* periods. Because of the Earth's motion about the Sun, these rotation periods differ slightly from the true, or *sidereal,* periods that would be observed from a fixed point in space.

11

The Births and Deaths of Comets

Early views conflict as to whether comets arrive from within the solar system or from interstellar space. Interstellar comets can become members of our solar system through repeated interactions with massive Jupiter, but the process is inefficient and captures are rare. A cloud of comets at the very edge of the solar system is a likely source of long-period comets. A belt of comets beyond Neptune may provide short-period comets. The birth and death of comets continue as topics of study.

ALTHOUGH THEIR MIDLIFE PHENOMENA are now at least partially understood, the birth and death processes of comets are still fraught with fundamental questions.

- Do comets originate within the solar system or do they initially arrive from interstellar space?
- Since comets disintegrate with time, by what processes are their numbers replenished?
- Do active periodic comets disintegrate completely into meteor streams or do they evolve into inactive asteriodlike objects?
- How often do comets collide with the planets, including the Earth, and of what significance are these events?

As if to distinguish themselves from the well-behaved members of the solar system, comets can travel in almost any type of orbit, from near-circular to near-parabolic, and with inclinations that show no preference for a particular value, either direct or retrograde. There are over 100 short-period comets of less than 200 years. With a few exceptions, they all have direct orbits. A second group, of long-period comets, have paths that differ very little from parabolas. They are only observed at one return, so their orbital periods are not well determined. For ease of computation, their orbits are often assumed to be parabolic, though when their true orbital shapes are carefully investigated, they are most often found to travel on very eccentric ellipses. The presence of two such diverse cometary populations provides an enormous challenge to theorists seeking a common origin for both groups. If one adopts an origin within the planetary system, explaining the existence of the long-period population is problematic, whereas an origin well beyond the planetary system—including interstellar space—presents difficulties in explaining the abundance of short-period comets.

The paucity of real data to guide theories of cometary births and deaths has led to a great number of unconstrained ideas in these areas. This is nowhere more evident than in the study of comet-Earth collisions. The frequency and ramifications of such events are uncertain, as is the role that cometary collisions have played in the origin and evolution of life.

Interstellar and Solar System Sources of Comets

Comet Formation—Inside or Outside the Solar System?

One of the few solid facts that bear directly upon the source of comets was stated in 1705 by Edmond Halley. After computing orbits for 24 well-observed comets, Halley noted that none of them had obvious hyperbolic motions. He believed that comets were bound to the Sun and return after very long periods of time. For more than two hundred years after Halley's observation, there was no solid additional evidence identified for the birthplace of comets. Various theories favored either a solar system or interstellar origin. These included those of Immanuel Kant (solar system origin), Pierre-Simon de Laplace (interstellar origin), Joseph Lagrange (solar system), Raymond A. Lyttleton (interstellar), and Sergey Konstantinovich Vsekhsvyatskij (solar system). In the next section, the 1950 work of Jan Hendrik Oort will be reviewed. Oort's work would focus attention rather strongly on the solar system origin of comets for reasons that included Halley's initial observation two and a half centuries earlier.

Fifty years after Halley's 1705 work, the German philosopher, Immanuel Kant (1724–1804), published his cosmology, in which the solar system was envisaged as part of a vast system of stars making up the Galaxy. Once created by God, the universe evolved according to the laws eventually discovered by Isaac Newton. The bodies of each solar system formed by condensing from diffuse, primordial material, and Newtonian forces were re-

302

Immanuel Kant. (*Yerkes Observatory photograph.*)

sponsible for the flattening of the systems into disks. The various bodies of each solar system lasted only as long as it took them to fall back into their central stars, and it was this infalling material that provided the fuel for the heat. Eventually, the central stars would explode into a diffuse cloud of material and the planet formation process would begin anew. This cyclical formation and destruction of solar systems continued for all time.

In Kant's cosmology, comets were assumed to have formed with the planets from the solar nebula, but at much greater distances from the Sun. While planetary orbital eccentricities and masses increase with distance from the Sun, in the most remote regions the low density of the nebular material and the lightness of the particles there slow the formation process. Comets would be composed of the lightest type of particles, which explained the cause of their vapor and tails. In the eighteenth century, Kant's book on cosmology was not well known, partly because many of the freshly printed copies were impounded after the publisher went bankrupt. However, Kant's work was important because it anticipated, in a qualitative way, the views of Pierre Laplace in the early nineteenth century. The nebular cosmology, as it has become known, is often referred to as the *Kant-Laplace theory*.

Although Laplace explicitly presented this cosmogonic hypothesis as a version of Comte de Buffon's work made more consistent with the truths of celestial mechanics, his ideas were closer to Kant's, whose work he may not have seen. In his *Exposition du système du monde*, published in 1796, Laplace suggested that the solar system formed when the rotating solar nebula flattened and formed rings of material as it contracted. These rings then coalesced into planets. The extended rotating atmospheres of the *protoplanets* themselves formed rings of material that subsequently became their satellite systems. Laplace's cosmology thus explained the fact that all planets and satellites then known rotated in the same direction and revolved about their primaries in nearly circular orbits, in the same plane and in the same eastward direction.

The discovery, in 1798, that two Uranian satellites, Titania and Oberon, move on retrograde orbits was an immediate embarrassment to Laplace's hypothesis, and one that was never mentioned in later editions of his work. Another problem was that the Sun's current slow rotation is inconsistent with planetary formation as outlined in the nebular hypothesis. However, the simplicity of Laplace's concept made it appealing to many, and despite its failings, modern variations on the Kant-Laplace hypothesis still form the basic paradigm for the origin of the solar system. Collisional processes and the magnetic braking of solar rotation are a few of the modifications that have been made to keep it viable.

Obviously influenced by the views of Descartes and Isaac Newton, William Herschel presented his cometary ideas in 1808 and 1812. According to Herschel, comets traveled through interstellar space collecting nebulous matter and transferring it to the stars they passed, thus replenishing the fuel used in making light. Older comets would have lost most of their nebulous matter and would produce more modest tails. With each repeated stellar passage, comets would become more consolidated and dense. Ultimately, an evolved comet would lose all of its nebulous material and form a planet.

In 1812, Joseph Lagrange began his discussion on the origin of comets by agreeing with Laplace's view that the primordial solar nebula could explain the near-circular orbits of the planets. However the very eccentric and highly inclined orbits of comets did not easily fit within the Laplacian hypothesis. Noting Wilhelm Olbers' idea that the asteroids originated from a fragmented planet between Mars and Jupiter, Lagrange set about investigating the conditions under which explosive events on the planets could produce bodies traveling on cometlike orbits about the Sun. Lagrange worked out formulae giving the velocity necessary to eject a body into a cometlike orbit at any given inclination. He showed that the maximum ejection velocity for a direct or retrograde parabolic cometary orbit would be 1.73 or 2.24 times the planet's orbital velocity. The ejection velocity would have to be larger if one were to include the additional velocity necessary to overcome both the parent planet's gravity, its escape velocity, and the resistance of the planet's atmosphere. For the Earth, Lagrange took these effects into account and computed that the velocity required for bodies to leave the Earth on direct parabolic trajectories would be up to 145 times the velocity of a speeding bullet, whereas velocities up to 180 times faster than that would be required for retrograde parabolic trajectories. Lagrange took the average speed of a bullet to be 70 times less than the Earth's orbital velocity, or about 430 meters per second. The process by which bodies could be ejected from the surface was thought due to extreme pressure from the Earth's hot interior. Lagrange ended his memoir by noting that his hypothesis for the origin of comets, together with Laplace's nebular hypothesis for the formation of the planets, could give a unified explanation for the origin of all solar system bodies.

Lagrange did not comment on the likelihood of observed long-period comets forming from planetary eruptions. However, François Félix Tisserand (1845–1896) further developed Lagrange's formulae in 1890 and concluded that in order for a long-period comet to pass within the Earth's zone of visibility, a comet would have to be expelled from a planet with well-defined direction and velocity, which was unlikely to occur very often. In view of the large observed number of near-parabolic comets, this eruption mechanism was considered unlikely.

In 1813, Laplace presented his own ideas for the origin of comets. His nebular theory could explain the central dominance of the Sun within the solar system and the fact that the planetary orbits were nearly circular and in the same plane. However, the cometary orbits were neither in the same plane nor near-circular. Like William Herschel, Laplace preferred to consider comets objects of interstellar origin, formed by the condensation of nebular material spread out in the universe, which traveled from solar system to solar system. He calculated that the Sun's sphere of influence was 100,000 AU, and outside this limit comets were assumed to exist in an interstellar field with velocities ranging from zero to infinity and with each velocity equally probable. As the field of interstellar comets moved past, the Sun attracted comets whose relative velocity was nearly zero. Those with considerable relative motions with respect to the Sun would rarely be drawn to it, and only when their initial motion was directed very close to it. Using probability arguments, Laplace showed that of 5713 comets that begin at 100,000 AU and drop into the Earth's zone of visibility, a perihelion distance less than 2 AU, only one would be expected to be strongly hyperbolic. The vast majority of comets would reach the Sun on nearly parabolic orbits. A few comets might have their orbital sizes decreased as a result of planetary perturbations, or perhaps a resisting medium. Here, Laplace had anticipated the mechanism by which comets' near-parabolic paths are converted into short-period orbits as a result of planetary perturbations.

Although Laplace's results agreed with the observed distribution of cometary orbits and seemed to validate the interstellar origin of comets, he

305

neglected the Sun's motion with respect to the interstellar stars and comets. In 1783, William Herschel analyzed several nearby stars that had been observed many years apart and noted changes in their positions. He deduced that they are in motion with respect to one another and that the Sun has a so-called *proper motion* with respect to these stars. He correctly surmised that the Sun was moving toward the stars in the constellation Hercules. The point toward which the Sun appears to be moving is termed the *solar apex*, with the opposite point being the *solar antapex*.

In 1866 and 1867, Giovanni Schiaparelli showed that if the Sun's proper motion were taken into account, the predicted number of observed hyperbolic comets would be far greater than Laplace's estimate. Moreover, no comets showing the expected departures from parabolic motion have been observed. Schiaparelli's work would have decisively ruled out the interstellar origin of comets, except that the predicted distribution of observed cometary orbital eccentricities depends rather critically upon the assumed, but unknown, relative velocities within the interstellar field itself. Thus, while it can be argued that the observed lack of hyperbolic orbits strongly suggests a solar system origin for comets, this argument is not conclusive.

Further evidence for the lack of hyperbolic cometary orbits was presented in 1894 by the German astronomer—and Catholic priest—Anton Karl Thraen (1843–1902) and 20 years later by Svante Elis Strömgren (1870–1947). They pointed out that some comets whose orbits were slightly hyperbolic would appear much less so if the planetary perturbations they suffered on their way into the inner solar system were taken into account.

The Cometary Capture Process

From the early 1800s through the mid-twentieth century, various scientists sought to verify either the interstellar origin of Laplace or the solar system origin of Lagrange. There were some vigorous supporters of the Lagrange hypothesis, particularly Sergey Konstantinovich Vsekhsvyatskij (1905–1984), and a few who objected to the lack of obvious hyperbolic orbits from interstellar space. The consensus opinion amongst scientists during most of this period, however, was that comets originate in interstellar space. The major issue discussed during this time was the process of capturing interstellar comets on parabolic orbits by planetary encounters and explaining the surprisingly large number of observed short-period comets.

The most influential early work on the planetary capture process was done by Hubert Newton in 1878 and 1893. His important conclusions included the result that planetary perturbations acting on comets with parabolic orbits of every inclination would eventually produce a family of short-period comets, which—as a rule—would move in moderately inclined orbits with direct motions. A comet in a low-inclination, direct orbit can overtake Jupiter and spend enough time in front of this giant planet to have its orbital energy decreased sufficiently to be captured. However, Newton demonstrated that this capture process was very inefficient. Of one billion, 10^9, parabolic comets coming for the first time into a sphere with a radius equal to Jupiter's orbit, 2670 had their periods reduced to less than twice Jupiter's orbital period, 839 had less than Jupiter's period, 12 years, and 126 had less than half Jupiter's period. Of the 839 short-period comets whose periods would be reduced to less than the orbital period of Jupiter, 257 would have orbital inclinations under 30 degrees but only 51 would have inclinations greater than 150 degrees. In addition, those comets on direct orbits would be far more likely to undergo additional perturbations, which would further reduce their orbital periods at subsequent returns.

Several authors have recently presented results—generated with the aid of high-speed computers—similar to those obtained analytically by Hubert Newton nearly a century before. For example:

- Jupiter is the dominant planet in the cometary capture process.
- The capture of a comet into a short-period orbit is not likely to result from a single close approach to Jupiter.
- As a result of Jupiter close approaches, long-period comets traveling on direct orbits with low inclinations are far more likely to have their periods shortened than are comets traveling on retrograde orbits with low inclinations.
- Repeated planetary encounters of long-period comets will ultimately result in a family of short-period, low-inclination comets that move on direct orbits.
- Short-period comets could result either from a solar system source or from the capture of comets on initially parabolic orbits as a result of successive encounters with the major planets. In the former case, cometary disintegration must be slow in order for comets to have existed within the planetary system since its formation.
- Repeated planetary perturbations would thoroughly mix the directions of the orbital *aphelia* for periodic comets so this information cannot be used to discriminate between interstellar and solar system sources.

Except for the short-period comets, which could have been formed either inside or outside the solar system, Newton felt that the weight of evidence—while not overwhelming—favored Laplace's interstellar origin for long-period comets. Like Laplace, Newton had neglected the proper motion of the Sun in his computations.

In 1920, the American astrophysicist Henry Norris Russell (1877–1957) extended Newton's work, and verified his result that captured short-period comets were a likely result of successive encounters with Jupiter. The orbit of every recently captured comet should pass close to the orbit of the planet that captured it. Since the observed distribution of short-period comet orbits did not reflect this, Russell concluded that these comets' orbits had been shifted in space by subsequent planetary encounters. Hence they were captured long ago. Russell also pointed out that comets whose aphelion distances were close to a major planet's orbit were not necessarily captured by it, so assigning certain comets to the families of Saturn, Uranus, and Neptune was not warranted. However, it was clear that Jupiter, as the solar system's primary perturbing body, did have a family of short-period comets. The capture of a comet from a parabolic orbit is a rare event, and the production of a short-period comet is rarer still. The large number of observed short-period comets, coupled with their short lifetimes, suggested to Russell that their numbers have been increased by the disruption of large comets into smaller ones as a result of Jupiter's tidal forces.

Once the generic relationship between comets and meteors had been generally accepted in the second half of the nineteenth century, most astronomers believed that the heads of comets were made up of individual particles orbiting the Sun as a group. English astronomer Richard Proctor was a leading exponent of this *flying sandbank* comet model. Because the Jupiter capture process was so inefficient, he favored the hypothesis whereby comets originate by being expelled from planets. In 1884 Proctor pointed out the incompatibility of his comet model with the Jupiter capture process. A flying sandbank comet could not hold together unless each individual particle suffered identical perturbations when passing close to Jupiter. The enormous size of comets would ensure differing perturbing circumstances for each particle, so the group of cometary particles would greatly expand subsequent to a Jupiter close approach. Since no such effect had been observed for comets that had passed close to Jupiter, Proctor questioned the reality of the Jupiter capture process. With hindsight, he should have questioned the flying sandbank model instead. In the twentieth century this model for the cometary nucleus was championed by Raymond Lyttleton.

The Theories of R.A. Lyttleton and S.K. Vsekhsvyatskij

In 1948 and 1951 English astronomer Raymond Lyttleton published one of the few theories that attempted to explain the formation of comets as well as their subsequent dynamic behavior. In his hypothesis, comets were formed when the Sun passed through an interstellar cloud of dust. The dust particles followed hyperbolic trajectories with the Sun at one focus and collided with one another along a line parallel to the relative velocity between the Sun and dust cloud. An observer moving through space in front of the Sun and in the direction of its motion would see the dust particles collide with each other behind it in the antapex direction. During the collision, the particles lose their relative velocities and then agglomerate to produce small discrete clouds of particles, which Lyttleton identified as the comets. Some of the particles' initial hyperbolic velocities were diminished, thus forming parabolic or elliptic orbits. Particles colliding too close to the Sun vaporized on impact, while those colliding too far from the Sun were lost to the system. Thus for particles with an initial radial velocity with respect to the Sun of 1 kilometer per second, Lyttleton estimated that comets could form between 17 and 1100 AU from the Sun in the direction of the solar antapex.

Although Lyttleton's theory was attractive in that it attempted to explain both how and where comets formed, it suffered from several severe problems. First, it was difficult to understand why the newly formed comets did not simply fall into the Sun, since they formed with no transverse motion, or angular momentum, with respect to it. How the formation process, which took place along a particular direction in space, could produce long-period comets, which were observed from all directions in the sky, was also problematic. Furthermore, why the observed long-period comets evolved into a population with very different values from their original semimajor axes was also difficult to explain. There were problems with the concept of a comet as a cloud of independent, small particles whose activity was assumed due to intergrain collisions that were particularly frequent near perihelion. Cometary nongravitational forces were difficult to explain, as was the fact that some comets survived close solar passages without being vaporized. Lyttleton's theory was finally abandoned in favor of the icy-conglomerate model for the cometary nucleus when images of comet Halley's coherent nucleus were returned during the spacecraft flybys in March of 1986.

Lyttleton's hypothesis for the origin of comets did not draw an immediate or large following. However, it was certainly preferred to the contemporary solar system origin hypothesis of the Soviet astronomer Sergey Vsekhsvyatskij. This hypothesis was motivated largely by an attempt to overcome the troublesome abundance of observed short-period comets.

By investigating the intrinsic brightness of 94 periodic comets, Nicholas Bobrovnikoff determined the rate at which comets became fainter with time. In his 1929 publication, he assumed that comets never got brighter than a certain level, absolute magnitude of 1, then determined that they cannot live longer than a million years. Hence, the relatively short lifetimes of comets were inconsistent with their forming along with the planets. Bobrovnikoff reasoned that if they formed with the planets, comets would have dissipated long ago.

Celestial Pinball: Catching Comets and Guiding Spacecraft

Long-period comets have enormous orbital sizes and can travel great distances beyond our planetary system. By comparison, few short-period comets ever travel farther from the Sun than the planets. They are said to have more binding energy than long-period comets, and this quantity is represented by the inverse semimajor axis of their orbit. The smaller a comet's semimajor axis, a, the larger its gravitational binding energy, 1/a, with respect to the Sun. A comet's binding energy contrasts with its orbital energy; the latter is directly, rather than inversely, proportional to its orbital semimajor axis. Thus long-period comets are said to have more orbital energy, but less binding energy, than short-period comets. When a comet passes close to a planet, it can either gain or lose orbital energy, depending on the circumstances. It will lose orbital energy if its initial parabolic path brings it close to a planet's front side and gain energy if it passes behind. If it gains too much orbital energy from one of these encounters, its orbit becomes parabolic or hyperbolic and it flies out of the solar system, never to return.

The same principles hold for a comet entering, rather than leaving, the solar system. A comet located in interstellar space that has no initial velocity with respect to the Sun would drop into the planetary region with a parabolic orbit. Its value of 1/a would be zero. However, even the slightest initial velocity of this same comet would give it a hyperbolic orbit and its 1/a value would be negative. In general, comets entering the solar system on strongly hyperbolic orbits cannot be captured and bound to the Sun. However, if a comet whose initial path is not strongly hyperbolic loses orbital energy during a planetary encounter, it can be captured into a shorter period elliptic orbit, with more binding energy. During each planetary encounter the comet's orbital energy gain or loss will be matched by that of the planet. However, since the planets are so much larger than the comets, they experience only imperceptible changes in orbital velocity. Crudely speaking, the situation is rather like a cometary spitball ricocheting off a planetary battleship.

These same principles of celestial pinball have been successfully employed to alter the courses of manmade spacecraft. Venus flybys were employed by the Soviet VEGA 1 and 2 to redirect their flight paths to fly closely past comet Halley in March 1986. NASA's Voyager 1 and 2, launched in 1977, flew past Jupiter and Saturn, and Voyager 2 went on to Uranus and Neptune. The Saturn, Uranus, and Neptune encounters were made possible because Voyager received energy boosts along the way from backside passages of Jupiter, Saturn, and Uranus respectively. Both spacecraft received enough orbital energy from planetary flybys that they are now on hyperbolic orbits with respect to the Sun. One day, thousands of years from now, they will fly through the Oort cloud of comets and leave the solar system forever.

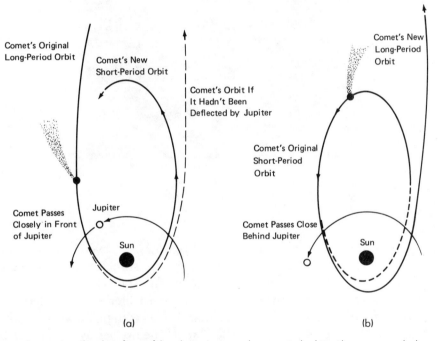

(a) (b)

a. By passing closely in front of the planet Jupiter, a long-period orbit with a near parabolic orbit, can lose enough orbital energy so that it is captured into a shorter period orbit about the Sun.

b. A short-period comet can gain enough orbital energy to become a long-period comet if it makes a close backside passage of Jupiter.

Nicholas Theodor Bobrovnikoff. (*Courtesy of the Mary Lea Shane Archives of the Lick Observatory.*)

After noting the decreasing intrinsic brightness of several short-period comets at successive apparitions, Vsekhsvyatskij estimated cometary lifetimes to be no more than a few thousand years.[1] Yet short-period comets are seen in abundance in the inner solar system. Short lifetimes, coupled with the inefficiency of capturing new interstellar comets with planetary perturbations, led Vsekhsvyatskij to suggest the planets themselves as a more fertile source of the short-period comets. Hence, he resurrected Lagrange's hypothesis. In 1930 and 1931 he argued that volcanic eruptions from the major planets were the source of comets. Twenty years later he advocated similar eruptions from the satellite systems of the planets—mainly Jupiter—as the source, because the escape velocity from these small satellites was only a fraction of that required to escape Jupiter, 60 kilometers per second. In Vsekhsvyatskij's view, short-period comets continually issued forth from satellite eruptions, while long-period comets formed millions of years ago during cataclysmic eruptions from the outer, major planets.

The ejection of high-velocity bodies from the planets or their satellites seemed such an unlikely physical mechanism that few were won over. It was pointed out that the Earth and Moon do not eject bodies into cometlike orbits. Planets cannot eject enough mass at sufficient velocities to explain long-period comets even in the unlikely event that explosive planetary events could be shown to result in cometlike objects.

In the mid-twentieth century, neither the interstellar origin hypothesis of Lyttleton nor Vsekhsvyatskij's ideas for a solar system source could easily explain the observed characteristics of cometary populations.

The Oort Cloud of Comets

In 1932 the Estonian born astronomer Ernst Julius Öpik (1893–1985) investigated the stability of distant meteors or comets to stellar perturbations and concluded that they would remain bound to the Sun at distances of one million AU. Even at this distance, roughly four times the range to the nearest star, Öpik argued that comets could remain connected to the Sun over the entire age of the solar system. He concluded that as a result of stellar perturbations, the perihelion distances of long-period comets would tend to increase with time, so their orbital distribution would eventually form a cloud, or shell, at the edge of the solar system. However, a small percentage of these objects, with large initial perihelion distances, could have their orbits modified to smaller perihelion distances. Öpik's work established the possibility that a cloud of comets surrounding the planetary system out to interstellar distances could remain bound to the Sun even when the perturbing effects of passing stars were taken into account.

The next significant work that led to the concept of the Oort cloud of comets, was published by the Dutch astronomer Adrianus Jan Jasper Van Woerkom in 1948. The importance of this work is the systematic way with which he detailed the many problems with both the solar system and interstellar origins of comets. Considering the solar system origin, Van Woerkom acknowledged that Vsekhsvyatskij's eruption mechanism might explain the orbital properties of the short-period comets, but not those of long-period comets. Following Tisserand's objection, he noted that unless planetary explosions are somehow favorably directed, eruptions would produce a ratio of 3 to 1 for direct to retrograde long-period comets, a result that was contrary to the observations. In addition, the eruption theory could not explain the nearly uniform distribution of orbital inclinations observed for long-period comets.

Ernst Julius Öpik

Van Woerkom also presented an in-depth analysis of several problems that had to be overcome before an interstellar origin of comets could be tenable. He identified three basic difficulties:

1. There are very few observed hyperbolic comets. In order for the majority of observed elliptic, long-period comets to have been captured from an interstellar field of comets, they would have to possess very small initial velocities, less than 1 kilometer per second, relative to the Sun. Passing stars and galactic differential rotation made this unlikely.

2. The mechanism of single-event capture of comets by Jupiter was too inefficient to explain the observed number of short-period comets. Van Woerkom began with Herbert Newton's result that only one in a million comets coming to perihelion with a parabolic orbit would be changed into an elliptic orbit with a period less than Jupiter's. The number of comets that came within Jupiter's distance was approximately 6 per year so the number of comets captured was then 6 x 10^{-6} per year. However, the number of newly discovered comets of Jupiter's group between 1850 and 1930 was 37, or 0.47 per year.

Hence, there were far too many observed short-period comets to be accounted for by single-event captures by Jupiter.

3. The distribution of reciprocal semimajor axes for the observed comets was not consistent with what would be expected from the capture process. Table 11.1 is a modified version of data that Van Woerkom presented. It shows, for each interval of reciprocal semimajor axes, the corresponding range in semimajor axes, a, the range in periods, and the number of observed comets that fell into each group. The vast majority of long-period comets, 177, possess semimajor axes so large as to put them into the first group. Most observed long-period comets have enormous orbital sizes.

Van Woerkom pointed out that the data were inconsistent with the picture of interstellar comets being captured into long-period comets by repeated Jupiter encounters. He determined that the principal effect of Jupiter perturbations on comets in near-parabolic orbits was not direct capture but small changes in 1/a of approximately 0.001 AU^{-1}, or a typical velocity change of less than 1 kilometer per second. The accumulation of small perturbations in 1/a could be modeled as a diffusion process whereby the long-period comets would gradually have their orbits modified and either be thrown into interstellar space or become short-period comets. Their 1/a values would diffuse into smaller or larger quantities respectively. After the first few returns to the Sun, the values of 1/a for these comets would no longer fall

Table 11.1 *Modified Version of Van Woerkom's Table Listing the Number of Observed Comets from 1850 to 1936 in Equal Intervals of Reciprocal Semimajor Axis, 1/a.*

Range of 1/a, 1/AU	Range of a in AU	Range of Periods in Years	Number of Comets
0.000–0.002	∞–500	∞–11,180	177.0
0.002–0.004	500–250	11,180–3,953	10.0
0.004–0.006	250–167	3,953–2,152	8.0
0.006–0.008	167–125	2,152–1,398	7.0
0.008–0.010	125–100	1,398–1,000	2.5
0.010–0.012	100–83	1,000–761	6.5
0.012–0.014	83–71	761–604	1.0
0.014–0.016	71–63	604–494	1.5
0.016–0.018	63–56	494–414	1.5
0.018–0.020	56–50	414–354	4.0
0.020–0.022	50–45	354–306	2.0
0.022–0.024	45–42	306–269	0.0

into the range from 0.000 to 0.002. Within approximately one million years, all long-period comets would disappear into the inner solar system or interstellar space. Yet the vast majority of observed comets on near-parabolic orbits do not seem to have done either. Hence, either they originated within the last million years or there is a source of comets far enough away that they are not affected by the perturbing action of the Sun and planets.

Capture of comets traveling on near-parabolic orbits by repeated Jupiter encounters would result in a flat distribution of reciprocal semimajor axes. If most observed long-period comets formed by this mechanism, the diffusion of 1/a values would cause roughly an equal number of comets in each interval of 1/a, including negative values of 1/a to account for the hyperbolic comets produced.

Van Woerkom reached the conclusion that, although the marked concentration of orbits near the parabolic limit appears at first sight to indicate an interstellar origin for comets, there seem to be insurmountable objections to such a scenario. In this classic work, he showed that Vsekhsvyatskij's solar system origin of comets could not explain the long-period comets, and orbital characteristics of observed comets were extremely difficult to reconcile with the capture of interstellar comets by Jupiter. None of the existing theories of cometary origin were viable!

In discussing the objections to the interstellar origin of comets, Van Woerkom mentioned in passing that a cloud of comets moving permanently with the Sun might be free from some of the objections he raised. This latter thought was developed into the *Oort cloud.*

The Dutch astronomer Jan Oort published a work in 1950 that was to bring nearly universal agreement about the source—if not the origin— of comets. Oort concluded that the observed distribution of cometary orbits could be explained by supposing a cloud of one hundred ninety billion, 1.9×10^{11} comets surrounds the Sun as far as 50,000 to 150,000 AU and it is affected from time to time by the perturbing effects of passing stars.

Like Van Woerkom before him, Oort constructed a table of 1/a values for several long-period comets. However, he was careful to include orbital information for only 19 comets with well-determined orbits that had been computed for a time before each comet entered the realm of the planets. Thus, his table represented these comets' original orbits before they had been perturbed into somewhat different paths by planetary encounters. Table 11.2 is a slightly modified version of the information that Oort presented in 1950. From the information that went into making his table, Oort concluded that the average value of 1/a for the 10 objects in the first interval is 0.000018, or the average semimajor axis is 55,555 AU. These comets would spend the majority of their time near aphelion, which, for these highly eccentric orbits, would be near 110,000 AU.

Jan H. Oort

Oort reminded his readers of Van Woerkom's observation that within one or two million years after their first perihelion passage, practically all of the long-period comets would have suffered planetary perturbations and either diffused into shorter period comets or diffused out of the solar system altogether. Thus, to explain why there are so many observed long-period comets, one had to suppose that comets either were younger than one million years or were continually brought into the inner solar system from a region so remote that they suffered little from planetary perturbations. Since there were no obvious hyperbolic cometary orbits to suggest a continuing interstellar source, Oort supposed that long-period comets arrived into the planetary region from a cloud of comets surrounding the Sun at a distance of between 50,000 and 150,000 AU. If one imagines that the planetary system, extending to the outermost orbit of Pluto, was somehow shrunk to the size of a dime, then Oort's cloud would extend from 11 meters to some 33 meters from the dime's center.

Once the Oort cloud was postulated, it was necessary to find a mechanism whereby comets in it could be perturbed into the Sun's immediate neighborhood so they could be seen from Earth. Van Woerkom pointed out that while planetary perturbations could alter orbital semimajor axes, these effects could not as easily reduce the comets' perihelion distances to bring

317

Fire and Ice: Comets Colliding with the Sun

More than 3 centuries ago Edmond Halley recognized that the comet of 1680 had passed within 0.006 AU of the Sun's center. Since the radius of the Sun's disk is 0.0047 AU, the question arose as to whether a comet can pass closer than this distance and actually collide with the Sun. This question was answered in the affirmative when the Solwind coronagraph on the U.S. Air Force space test program satellite P78-1 discovered six comets that passed close to, or collided with, the Sun. These six discoveries were made between the time of the satellite's launch in 1979 and its deliberate destruction in 1985 during a Department of Defense antisatellite weapons test. Images of the comets passing into the Sun's atmosphere were only possible because a special occulting shield blocked out the solar light, allowing the much fainter comets to be seen nearby. None of these comet's survived their solar encounters.

These kamikaze comets were named SOLWIND 1 through 6. None were noticed by ground-based observers but the coronagraph on board the Solar Maximum Mission satellite, SMM or Solar Max, also detected SOLWIND 5. Launched on February 14, 1980, Solar Max was not available to observe SOLWIND 1 and blown fuses disabled the coronagraph and several other instruments until space shuttle astronauts repaired them in April 1984. Solar Max then observed SOLWIND 5, and by the time it was destroyed during its reentry into the Earth's atmosphere in early December 1989, it had discovered 10 of its own sungrazing comets, SMM 1 through 10.

Sequence of photos showing the collision of comet Solwind 1 with the Sun on August 30 and 31, 1979. The comet is seen approaching the Sun's occulted disk from the right. After colliding with the Sun, the comet apparently disintegrated completely. The Greenwich times of the photographs are given in the lower left corner of each image. (*Photographs courtesy of the Naval Research Laboratory.*)

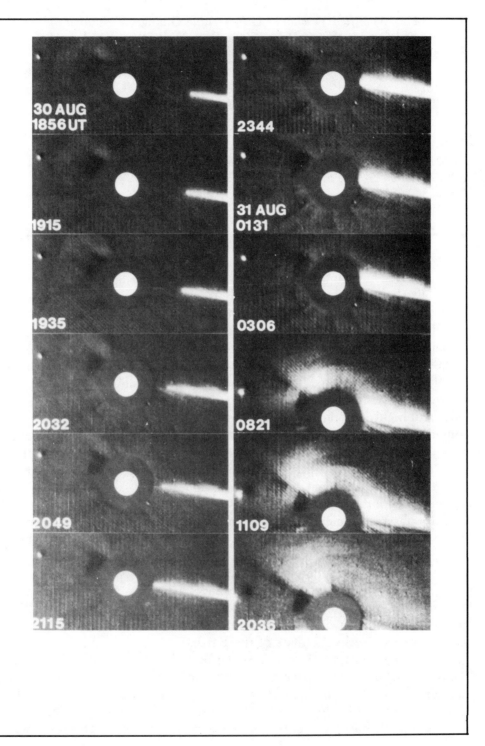

Table 11.2 *Modified Version of Oort's Table Giving the Number of Long-period Comets Whose Orbits Have Been Well Determined as a Function of their Reciprocal Semimajor Axis*

Range of 1/a, 1/AU	Range of a in AU	Range of Periods in Years	Number of Comets
< 0.00005	> 20,000	> 2,828,427	10
0.00005–0.00010	20,000–10,000	2,828,427–1,000,000	4
0.00010–0.00015	10,000–6,667	1,000,000–644,331	1
0.00015–0.00020	6,667–5,000	644,331–353,553	1
0.00020–0.00025	5,000–4,000	353,553–252,982	1
0.00025–0.00050	4,000–2,000	252,982–89,443	1
0.00050–0.00075	2,000–1,333	89,443–48,686	1
> 0.00075	< 1,333	< 48,686	0

them close to the Sun. Oort reasoned that stars passing near the distant cloud, like bullets through a cloud of gnats, would alter the course of nearby comets and send a few racing toward the distant Sun.

Under the perturbing influence of passing stars, Oort determined that comets could remain in stable orbits, bound to the Sun, to distances of some 200,000 AU, a figure five times smaller than the result Öpik had determined in 1932. Oort attributed the difference to Öpik's failure to account for Jupiter's perturbations and less-elongated cometary orbits. For example, circular and highly elliptical orbits can have the same semimajor axis, a, yet a comet on a highly elongated orbit can travel nearly twice as far from the Sun as one on a circular orbit. Oort determined that stellar perturbations acting over the lifetime of the solar system would completely rearrange the orbital orientations into a random distribution, even if their initial orbits began with a preference for the ecliptic plane. Thus the wide variation of orbital inclinations for observed comets with nearly parabolic orbits was explained in a natural way. Oort then asked how many comets would be required in the cloud to explain the observed flux of one dynamically new comet passing each year inside a Sun-centered sphere with a radius of 1.5 AU. Assuming only the effects of stellar perturbations, he determined that some 1.9×10^{11} comets were required in the cloud. Assuming an average mass of a comet to be 10^{16} grams, the total mass of the Oort cloud of comets would equal about 10^{27} grams or a few tenths the mass of the Earth.

Oort then considered whether stellar perturbations acting on comets in the distant cloud could explain the distribution of 1/a values. Van Woerkom had already pointed out that comets repeatedly perturbed by Jupiter would show a flat distribution in 1/a so that there would be roughly equal numbers of comets in all the tabular intervals, both positive and negative. This result was completely inconsistent with the observed data in Oort's table. However, for a cloud of comets traveling through space with the Sun, Oort concluded that a comparison of the observed distribution of 1/a values in his table satisfactorily agreed with the distribution predicted from his comet cloud hypothesis. The single exception was the overabundance in the first interval of 1/a.

Oort's comet cloud model predicted six times fewer of the very longest period comets than were actually observed. This discrepancy could be explained by assuming that new comets, fresh from the Oort cloud, were intrinsically brighter on their first return to the Sun and were not rediscovered at subsequent returns, when their 1/a values would be larger. Together with Maarten Schmidt in 1951, Oort estimated that approximately 80 percent of Oort cloud comets making their first, or first few, passages by the Sun should be no longer discoverable at subsequent returns. Since none of the comets in their sample exhibited disruptions and disappeared, cometary disintegration could not explain the paucity of the theoretical long-period comets in their model results. They concluded that new comets from the Oort cloud were brightest on their first few returns to the Sun, but become too faint to be seen on subsequent returns, when planetary perturbations had reduced their semimajor axes or increased their 1/a values. Thus the first interval in Oort's table is overpopulated with respect to theoretical predictions.

Beginning with the groundwork that Van Woerkom provided, Oort's cloud could explain the source of the long-period comets. Perturbed out of the cloud by stellar perturbations, the comets arrived in the inner solar system with random inclinations and near-parabolic eccentricities.

With its publication in 1950, Oort's cloud quickly became the favored model for the source of comets. However, his suggestion that the origin of comets and asteroids was due to the breakup of an Earth-sized planet between Mars and Jupiter was not as well received. In addition, Oort considered his comet cloud model only in light of the long-period comets and those with periods down to about 100 years. He did not offer a solution to the problem of comets whose periods were less than Jupiter's twelve-year period. There were still far too many of these comets to be explained by available theories.

Further Refinements and Alternatives to the Oort Cloud

Much of the work on the source of comets since Oort's paper of 1950 has supported his basic ideas concerning the provenance of long-period comets. Many of the changes are only refinements to the original theory. For example, the inner region of Oort's original cloud has become more populated and the important perturbative effects of the galactic nucleus, and particularly the galactic plane, have been recognized. It has been suggested that occasional passages of the solar system through giant, interstellar molecular clouds may strip existing long-period comets out of the Oort cloud and possibly replenish the supply with new comets at the same time. To help repopulate the outer Oort cloud and explain the population of short-period comets, a massive inner comet cloud and a comet belt beyond Neptune's orbit were suggested. The theoretical picture of the comet cloud that emerged in the late twentieth century is that of a flattened disk of comets beyond the orbit of Neptune that gradually spreads out into an inner cloud and finally forms the roughly spherical Oort cloud itself. However, this concept may not maintain its scientific consensus. With characteristics of the observed cometary populations acting as the only constraints on theoretical constructs, progress in this field does not seem to occur in a straightforward manner and a complete convergence of views has not yet occurred.

Updating and Verifying the Oort Cloud and the Capture Process

In 1978, Brian Marsden, Zdenek Sekanina, and Edgar Everhart (1920–1990) extended a 1973 work by the first two authors and presented near-parabolic orbits for 200 comets. Each orbit was computed with respect to the solar system's center of mass, and original and future values of $1/a$ were determined corresponding to times when the comet was 60 AU from the Sun on the inbound and outbound paths. Thus these $1/a$ values represented the original orbital shapes before encountering the planetary system and the future shapes after having been affected by planetary perturbations. If they used only the best determined, original orbits and eliminated comets that came within two AU of the Sun—to reduce the effects of outgassing on their orbital parameters—the authors found that the average value for orbital semimajor axes was about 22,000 AU. Thus the average extent of the Oort cloud is twice this distance, or about 44,000 AU. If the effects of planetary perturbations and outgassing on the orbits of near-parabolic comets are removed, these objects appear to arrive from an average distance of about 44,000 AU from the Sun. Thus Oort's original estimate of the cloud's extent was reduced. They also confirmed Oort's suggestion that a large number of comets on nearly parabolic orbits are observed during their first approaches to the Sun and that fading tends to prevent them from being rediscovered on subsequent returns.

In a series of papers from 1969 through 1977, Edgar Everhart used computer simulations to mathematically follow the paths of thousands of simulated comets, each making thousands of revolutions about the Sun. These numerical experiments were designed to investigate long-term

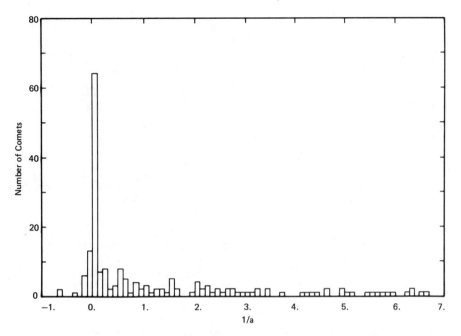

Distribution of the inverse semimajor axes for about 200 observed long-period comets. The number of comets is given for each interval of $1/a$, where the units printed along the horizonal axis of the graph are $1/(1000 \text{ AU})$. The vast majority of comets from the Oort cloud have values of $1/a$ near 0.00005, a figure that corresponds to semimajor axes of approximately 22,000 AU. Based on the 1978 work of Brian Marsden, Zdenek Sekanina, and Edgar Everhart.

changes in cometary orbits as they evolved away from their initial parabolic paths under the influence of planetary perturbations. Many of Everhart's numerical experiments verified the analytic results obtained by Hubert Newton in 1878 and 1893 and by Van Woerkom in 1948. Just as Newton pointed out, Everhart determined that if short-period comets in Jupiter's family could be captured with a single, very close passage of the giant planet, then one would expect a sizable fraction of all short-period comets to be retrograde and there are no observed retrograde members of Jupiter's family.

Everhart's results suggested that Jupiter could produce the observed short-period comets from a flux of near parabolic comets. Rather than requiring one very close passage, however, the process required many hundred not-so-close passages near Jupiter to effect the captures. Some 90 percent of Jupiter family comets could have evolved from orbits that were originally parabolic, assuming they had low inclinations, less than 9 degrees, and perihelia near Jupiter's orbit between 4 and 6 AU. Using a sophisticated numerical integration, Everhart followed the evolution of a large sample of orbits with

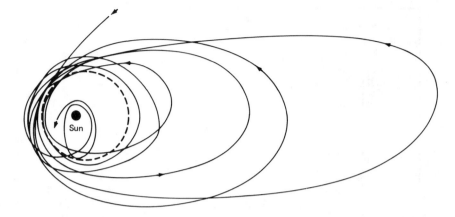

A comet whose long-period orbit has a perihelion near Jupiter's orbit (dashed circle) can be transformed into a short-period comet by repeated Jupiter encounters. Once the transformation is complete, the comet's aphelion, rather than its perihelion, is located near Jupiter's orbit. This diagram is simplified in that the transformation would take hundreds of orbits.

these characteristics for up to 2000 revolutions about the Sun. Something like 1 percent of these objects did attain revolution periods shorter than Jupiter for a time.

In a series of papers published during the 1960s and 1970s, the Soviet astronomer Elena I. Kazimirchak-Polonskaya demonstrated that comets on low-eccentricity orbits in the neighborhood of Neptune could have their orbits modified to cross the orbit of Uranus. Uranus, in turn, could modify the orbit so it crossed Saturn's. In effect, the perturbations of the outer planets can modify a comet's orbit so as to hand it over from one planet to the next inner planet.

Everhart demonstrated that the outer planets can play the same role in capturing comets on near-parabolic orbits into short-period orbits. Comets entering the zone defined by the orbits of Neptune and Uranus could have their perihelia passed down into the Saturn, then the Jupiter zone. Everhart concluded that comets with low-inclination orbits and initial perihelia up to 34 AU could ultimately be captured into the orbits of visible short-period comets by diffusing down from one planet to another. Although the process is not very efficient, short-period comets could evolve from comets on near-parabolic orbits, but only if they were nearly in the ecliptic plane. If this type of comet began with a large perihelion distance, it could evolve into a short-period comet with a small perihelion distance. However, it was not likely to evolve into a long-period comet with a small perihelion distance. Similarly, if this type of comet began with a small perihelion distance, it was difficult for

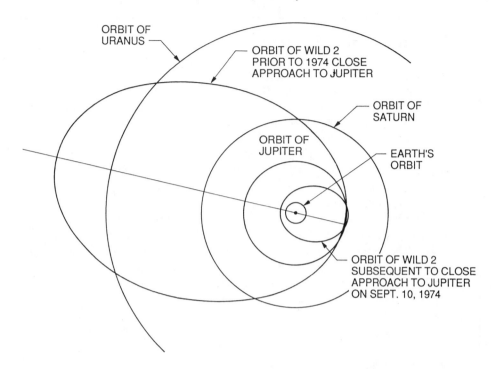

ORBIT OF
URANUS

ORBIT OF WILD 2
PRIOR TO 1974 CLOSE
APPROACH TO JUPITER

ORBIT OF
SATURN

ORBIT OF
JUPITER

EARTH'S
ORBIT

ORBIT OF WILD 2
SUBSEQUENT TO CLOSE
APPROACH TO JUPITER
ON SEPT. 10, 1974

Orbit of periodic comet Wild 2 before and after passing within 0.006 AU of Jupiter in September 1974. The Jupiter encounter caused the comet's orbital size to shrink and its perihelion distance to decrease.

it to evolve into a short-period comet with a small perihelion distance. In other words, observable long-period comets rarely evolve into observable short-period comets. Moreover, if a comet source near Jupiter and Saturn provided the observable long-period comets, they must first be ejected to the distant Oort cloud. Thus, after examining the results of his numerical experiments, Everhart concluded that the evolution of long- and short-period comets from the Oort cloud was the most likely scenario, but he did not rule out a Jupiter-Saturn belt of comets as a possible source of the short-period comets. As shown in the next section, an outer solar system belt of comets is now considered a likely source of short-period comets.

Everhart assumed that the only loss mechanism for comets was their hyperbolic ejection from the solar system as a result of planetary perturbations. In subsequent numerical experiments, Paul R. Weissman allowed several loss mechanisms. Weissman found that a good fit to the orbital characteristics of observed comets was obtained if he assumed that 65 percent of the observed new Oort cloud comets are lost from the solar system

The Tunguska Event

On June 30, 1908, just after 7 A.M. local time, a colossal midair explosion occurred over the Podkamennaya Tunguska River basin in Siberia. Although only several unsuspecting reindeer were close enough to be killed by the blast, it was powerful enough to knock people off their feet some 60 kilometers away. Eyewitness accounts reported an extremely intense bluish-white streak appeared in the sky and tremendous crashes and roars were heard. Seismic disturbances were recorded over a large area of what is now the Soviet Union and barometric anomalies were noted around the globe.

The first scientific party to reach the sparsely inhabited area was led by Soviet scientist Leonid A. Kulik in 1927, 19 years after this strange event. It was only then that the extent of the blast damage was known. Within 30 kilometers of the blast area, most trees had been knocked down, their tops pointing radially away from the blast center. As a result of such an enormous explosion, Kulik expected to find a sizable crater in the Earth's surface and yet he found none—just the charred remains of fallen trees. In 1978, Lubor Kresák presented the possibility that the event was caused by an errant fragment from periodic comet Encke, since the Tunguska object fell at a time when the Earth was passing through the comet's debris and from a direction that coincided with the motion of the debris. In 1983 Ramachandran Ganapathy analyzed several submillimeter-sized metallic spheres extracted from the Tunguska region and found them to be of extraterrestrial origin because of their high content of iridium and other metals, such as cobalt and nickel, that are abundant in meteorites. Ganapathy concluded that the blast was due to an above ground, explosive disruption of a stony meteorite some 100 meters in diameter. Zdenek Sekanina came to a similar conclusion after noting that it was extremely unlikely that a relatively fragile comet could have survived the aerodynamic pressures on an object that followed the Tunguska object's flight path. What little evidence there is points toward either a comet or a large stony meteorite as the source of the Tunguska event.

With very little evidence to constrain the possible explanations of this mysterious event, an extraordinary variety of speculations have been

due to planetary ejections. Some 27 percent are lost to random disruptions such as splitting, and another 7 percent are lost as a result of their losing the ability to outgas—they become asteroidlike bodies. The remaining comets are lost by either planetary collisions or tidal disruptions when they pass too close to the Sun or a major planet.

reported in the literature. These range from the highly speculative—antimatter explosions or a collision with a small black hole—to the ridiculous—the jettison of a UFO's nuclear propulsion system—and on to the sublime. In this latter category must fall the idea put forward by Soviet science writers in 1964. This scenario begins in 1883 with the volcanic eruption of Krakatoa. The volcano allegedly generated strong radio signals and intelligent beings on a planet orbiting the star 61 Cygni considered these signals a message from Earth. Proper interstellar etiquette required that they send a return message, which was accomplished with a powerful laser. Unfortunately, the 61 Cygnians overestimated the necessary broadcast power and, rather than sending a coded message to Earth, they zapped the Tunguska basin instead.

This photograph, taken 19 years after the Tunguska explosion, shows the charred and fallen trees near the blast site. (*Photograph courtesy of E.L. Krinov, Soviet Academy of Sciences.*)

Galactic Perturbations and the Effects of Giant Molecular Clouds

After the introduction of the Oort cloud in 1950, Everhart, Weissman, and others made efforts to clarify the capture process that becomes important once comets have entered the planetary realm. There has also been im-

327

portant work on the perturbations suffered by comets while they are still in the distant cloud. In addition to the stellar perturbations suggested by Oort, galactic effects have been recognized as important and possibly also the perturbative effects of interstellar molecular clouds of hydrogen through which our solar system passes from time to time.

One of the first to consider the effects of the Milky Way on the Oort cloud comets was the Soviet astronomer Gleb A. Chebotarev in the mid-1960s. He considered the perturbing effects of the galactic nucleus on Oort cloud comets and estimated the maximum distances at which they could remain bound to the Sun would be 230,000 AU on circular orbits. The 1972 work of Vadim A. Antonov and I.N. Latyshev considered the perturbing effects of the galactic plane as well as the galactic nucleus and estimated that the largest dimension of the Oort cloud would be 293,000 AU from the Sun in the direction of the galactic center.

Robert S. Harrington published a work in 1985 that showed the impact of galactic disk perturbations on Oort cloud comets. These effects, which are at least as effective as stellar perturbations in removing comets from the solar system, are due to an unequal gravitational attraction of the disk on the Sun and a particular comet. These galactic tidal perturbations are so effective they can alter a comet's perihelion distance during a single orbit, so as to completely remove it from the planetary region. Harrington also demonstrated that the effect was not operative on comets whose aphelia were located along the line between the Sun and the galactic center. One might expect a possible concentration of orbital semimajor axes aligned in this "protected" direction. Harrington concluded that this effect may have been interpreted by some as indicative of a departure of the Oort cloud from a random spherical distribution.

Julia Heisler and Scott Tremaine extended Harrington's results, noting that galactic tidal effects would be up to twice as effective as stellar perturbations in removing comets from the outer Oort cloud, either by transforming them into shorter period comets or by ejecting them into interstellar space. Thus, not only would the galactic tidal perturbations be effective in removing comets, they would also cause Oort cloud comets to drop into the planetary region. The importance of these effects is due to the facility with which the component of this tidal force that is vertical, or perpendicular, to the galactic plane can continually alter the direction of a comet's rotational velocity—modify its angular momentum—with respect to the Sun. As soon as a comet loses enough angular momentum, its velocity component perpendicular to the Sun-comet line can be reduced to a point where it can drop very close to the Sun.

In his 1983 work, John Byl examined the effect of galactic perturbations on the motions of long-period comets and noted that the changes in

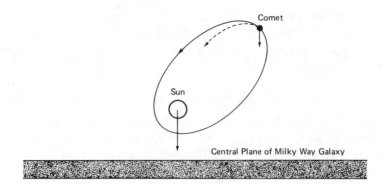

Schematic illustration of the galactic tide. The Sun and its family of comets are located somewhat above the central plane of our Milky Way Galaxy's massive disk of stars. Since the Sun and any of its long-period comets will most often be at different distances from the galactic plane, the gravitational force exerted by this plane on the Sun will differ from that exerted on the comet. The difference between these two forces represents the galactic tidal effect. Depending on the relative positions of the comet and Sun with respect to the galactic plane, this tidal effect can either increase the comet's orbital perihelion distance or decrease it, as in this diagram.

initial perihelion distance are greatest when the aphelion is located at a galactic latitude of approximately 40 degrees. These comets should be—and apparently are—most evident in the observed sample. Four years later, Armand Delsemme analyzed the orbits of 152 comets that arrived from the Oort cloud and noted a paucity of aphelia near the galactic poles and equator. The vertical galactic tidal force, which he assumed responsible for bringing these comets into the observable region, is least effective when the comets' orbits are located either in the galactic plane or at the galactic poles. Hence, Delsemme claimed observational evidence for the effect of the vertical galactic tide.

It seems likely that the tidal forces created by the massive galactic disk dominate the dynamical evolution of comets in the outer regions of the Oort cloud. Apparently, the effects of occasional stellar passages near the cloud, while still significant, are only of secondary importance. Moreover, with the discovery of giant, interstellar molecular clouds of hydrogen in the galactic spiral arms, it appears possible that Oort cloud comets might also suffer substantial perturbations as a result of the solar system passing near these collections of hydrogen.

The Sun and its attendant planets and comets reside within the Milky Way Galaxy, about halfway between the massive central nucleus region and the edge of the planar disk that extends out to 50,000 *light years*.[2] There are perhaps 20 times more comets in the Oort cloud than stars within the Galaxy. However its total mass is about 2×10^{11} solar masses, whereas the Oort cloud is far less than one solar mass. Thus, the numerous, but tiny, Oort cloud comets are perturbed by both the galactic nucleus region and the closer stars within the planar disk. As the Sun revolves about the galactic center once every 250 million years, it does so in a wavelike motion through the central portion of the flat galactic disk. Its motion is up and down, like a porpoise in water. In so moving, it occasionally passes near giant molecular clouds within the galactic disk. These clouds of mostly neutral hydrogen contain enough mass to form up to a million suns, and upon encountering the Oort cloud comets, they might also be expected to provide substantial perturbations.

In 1978, Ludwig Biermann and Reimar Lüst suggested that encounters between the solar system and interstellar, giant molecular clouds might lead to the ejection of large numbers of comets from the Oort cloud. Victor Clube and William Napier then suggested, in 1982 and 1984, that the Oort cloud has been disrupted by approximately 10 passages through these clouds over

Ludwig Biermann (1907–1986). (*Photograph courtesy of Fred L. Whipple.*)

the age of the solar system, and that each encounter was capable of stripping away that portion of the cloud beyond 1000 AU from the Sun. At the same time, the solar system was supposed to capture some 10^{11} new comets from the same giant molecular cloud encounter. Thus these authors did not consider the comets in the Oort cloud as primordial objects, since they were bound to the Sun only until they passed through another giant molecular cloud, when they were stripped away and replaced by an entirely new supply.

Piet Hut and Scott Tremaine pointed out that the presence of large numbers of wide binary stars, with stellar separation distances comparable to Oort cloud dimensions, implied that the perturbations from giant molecular clouds have not disturbed these star pairs. Hence, these perturbations may not have decimated the Oort cloud population as Clube and Napier suggested.

The Distribution of Cometary Orbital Aphelia

The English astronomer Richard Christopher Carrington (1826–1875) is best known for his determination of the solar rotation properties he deduced from years of sunspot data. It was Carrington who first pointed out in 1861 that if comets were captured from interstellar space, one would expect them to enter the solar system from the direction toward which the Sun was moving. Thus the orbital semimajor axes of comets should roughly align with the direction of the Sun's motion through space. However, after analyzing 133 parabolic and hyperbolic cometary orbits, he could find no clustering of orbital aphelia in either the solar apex or antapex direction. Though the distribution of cometary orbits did not seem to be completely uniform over the sky, Arthur Stanley Eddington reached a similar conclusion in 1913.

Ichiro Hasegawa, in 1976, examined the aphelia of long-period comets observed since 1800 and noted a concentration near the solar antapex direction as well as a tendency for them to lie in the galactic plane. Using 542 near-parabolic cometary orbits, Richard S. Bogart and Peter D. Noerdlinger also found, in 1982, a modest concentration of aphelia in the solar antapex direction.

The next year, Ludwig Biermann and his colleagues analyzed the orbits of comets fresh from the Oort cloud. They also found a clustering of aphelia near the Sun's antapex direction for 17 of the 80 comets in their sample. However, they concluded that these comets had similar aphelia directions, not because of the Sun's motion through space, but because they all had been perturbed by the same star passing through the Oort cloud a few million years ago. By removing these comets from their sample, there remained no trace of a preference for the direction of the Sun's antapex.

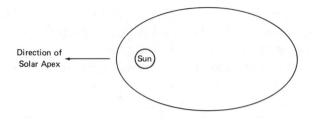

Schematic illustration of Sun and long-period comet orbit travel-ing toward solar apex. If a comet were to be captured from interstellar space, one would expect its orbital semimajor axis to be aligned with the direction of the Sun's motion through space—that is, aligned with the direction of the solar apex.

A Possible Inner Oort Cloud and a Comet Belt Beyond Neptune

The Inner Oort Cloud

An inner comet cloud and a comet belt beyond Neptune have emerged as theoretical constructs made necessary to resupply the outer, classical Oort cloud and to provide the observed numbers of short-period comets. The inner cloud comets would be far more numerous than the outer population and could serve as a reservoir to resupply it as it is depleted, particularly by perturbations by the galactic plane, passing stars, and giant molecular clouds. The inner cloud would be dynamically more stable than the outer cloud because the comets there are more tightly bound to the Sun and less prone to outside perturbing effects. The inner cloud may begin just beyond the orbit of Neptune and merge into the Oort cloud at approximately 10,000 to 20,000 AU from the Sun. Inner cloud comets may be in orbits that lie near the ecliptic plane, but those in the outer cloud have random orientations be-cause they have been frequently perturbed by galactic tidal effects, passing stars, and giant molecular clouds.

The concept of an inner comet cloud was put forward by Jack G. Hills in 1981. Although Oort originally considered the possibility of such an inner cloud of comets, his preferred model was a thick shell of comets surrounding the Sun with an outer and inner boundary. Hills pointed out that the appar-ent inner boundary, near 20,000 AU, was strictly an observational artifact and one would expect there to be comets closer to the Sun. The inner cloud comets are infrequently perturbed either into the planetary region where they can be seen or out to replenish the outer comet cloud. Hills suggested that if outer cloud comets originated in the inner cloud and were later ele-vated into their present orbits by the perturbations of passing stars and Jupiter, the total mass of the inner cloud would have to be about 200 times greater than the outer one. Because there are likely to be so many more comets in the inner cloud than in the Oort cloud, one might expect, in the event of a very close stellar passage, that these comets would arrive into the inner solar system in showers. During one of these rare shower events, the comets might cause considerable planetary cratering and possibly contribute to biological extinction events.

Because they have smaller semimajor axes, inner cloud comets that have been perturbed into the planetary region would return to the Sun's neighborhood more frequently than comets arriving from the Oort cloud. Hence they are lost to the planetary region more quickly, either being ejected into interstellar space by planetary perturbations or disintegrating in the solar neighborhood. For most of the time between the close stellar passages that repopulate their numbers, there would be no surviving, observed examples in the inner solar system. Oort cloud comets, at distances greater than 20,000 AU, arrive and are observed more or less continually in the inner solar system.

Mark E. Bailey demonstrated in 1986 that out to 30,000 AU from the Sun, the galactic tidal effects perturb far more comets into the planetary region than do stellar encounters. Using only galactic tide perturbations, Bailey showed that most short-period comets originate in orbits near the cloud's inner boundary, while the new comets arrive from the outer Oort cloud. Thus the short- and long-period comets may originate in the inner and outer comet clouds respectively. As a result of the galactic tidal effects, comets from the inner cloud drift into the planetary system and are first captured by Uranus, then passed from planet to planet until they become observable short-period comets.

Martin Duncan, Thomas Quinn, and Scott Tremaine assumed that comets formed in the region of the outer planets and, using numerical experiments, then investigated the process by which cometary orbits evolve as a result of planetary perturbations, stellar encounters, and the galactic tide. Their principal goal was to determine the relative efficiency of populating the Oort and inner clouds and hence derive their relative populations. Their results, published in 1987, indicated that the formation of the current cloud is driven by an interaction between planetary perturbations and those of the galactic tide. The inner cloud contains five times as many comets as the Oort cloud, a figure substantially less than that determined by Hills, who considered only stellar perturbations. They pointed out that comet showers would be triggered about every 100 million years, when a star passed closer than 10,000 AU to the Sun. During these showers, which would last about three million years, the rate at which comets entered the solar system might be as much as 20 times higher than during the rate experienced now.

Because of their protected locations, comets cannot be directly observed arriving from the inner comet cloud—unless we are unlucky enough to witness a rare comet shower from this region, since life on Earth might not survive direct hits from a cometary bombardment. However, in addition to the theoretical justification for this inner comet cloud, there are also some observational hints for its existence.

Mark E. Bailey suggested in 1983 that the inner cloud might be observed from the infrared radiation, or reradiated sunlight, the cloud could be expected to emit. After analyzing data from the Infrared Astronomical Satellite, IRAS, which operated from January to November 1983, a team led by Frank J. Low announced they had detected infrared radiation near 100 microns that could be attributed to cold material at the edge of the solar system. Hartmut H. Aumann and his colleagues reported that the IRAS data also showed evidence of a cloud of dust particles, out to about 85 AU, surrounding the bright star Vega. Paul R. Weissman and Doyal A. Harper and his colleagues suggested that the source of this dust might come from an inner cloud of comets surrounding Vega. Sublimation of cometary ices would release the embedded dust particles and maintain the supply of dust surrounding the star.

It must be stressed that the relative populations of the inner and outer comet clouds are very uncertain and there is little more than hints that the inner cloud exists at all.

A Comet Belt Beyond Neptune

The solar system origin theories of Gerard P. Kuiper (1905–1973) and Alastair Graham Walter Cameron (discussed in the next section) both suggested the existence of small solid debris remaining in the outskirts of the solar system. Fred Whipple in 1964 questioned whether this cometlike material might be the answer to some differences between the observed and predicted motion of Neptune. If a comet belt does exist beyond Neptune and the mathematical model used to predict the planet's motion did not include the belt's perturbative effects, then the actual and predicted motions would be discordant. The mass of the outermost planet, Pluto, was insufficient to perturb Neptune in any sensible way, and Whipple concluded that the discrepancies between Neptune's observed and predicted latitude positions could be reduced if one assumed that a properly oriented comet belt of 10 to 20 Earth masses exists at a distance of 40 to 50 AU from the Sun. Four years later, S. El Din Hamid, Brian Marsden, and Fred Whipple noted the lack of any perturbative effects on the motion of comet Halley and placed an upper bound on the comet belt's mass of 1.3 Earth masses if it was located at 40 to 50 AU from the Sun.

Mark Bailey, Julio Fernández, and George Wetherill

The differences noted in the observed and computed positions of Neptune may well be due to observation inaccuracies, so there is no clear need to invoke the perturbative effects of an unobserved solar system body or bodies to explain its motion. However, the existence of a belt of comets beyond Neptune is still an attractive idea, since it might easily explain why so many short-period comets are observed. It is far easier to perturb comets into short-period orbits from a near-circular belt region beyond Neptune than to attempt to capture them from near-parabolic orbits as they drop in from the Oort cloud.

In 1980, Julio A. Fernández suggested a comet belt between 35 and 50 AU from the Sun as a source of the observed short-period comets. Assuming there are about 73 observed short-period comets and each one has a lifetime of about 1500 years, Fernández determined that 73/1500, or 0.05, new comets per year are required to maintain the current population. Given the loss mechanism of planetary perturbations, some 300 new Oort cloud comets each year would be required to maintain the observed population of short-period comets. However, if these objects originate within the comet belt, the losses due to planetary perturbations would be far less; only one belt comet each year would be required to maintain the population. While only about 6 percent of belt comets that crossed Neptune's orbit would evolve to become observable short-period comets, this was still 300 times more efficient than bringing them in directly from the Oort cloud. In addition, the increased efficiency of the process reduced the required total mass of the comet belt. It need not be larger than one Earth mass. Fernández's hypothesis provided a source for the observed short-period comets in the comet belt. The source for the long-period comets remained in the Oort cloud.

The 1988 publication by Martin Duncan, Thomas Quinn, and Scott Tremaine raised serious questions about the capture of comets on initial long-period orbits into those with short-period orbits. Although their techniques were an extension of some of Edgar Everhart's earlier numerical experiments, their main conclusion was opposite to Everhart's. Whereas Everhart concluded that planetary captures can transform Oort cloud comets into the observed population of short-period comets, Duncan, Quinn, and Tremaine concluded this was unlikely. Using their mathematical computer model, they followed the orbital evolution of several thousand comets on near-parabolic paths until the comets were either captured by the major planets into short-period orbits or ejected from the solar system altogether. Near-parabolic comets, after evolving into short-period comets during the computer simulations, approximately retained the orbital inclinations they had upon leaving the simulated Oort cloud. Since the observed sample of short-period comets shows a strong preference for low-inclination direct orbits, the authors concluded that these comets originally had direct orbits of low inclination. In other words, most observed short-period comets could not have come from the nearly spherical Oort cloud.

When the authors moved the source of their simulated comets from locations in the Oort cloud to positions in a flattened comet belt beyond Neptune, they could reproduce the observed orbital characteristics of the short-period comets. Just like those of the observed short-period comets, the ones simulated in the computer model had their aphelia near Jupiter's orbit and crossed the ecliptic plane near their aphelion and perihelion points. Some 17 percent of the comets that cross Neptune's orbit will eventually become observable short-period comets. This increased efficiency reduced to only one-fiftieth of an Earth mass the amount of comet belt material necessary to maintain a steady supply of short-period comets over the age of the solar system.

The problem of too many observed short-period comets may have been solved with a very modest amount of cometary material beyond Neptune. A remaining fly in the ointment was how to perturb the belt comets, whose orbits were nearly circular, into Neptune-crossing orbits. Fernández suggested one possibility: a few larger than normal comets could perturb enough smaller belt comets across Neptune's orbit to begin the journey toward the Earth's region of the solar system. Alternatively, Martin Duncan and colleagues suggested that comets between Uranus and Neptune could achieve planet crossing orbits as a result of planetary perturbations over long periods of time.

The 1989 work of Chris R. Stagg and Mark Bailey questioned the validity of the mathematical shortcuts used by Duncan and his colleagues to represent the capture of long-period comets with perihelion distances larger than about 15 AU. They argued that the source of the short-period comets could be the inner comet cloud. The greater time required to capture inner cloud comets on high-inclination orbits would allow them time to completely de-gas before being captured into short-period orbits and they might be masquerading as inactive, asteroidlike objects that have escaped detection. Thus the inner comet cloud cannot be ruled out as the source of short-period comets.

Cometary End States

In the late twentieth century, there seems to be a scientific consensus building for the source of long-period comets being in the Oort cloud with either the inner cloud or the comet belt providing the short-period comets. While the extent of the Oort cloud and the number of its members are still open issues, its existence is not seriously questioned. However, the inner cloud and comet belt are not proven entities, and the birthplace and ultimate fate of comets are still very much in question. Their end states—birth and death processes—are far from certain.

The Birth of Comets in Interstellar Space

Oort's 1950 work did not establish a complete adherence to his theory for the solar system origin for comets, and there were noteworthy attempts to devise viable theories for an interstellar origin. William Hunter McCrea proposed that comets could be formed in relatively high-density interstellar clouds that form in the Galaxy's inner spiral arms. When the Sun, in its journey about the galactic center, ran through one of these comet formation regions, it captured many of them. The solar system could be expected to pass through a galactic spiral arm once every one hundred million years, or about 50 times during the solar system's lifetime. The Sun recently passed through a spiral arm, so if McCrea's mechanism is valid, we are currently privileged to see comets.

According to the hypothesis put forward by Victor Clube and William Napier mentioned earlier in this chapter, comets already existing in the Oort cloud can be effectively stripped away as the Sun passes through a giant molecular cloud. Whether this loss decimates the Oort cloud population or merely reduces its numbers is not altogether clear. In either case, the Oort cloud must be resupplied in some fashion. Clube and Napier suggested that the same gas cloud passages that strip the comets away would also repopulate the Oort cloud. Alternatively, the outer Oort cloud could be resupplied with the far more numerous inner cloud comets via planetary perturbations.

The problem with all interstellar origin theories is the very low likelihood of their being captured by the solar system. In 1982 Mauri J. Valtonen and Kimmo A. Innanen computed that the capture of interstellar comets would only be possible if the comets had a relative velocity, less than 0.4 kilometers per second, with respect to the solar system. These low velocities could only be achieved for interstellar clouds that move in very nearly the same direction and velocity as the Sun and this is exceedingly unlikely. The Sun's velocity relative to neighboring stars is approximately 17 kilometers per second so that the interstellar origin hypothesis is difficult to defend.

The Birth of Comets Within the Solar System

An hypothesis for the cometary birth process was put forward in 1951 by Gerard P. Kuiper. The year before, J.H. Oort suggested that comets may have formed with the asteroids during the breakup of a planetary body between Mars and Jupiter. Kuiper pointed out that comets were likely composed of icy materials and ice would not be a major constituent of comets if they formed as close to the Sun as the asteroid belt. He proposed that comets formed between 35 and 50 AU from the Sun as the condensation products of the outer solar system. During the breakup of the primordial solar nebula, the icy material and dust of the outer solar system first settled toward the equatorial plane of the rotating nebula to form a so-called *accretion disk*. When the density was sufficient, irregularities in the flattened accretion disk formed into clumps of dirty ice by gravitational attraction. Over a thousand million years, these clumps agglomerated into larger kilometer-sized objects, which Kuiper identified with comets. Much of this material continued to agglomerate into the major outer planets, which grew until exhausting all cometary material within their gravitational reach. Comets that did not become building blocks for one of the major planets either remained in the comet belt beyond Neptune or were perturbed by the planet Pluto into orbits that crossed Neptune's or another outer planet's orbit. Eventually these comets passed close enough to a major planet to be thrown either into the inner solar system, completely out of it, or into the Oort cloud. Once in the Oort cloud, a comet could be perturbed by a passing star and returned to the planetary system from whence it came.

Though apparently developed independently, Kuiper's view of a cometary nucleus was very similar to Fred Whipple's icy-conglomerate model of the year before. In a note added while his own paper was being proofed, Kuiper mentioned the recent appearance of Whipple's work. His hypothesis fits very nicely both with Whipple's model of the cometary nucleus and with Oort's source of comets in a circumsolar cloud. Since Kuiper's paper appeared in 1951, the estimated mass of Pluto has been revised downward to such an extent that its perturbative effects on Kuiper belt comets would not be sufficient to move enough of them into Neptune-crossing orbits. Another perturbation mechanism has to be substituted.

According to Victor S. Safronov, the accretion time for Uranus and Neptune would be 10^{11} years if their formation zones contained only enough material to accrete the planets themselves. This was far longer than the solar system's estimated lifetime. By postulating that the original protoplanetary cloud contained 6 to 7 times the current mass, the planets could be formed rapidly enough to escape this paradox. Much of the mass remaining after the planetary formation process was ejected from the solar system altogether by Jupiter perturbations; the more gentle Neptune perturbations would result in comets being placed into the Oort cloud.

Alastair Graham Walter Cameron pointed out that many solar type stars, early in their lifetimes, undergo a phase whereby they explosively eject material. These stars are referred to as *T Tauri* stars after a prototype variable star in the constellation Taurus. If our Sun underwent such a phase, the resultant explosion might have swept away the gas and dust of the early solar nebula. Thus the planets must have formed quickly, before their building material was eliminated. To do this, far more than the current solar system mass would be necessary initially to provide the requisite density of building material. Surrounding the massive main planetary nebula, Cameron envisaged 2 to 4 satellite nebulae with masses of a few tenths of a solar mass. Each would have sufficient density to allow the accretion of comets but not enough to form the larger planets. Already on the outskirts of the planetary region, these comets would orbit the Sun in very long-period, near-circular orbits. If the Sun were to lose a portion of its mass during a T Tauri phase, the most distant comets would become gravitationally unbound to the Sun. Their aphelion distances would move out farther and farther as more and more Solar mass was lost. However, those closest to the Sun would be given less of an energy boost than the more distant ones, and they might remain bound to the Sun in the distant Oort cloud. Cameron's ingenious method for removing comets from their birthplace at a few hundred AU from the Sun to the Oort cloud at 100,000 AU avoided the very inefficient process of gravitationally flinging them out of the planetary system using strong perturbations from the outer planets.

In 1982 Jack G. Hills also suggested that comets formed beyond the existing planetary system. In his hypothesis, they formed in the collapsing layers of the proto-Sun at distances of 1000 to 5000 AU from the Sun. Radiation pressure from the Sun and neighboring protostars may have forced icy dust grains together. Most of the dust where the density was highest would be unaffected because of shielding by neighboring particles. The greatest pressure would be applied to dust grains in regions where there was little dust. The total effect was to cause dust to drift toward regions where the dust concentration was already high. Having formed in the inner comet cloud, rare stellar passages would then perturb the young comets into the planetary system and subsequent outer planetary perturbations would fling them into the Oort cloud. If the outer Oort cloud was populated in this inefficient fashion, Hills concluded there must be at least one hundred times more comets in the inner cloud than in the outer one.

All of the foregoing scenarios for the solar system formation of comets assumed that a protoplanetary disk of material once surrounded the Sun during the planetary formation process. An interesting verification that another star currently has such a disk of material came with the announcement of what appears to be an edge-on accretion disk surrounding the star β Pictoris to a distance of about 400 AU on either side of it. Alerted by results from the IRAS satellite that suggested a possible dust disk surrounding the star, the discovery observations were made by Bradford A. Smith and Richard J. Terrile in April 1984.

Since the introduction of the Oort cloud concept in 1950, most scientists have generally favored the hypothesis whereby the populations of both the long- and short-period comets accreted in the outer planetary regions. Formation in this region, rather than at Oort cloud distances, is favored because the higher densities in the inner protosolar nebula would allow comets enough time to form by accretion. However, this area of study changes rapidly and today's favored hypotheses are often tomorrow's historical footnotes.

One such footnote was the exploding planet theory put forward by Thomas Van Flandern in 1978. Reminiscent of Lagrange's solar system origin hypothesis, Van Flandern considered that an explosion of an enormous planet of 90 Earth masses some five and one-half million years ago resulted in the formation of both asteroids and comets. Apparent Oort cloud comets were fragments ejected on very long-period ellipses, making their first return to the planetary region at the present time. Mathematically following these comets back one revolution, Van Flandern found that many of the orbital perihelia were clustered at a particular position on the celestial sphere. This would be expected if comets had an exploding planet as a common origin. However, Robert Harrington, in his 1985 work, pointed out the clustering direction noted by Van Flandern was very close to the galactic center. This effect should be expected due to prolonged perturbations of Oort cloud comets by the galactic plane. Those whose perihelia were pointing toward the galactic center would escape the perturbations, which were capable of raising their perihelia entirely out of the inner solar system. Harrington concluded that because of galactic perturbations, the current distribution of comet orbit orientations can tell us nothing about their original distribution. In addition, there is no known mechanism for making a planet explode and no planetary cratering evidence for the predicted heavy bombardment of planetary surfaces five and one-half million years ago.

Photograph of dust cloud surrounding the star β Pictoris. An edge-on disk of dusty material extending to about 400 AU on either side of the star is evident. This disk material is only visible because the light from the relatively bright, central star was masked by an occulting shield within the telescope. The dark cross hairs were caused by light scattering from the occulting shield supports within the telescope itself. This photograph was taken in February 1985 by Richard J. Terrile and Bradford A. Smith at the Cerro Tololo Inter-American Observatory in Chile.

Overview of Current Ideas

The following outline presents a general overview on the origin of comets, the development of the comet clouds, and the delivery of comets to the inner solar system. These points are representative of ideas held by many—but not all—cometary authorities in 1990 and are listed here only to clarify the concepts presented in the foregoing pages. As future research on cometary origins is completed, the following scenario will have to be modified accordingly.

A. Formation of comets
   1. Comets form in the region of Uranus and Neptune as the icy debris from the outer planet formation process. Originally on near-circular orbits, these primordial comets comprise the Kuiper belt near and beyond Neptune's orbit.
   2. Long-term perturbations by the forming protoplanets increase the eccentricities and semimajor axes of these cometary orbits.
   3. After the orbital sizes have increased sufficiently, galactic plane perturbations raise their perihelia well beyond Neptune's orbit. They are then placed into the inner comet cloud, where they are immune from further planetary perturbations. Continued galactic plane perturbations eventually lift some inner cloud comets into the outer Oort cloud, thus resupplying the cloud, which is continually losing comets to interstellar space. The inner cloud extends to approximately 10,000 AU from the Sun and contains perhaps five times the number of comets in the outer Oort cloud.
   4. While in the outer Oort cloud, comets can be stripped away from the Sun through perturbations by the galactic tide and close stellar passages. More than 50 percent of the comets present when the solar system formed have been lost from the Sun's realm. The outer Oort cloud now has a total population of approximately 10^{11} to 10^{12} comets.
   5. Perturbations resulting from encounters with passing stars and giant molecular clouds are also important for kicking comets into interstellar space. It is these latter two mechanisms that are primarily responsible for the reorientation of the orbital inclinations of Oort cloud comets. After 4.5 billion years—the current age of the solar system—comet orbits initially in the plane of the planets have been shuffled into a *prolate spheroid* distribution about the Sun—roughly resembling a fat football. The long axis of the outer Oort cloud, which is pointed toward the galactic center, is roughly 200,000 AU from end to end, while the shorter axis is approximately 160,000 AU wide.

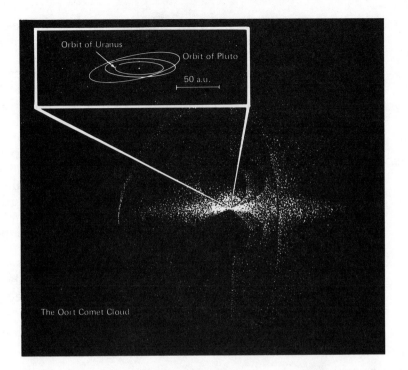

Illustration of inner and outer Oort cloud of comets. Beginning with a flattened disk of comets just beyond the planetary region, the inner cloud extends to approximately 10,000 AU from the Sun, then broadens into the roughly spheroidal outer Oort cloud. The latter cloud is shaped like a fat football with its longest axis, approximately 200,000 AU, directed toward the center of the Galaxy. The shorter axis is perpendicular to the long axis and approximately 160,000 AU in extent.

B. The return of comets from the inner and outer Oort clouds
   1. The major source of perturbations for comets already in the Oort cloud is the galactic tidal effect. This galactic tide is effective in changing the perihelia of comets in a continual, stepwise fashion, whereas passing stars and giant molecular cloud perturbations act more or less randomly. Thus, while these latter two effects are important for stirring up orbits in the Oort cloud, the galactic tide is primarily responsible for raising comets from the inner to the outer cloud and dropping comets from both clouds back into the planetary region. The galactic tide acts most effectively on comets whose orbital inclinations are near 45 degrees on either side of the galactic plane. They are most often perturbed inward toward the planetary region, where they can be observed or kicked out of the solar system entirely.

2. Comets dropped in toward the planetary system from the outer Oort cloud are said to have long-period, or nearly parabolic, orbits.

3. Comets with short-period orbits of the Jupiter group have the inner comet cloud or the Kuiper comet belt as their source. Planetary perturbations, over long periods of time, elongate the orbits of Kuiper belt comets, allowing them to eventually pass within the region where they can be captured by the major planets. A typical capture might first invoke relatively gentle perturbations by Neptune to lower the perihelion of a comet to Uranus' orbit. Uranus' perturbations then lower the perihelion to the region of Saturn, then onto Jupiter, where the inward migration continues into the inner solar system. Jupiter's family of short-period comets generally have their aphelia near the planet's orbit.

4. At present, the comet belt, as well as the inner and outer comet clouds, have likely merged to form one continuous distribution of comets extending beyond Neptune's orbit to the very edge of the solar system.

The Aging of Comets

Paul R. Weissman's numerical studies on the evolution of Oort cloud comets allowed him to estimate the relative efficiencies of various cometary loss mechanisms. He calculated that 65 percent of all comets that disappear are ejected from the solar system as a result of strong perturbations by Jupiter. This being the case, there must be some true interstellar comets. Based on the fact that no comet with a strong hyperbolic orbit has yet been observed, Zdenek Sekanina determined an upper limit upon their density in interstellar space. In his 1976 study, he concluded there is no more than one interstellar comet in a volume defined by a sphere whose radius is 12 AU. Interstellar comets certainly exist, but their density in space is far too low to supply those we actually observe.

The next most efficient loss mechanism is random cometary disruption, which accounts for roughly 27 percent of cometary disappearances. Lubor Kresák noted that aging among observed comets is evident from the escaping gas and dust that form the coma and tail, from splitting and flaring, and from differences in intrinsic brightness between new arrivals from the Oort cloud and those that have evolved to shorter period orbits. In fact, almost everything we observe of comets is related to their disintegration. There have been several attempts to determine the lifetimes of comets by noting their fading with time. However, Kresák concluded that while periodic comets decrease in mean intrinsic brightness with time, the average change is

Paul R. Weissman

only 0.1 magnitude per century. Since the variation in a particular comet's absolute magnitude from one apparition to the next can easily change by a magnitude, or even enter a dormant phase, fading is not a good indicator of a particular comet's aging.

Although there are a very limited number of examples, Kresák considered complete disintegration as a better aging indicator. The few periodic comets actually observed to disintegrate and disappear are comets Biela, last seen in 1852; Brorsen, 1879; Westphal, 1913; and probably Neujmin 2, 1927. For the long-period comets observed since 1840, Kresák found approximately eight whose disappearance could be attributed to their final extinction. Based on these few examples, he estimated that comets whose perihelion distances are equal to 1 AU can be expected to make 300 revolutions before disintegrating if their periods are less than 20 years, but only 20 revolutions if they are long-period comets. In each case, comets with smaller perihelion distances would age more rapidly than those with larger perihelion distances. These computations were meant only to be approximate averages and cannot be used to compute realistic lifetimes for individual comets. For example, from Kresák's analysis, one would determine the lifetime of comet Halley to be approximately 77 revolutions. However, Anton Hajduk derived a current age of 2300 revolutions for comet Halley by comparing its estimated mass loss per revolution with the total mass estimated to reside within its two meteor stream complexes, the Orionid and η Aquarid streams.

Two of the most obvious cometary aging phenomena are flaring and splitting. David W. Hughes concluded that cometary flaring does not have any obvious connection with a comet's heliocentric distance, so the mechanism responsible for outbursts is not triggered by increased solar heat as a comet approaches the Sun. The spectral features of cometary flares indicate only solar continuum radiation, so outbursts apparently result in clouds of dust surrounding the comet. Comet Tuttle-Giacobini-Kresák was an exception to this general rule when its extraordinary 1973 flare of 10 magnitudes exhibited strong molecular bands of C_2, C_3, and CN overlying the continuum. Lubor Kresák analyzed the circumstances of cometary splitting and estimated that, on average, one split would be observable for every 12 revolutions of a long-period comet, whereas short-period comets have a lower rate of one split per 90 revolutions. In addition, there is no strong correlation between major brightness flares and splitting events, suggesting that the separation phenomenon is not a violent process.

Zdenek Sekanina studied 21 comets that were observed to split, forming two, three, four, or five separate fragments. Apart from solar tidal stresses operating on the two sungrazing comets 1882 II and 1965 VIII Ikeya-Seki and similar stresses that might have acted on periodic comet Brooks 2 during its 1886 close approach to within 0.000955 AU of Jupiter, splitting seems to occur at random. Shorter lived fragments separate from the parent comet with higher accelerations, suggesting that they are low-mass fragments.

Sekanina suggested that possible mechanisms for splitting included the separation of a portion of the comet's crust from a rapidly rotating nucleus or the breaking apart of a rapidly rotating object whose tensile strength and density were quite low. As a rule, the secondary fragments of a split nucleus do not last more than a few months. Occasionally, a more persistent fragment, like that of comet Biela, can last a few years. During their study of the long-term motion of all short-period comets, Andrea Carusi and his colleagues noted that when they traced the motion of periodic comets Neujmin 3 and Van Biesbroeck before their close Jupiter approaches in 1850, both comets had nearly identical orbits. Possibly, they both resulted from a common parent object that split prior to 1850. The orbital similarities of long-period comets 1987 XXX Levy and 1988 III Shoemaker-Holt were noted by Conrad Bardwell. These two comets may have originated with the same parent object as a result of a split many years ago.

The only clear case of a family of comets resulting from a common parent object is the so-called *Kreutz sungrazers*. Apparently, it was Daniel

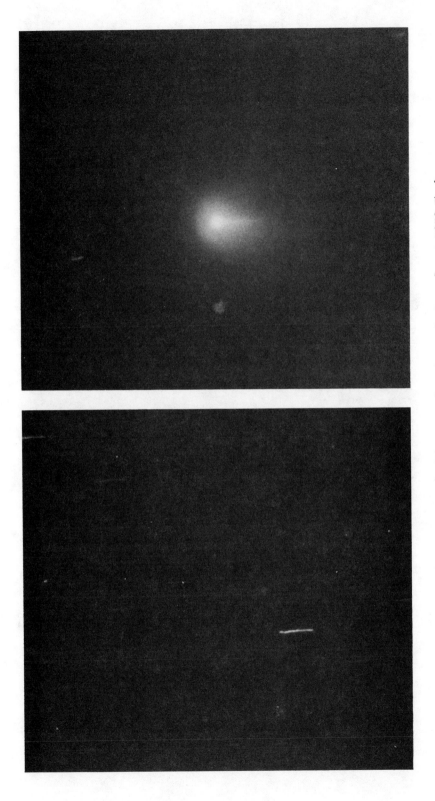

Periodic comet Tuttle-Giacobini-Kresák experienced two dramatic brightness flares in 1973. The first event occurred on about May 25 to 27, when the comet brightened by a factor of approximately 9500, or about 10 magnitudes. After fading to its brightness level before the first flare, the comet experienced a similar explosion on July 4, 1973, as a barely perceptible bit of fuzz in the center of the image. The left photograph shows the comet on July 4, 1973, as a barely perceptible bit of fuzz in the center of the image. The right image shows the comet four days later after it had already peaked in brightness. (Photographs courtesy of Richard E. McCrosky, Oak Ridge Observatory.)

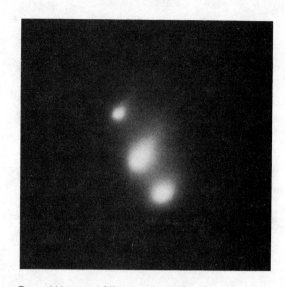

Comet West, or 1976 VI, split into four fragments as it rounded perihelion in February 1976. On May 26, 1976, when this photograph was taken, three of the four pieces were still apparent. (*Photograph courtesy of Henry L. Giclas, Lowell Observatory.*)

Lubor Kresák

Heinrich Carl Friedrich Kreutz. (*Courtesy of the Mary Lea Shane Archives of the Lick Observatory.*)

Kirkwood who first suggested that the sungrazing comets 1843 I and 1880 I formed a related pair with their possible parent being the comet seen splitting into two pieces by Ephorus in 372 B.C. Heinrich Carl Friedrich Kreutz (1854–1907) studied many of the sungrazing comets, noted that several had similar orbital elements, and suggested their common origin might have been due to the breakup of some primordial comet near perihelion. In honor of his extensive work, the members of this group are referred to as Kreutz sungrazers.

In his 1967 and 1989 studies, Brian Marsden identified eight acknowledged members of the group—1843 I, 1880 I, 1882 II, 1887 I, 1945 VII, 1963 V, 1965 VIII, and 1970 VI—as well as three uncertain members. The latter objects are the comets that reached perihelion near March 1, 1668, February 15, 1702, and May 17, 1882, seen only during the May 1882 solar eclipse. Several comets discovered using the Earth-orbiting SOLWIND and Solar Maximum Mission satellites are also likely to be Kreutz sungrazers. Marsden considered that almost all of the Kreutz group separated from one

Horrific Missiles or Life-Giving Providers?

One of the most persistent attitudes toward cometary apparitions has been the notion that malefic events follow in their paths. Particularly persistent is the concept that pandemic diseases somehow result from comet-Earth encounters. After the London plague of 1665, the comets of 1664 and 1665 were widely believed to have been the causal agents.

In an effort to avoid the London plague, Isaac Newton fled to the English countryside and developed many of the profound ideas that would later be published in his masterwork, the *Principia*. Included within this work is the notion that cometary material in space is accreted by the planets, thus resupplying them with the fluids spent upon "vegetation and putrefaction." Newton considered the continual arrival of cometary material to be essential for life on Earth.

In their 1979 book entitled *Diseases from Space,* English astrophysicist Fred Hoyle and Sri Lankan scientist Nalin C. Wickramasinghe resurrected the notion of cometary diseases. These two respected scientists suggested that life may have evolved from carbon-rich, organic molecules in the inner regions of cometary bodies. Protected from radiation and warmed to about 10 degrees centigrade, or 50 degrees Fahrenheit, by chemical processes, primitive life forms like bacteria and viruses might have evolved. Shed into space as tiny particles, these life forms presumably initiated biological life as we know it when the Earth, in its path around the Sun, ran into these particles and scooped them out of interplanetary space. Hoyle and Wickramasinghe further asserted that the process continues today and that viruses, particularly those of influenza and the common cold, rain down upon the Earth to cause epidemics of these diseases.

There are a host of objections to life and diseases from comets including no obvious connection between the arrival of large comets in the Earth's neighborhood and worldwide epidemics. Influenza viruses, for example, are parasitic and require host cells from specific animals to thrive. Presumably, the necessary animals do not reside within comets.

Although the delivery of life seeds to the Earth via comets seems most unlikely, comets may well have provided the Earth with the basic materials necessary for life to form. In 1961, the biochemist John Oró suggested that collisions of comets and the accretion of cometary matter to the Earth provided a plausible mechanism for supplying part of the Earth's early atmosphere. These same collisions might form local concentrations of organic molecules and liquid water on the Earth's surface where, in the presence of solar ultraviolet radiation, amino acids and other biochemical compounds could form spontaneously.

Comets may have provided much of the Earth's water, volatiles, and organic molecules, thus depositing some of the biogenic material from which primitive life ultimately formed. Subsequent impacts could easily have erased many of Earth's life forms, leaving only the most adaptable to develop further. This duality of comets, as both life threatening and life giving, is a theme that began with Isaac Newton and Edmond Halley in the late seventeenth century. In a more modern format, this theme saw a revival in the late twentieth century. Recent ideas suggest comets' reputation as life-threatening, horrific missiles may have to be modified to include our indebtedness for their supplying the building blocks of life itself.

The Earth is destroyed by a comet in this nineteenth century French cartoon.

of two comets that appeared around A.D. 1100. One of these progenitors might have been the comet seen near perihelion on February 5, A.D. 1106.

The Possible Evolution of Comets into Asteroids

Using some definitions, the Earth itself might be considered a comet since it has vaporizing ices, a large atmosphere, and a tail of ions in the antisolar direction. To simplify the following discussion, definitions provided in 1987 by William K. Hartmann and his colleagues are used. A comet is defined as a body formed in the outer solar system containing volatiles in the form of ices and capable of developing a coma if its orbit brings it close enough to the Sun. An asteroid, or minor planet, is defined as an interplanetary body that formed without appreciable ice content and thus never had, or can have, cometary activity.

There are a few hundred known near-Earth asteroids whose orbits are sufficiently eccentric that they can pass close to, or cross over, the Earth's orbit. They've been given the names of typical members of their group and each classification is based on their orbital characteristics with respect to the Earth's perihelion and aphelion distances, 0.983 and 1.017 AU respectively. *Aten asteroids* have semimajor axes less than 1 AU, and their aphelion distances are larger than 0.983 AU. *Apollo asteroids* have semimajor axes greater than or equal to 1 AU and perihelion distances less than 1.017 AU, so they can cross the Earth's orbit. The perihelion distances of *Amor asteroids* are less than or equal to 1.3 AU, but greater than 1.017 AU; they can approach, but not cross, the Earth's orbit.

Many near-Earth asteroids have orbital characteristics strikingly similar to short-period comets. Three examples are asteroids 944 Hildalgo, 2201 Oljato, and 3200 Phaethon. When Fred Whipple pointed out in 1983 that the Geminid meteor stream has orbital characteristics nearly identical to those of 3200 Phaethon, the association of this object with an extinct comet became particularly strong. Jack D. Drummond and Duncan I. Olsson-Steel also pointed out many minor meteor streams that could be associated with asteroids.

William Hartmann and his colleagues studied the outermost asteroid 2060 Chiron and found that it shows cometlike behavior, such as variable brightness and some indication of its becoming brighter as it approaches its 1996 perihelion date. This object is in an eccentric orbit, with perihelion and aphelion distances of 8.5 and 19.1 AU, and thus spends most of its time between Saturn and Uranus. In April 1989, Karen J. Meech and Michael J.S. Belton reported their observations of a coma surrounding 2060 Chiron. Although classified as an asteroid, it has many of the characteristics of a large comet.

As well as the notion that some asteroids look suspiciously like extinct comets, there is abundant evidence that comets become less active with age. Periodic comet Neujmin 1 and Arend-Rigaux are two comets that appear to be in the last stages of activity, perhaps because their orbits prevented them from making the close Jupiter approaches necessary to move them farther from the Sun. There is also circumstantial evidence that comets develop an inactive crust as they age. In introducing the icy-conglomerate model for the cometary nucleus, Whipple mentioned that comets might develop inactive crusts with age. The spacecraft images of comet Halley's nucleus revealed that most of its surface is covered with a black, apparently inactive crust. In fact, in the few cases where reflectivities and active areas of short-period comets have been estimated, the nuclei seem to be jet black, with only a portion of their surface areas remaining as sublimating ices.

Ernst Öpik first pointed out in 1961 that there are problems in explaining the number of asteroids that cross the Earth's orbit if one assumes they all originated in the asteroid belt between Mars and Jupiter. The mechanism of planetary perturbations, as then understood, was seriously inadequate to modify their original orbits to provide the observed number of Earth crossing asteroids moving on eccentric orbits. Öpik suggested that a number of inactive comets might be masquerading as Earth crossing asteroids. However, Jack Wisdom showed in 1985 that asteroids located in slightly eccentric orbits, $e = 0.2$, with semimajor axes of about 2.5 AU could evolve that way within a few hundred thousand years. Asteroids originally located in this zone of the asteroid belt would have a period one-third that of Jupiter and would periodically suffer the Jupiter perturbations necessary to make their motions chaotic. That is, their eccentricities could be enlarged sufficiently to take them into Earth crossing orbits. George W. Wetherill then concluded that this additional mechanism would greatly alleviate the problem of too many observed Earth crossing asteroids, so that the population of Aten, Apollo, and Amor asteroids might have had an asteroidal parentage. However, under the influence of planetary perturbations, some comet orbits could evolve so as to be completely interior to Jupiter's orbit and hence be safe from further perturbations by Jupiter. These comets would then be traveling on asteroidlike orbits and would eventually lose their ability to outgas. Hence, Wetherill considered it likely that extinct comets can still provide a significant fraction of the Earth crossing asteroid population.

The End of the World—Again

The fact that Apollo asteroids and comets can cross the Earth's orbit makes these objects the obvious candidates for forming impact craters on the Earth and Moon. Because it lacks the erosion processes present on Earth, the Moon retains many more of these ancient scars. The notion of cometary collisions causing major geological and biological effects on the Earth dates back to at least 1688, when Edmond Halley suggested that the deluge causing the Biblical flood might have been due to a *casual shock* of a comet. In an address to the Royal Society of London on December 12, 1694, Halley pointed out that a cometary collision might well cause a major extinction of species, but the fine debris would then settle onto the Earth's surface and render the soil more suitable for vegetable production and animal life. In 1750 Pierre Louis Moreau de Maupertuis (1698–1759) wrote that comets could certainly crash into the planets, and the resultant heat and contamination of atmosphere and water would lead to mass extinctions. The confused strata of the Earth's surface seemed to suggest that such collisions had already taken place.

A work published in 1980 by the father and son team of Luis W. and Walter Alvarez, together with Frank Asaro and Helen V. Michel, seems to have been most influential in renewing general interest in cometary collisions as the cause of historical extinction events. They sought to explain the great enhancement of the heavy metal iridium in limestone sediments corresponding to the time of the Cretaceous-Tertiary extinction event, 65 million years ago. Because it has long since settled into the Earth's core, iridium is relatively depleted in the Earth's crust, so an extraterrestrial origin seemed likely. Alvarez's team suggested that the impact of a 10-kilometer-sized asteroid might have delivered the iridium to the Earth's surface, at the same time triggering events that led to the death of many plant and animal species, including the dinosaurs. Upon impact, there would be an explosive release of some 60 times the asteroid's mass into the Earth's atmosphere in the form of pulverized rock. A fraction of this dust would remain in the atmosphere for several years and the resulting darkness would suppress photosynthesis, killing plants and the animals that fed on them.

Although an asteroid was initially identified as the agent of destruction, it was soon realized that a comet impact could provide an equally viable explanation for the observed iridium excess. David M. Raup and J. John Sepkoski, Jr. suggested that there was a mean interval of 26 million years between major extinction events they identified over the last 250-million-year period. This suggested periodicity of Earth impacts spawned a rash of hypotheses attempting to provide a mechanism for delivering showers of comets to the Earth's neighborhood every 26 million years. Although these latter hypotheses seem to be peripheral to the history of comets, it should be noted that Antoni Hoffman reexamined the data and found no periodicity by using a different, but equally acceptable, geologic time scale. Showers of comets might be expected to strike the Earth as a result of a star passing near the

The effects of the comet of 1811 on the Earth's environment were held responsible for the unusually good vintage grapes that year. For years afterward, the comet wines of 1811 were acclaimed on advertisements.

solar system's inner comet cloud, but they would be random, not periodic, events. In 1984, Zdenek Sekanina and Donald Yeomans analyzed the observed comets that have passed within 2500 Earth radii of the planet and concluded that a fair-sized, active comet could be expected to strike the Earth once every 33 to 64 million years, on average.

Summary

Although there has yet to be a convergence of views regarding the birth and death of comets, some important points have emerged.

1. No definite hyperbolic cometary orbits have been observed, which strongly suggests that comets are not now arriving from interstellar space. The capture of significant numbers of interstellar comets by planetary perturbations into the solar system requires a relative comet-Sun velocity typically less than 1 kilometer per second, which seems exceedingly unlikely.

2. The major outer planets control the capture of comets on near-parabolic orbits and the subsequent evolution of cometary orbits in the solar system.

3. Observed long-period comets do not evolve into observed short-period comets and vice versa.

4. The primary loss mechanism for comets is their ejection from the solar system by perturbations from the major planets, passing stars, the galactic plane, and giant molecular clouds.

5. The Oort cloud supply of comets must be replenished, possibly by a much more populous, but as yet undetected, inner cloud of comets.

6. The source of short-period comets may be a flat belt of comets just beyond the planetary system. The long-period comets arrive directly from the outer cloud that Oort originally envisaged.

7. The ultimate fate of the observed comets is not well understood. Many long-period comets are thrown into interstellar space and some short-period comets disintegrate into dusty debris called meteor streams, but whether or not these comets eventually become inactive asteroidlike bodies is not clear.

8. Comets could have struck the Earth, with enormous ramifications for developing life forms.

NOTES

1. In 1986 Lubor Kresák concluded that changes in a comet's intrinsic brightness with time were not a good indication of its aging process.
2. A light year is defined as the distance that light travels in one year, or approximately 9.46×10^{12} kilometers.

Epilogue

If there is a central theme that runs throughout the history of comets, it must be the public concern they have commanded—concern completely disproportionate to their infrequent visits, subtle radiance, and modest sizes. Before the seventeenth century, comets were considered portents—warning shots fired at a sinful Earth from the right hand of an avenging God. In the post-Newtonian era, when their paths were understood to intersect that of the Earth, they were considered actual agents of destruction. At one time or another, they have been blamed for presaging war and pestilence and held responsible for the deaths of great men and the birth of good wine, for periods of drought and Noah's flood, for severely cold weather and the London fire of 1666. They have been described as the carriers of both life-seeds to the early Earth and horrific missiles that will one day snuff out life as we know it.

The range of phenomena attributed to comets is extraordinary—some of it true, much of it nonsense. But all of it adds to their considerable mystique and perhaps explains the universal interest shown in these, the solar systems' smallest bodies. They are currently thought to be the building blocks of the major planets and sources for some of the Earth's water, volatiles, and organic molecules. Cometary impacts on Earth have deposited some of the biogenic material from which primitive life may have ultimately formed. Subsequent impacts, however, could have easily erased many of Earth's life forms, leaving only the most adaptable to develop further.

During their infrequent encounters with Earth, comets have directly influenced the evolution of life itself. Within the solar system, the diminutive size of comets is in no way proportional to their importance. Next to the Sun itself, theirs is the most important realm.

Appendix

Naked-Eye Comets Reported Through A.D. 1700

This appendix provides a compilation of the naked-eye comets until A.D. 1700. The date of the comet's first appearance is given followed by the country or countries from which useful records exist. If the comet was observed well enough so that an orbit has been determined, additional information appears in parentheses: the perihelion passage time, P; the minimum Earth-comet distance, d, in astronomical units; and the date on which this close approach occurred. One astronomical unit, AU, is approximately 93 million miles or 150 million kilometers. In many cases, brief notes are provided on the comet's motion and physical behavior with question marks (?) denoting possible uncertainties in the text descriptions.

At the end of each entry, the relevant sources are provided. Often, these sources can provide additional information on a particular apparition. Commonly used sources, such as Ho Peng Yoke and Alexandre Guy Pingré, have been abbreviated. For example Ho (5) refers to the fifth comet mentioned in Ho Peng Yoke's 1962 catalog and P500 refers to page 500 in Pingré's work of 1783–84. Primary sources are listed after each cometary apparition; other sources are mentioned only if they provide additional information. The following catalog of cometary apparitions is meant to provide a guide to where comprehensive observations were made and what records are extant. For the early apparitions of comet Halley, we have relied on the comprehensive observation summaries given by Stephenson and Yau (1985).

Chinese observers had different names for different cometary forms. One of these is the *po*, or bushy star comet, signifying a symmetric, diffuse image without a tail. However, the term *bushy star comet* was sometimes used to describe a comet with a tail as well. A *hui* comet, or *broom star*, is one with a tail. We have adopted the bushy star and broom star designation. When noting the angular length of a comet's tail, the Chinese usually used a linear unit of measure termed a *chi*, or foot. We have assumed that 1 chi is approximately 1.5 degrees, as determined by Tao Kiang in 1972, and this seems to be a good approximation when the tail lengths are fairly modest. However when the object is listed in the Chinese sources as tens of chi, the 1.5 degrees per chi conversion factor is no longer accurate, and longer tail lengths are often recorded by assuming that 1 chi is approximately 1 degree. The estimates given for the longer tail lengths must be considered very approximate.

Dates prior to October 1582 are referred to the Julian calendar. Although the perihelion passage times and Earth close approach times are Greenwich mean times, the Chinese dates of observation have been left in terms of the reported local times.

In general, Chinese records denote the particular asterism or lunar mansion where a particular comet was sighted on a given date. An *asterism* defines a group of stars that neighbor one another and each of the 28 lunar mansions defines a range of right ascension on the celestial sphere. Ho's identifications of the asterisms and lunar mansions with the more familiar constellations have been used throughout this catalog. Uncertainty in a comet's position is sometimes introduced because the original record mentions a lunar mansion to specify a comet's longitude only, whereas Ho generally identifies a specific constellation for each lunar mansion, thus specifying both the comet's longitude and latitude.

Physical and orbital information for cometary apparitions after 1700 are provided in the works by Kronk (1984) and Marsden (1989).

11th century B.C. **(perhaps about 1059** B.C.**); China.** When King Wu-Wang waged a punitive war against King Chou, a broom star comet appeared with the handle of the broom star pointing east. Ho (2), Pankenier (1983).

1002 B.C. A comet was seen in Leo (uncertain event), P251.

974 B.C. **Spring; China.** A bushy star comet appeared in the north polar region. Ho (3). Pankenier (1983) notes that Ho's date should read 963 B.C. and that a systematic four-year error in the reporting of events at that time assigns this event to 959 B.C.

633 B.C.**; China.** A broom star comet appeared in Auriga with its tail pointing toward Chhu State. Ho (4).

613 B.C. **Autumn; China.** A broom star comet entered the constellation of the Great Bear. This is probably the first comet for which a verifiable record exists. Ho (5).

532 B.C. **Spring; China.** A new star was seen in Aquarius. Ho (6).

525 B.C. **Winter; China.** A bushy star comet appeared in the winter near Antares. Ho (7).

516 B.C.; China. A broom star comet appeared. Ho (8).

500 B.C.; China. A broom star comet was seen. Ho (9).

482 B.C. Winter; China. A bushy star comet appeared in the east. Ho (10).

481 B.C. Winter; China. A bushy star comet was seen. Ho (11).

480 B.C.; Greece. At the time of the Greek battle of Salamis, Pliny noted that a comet, shaped like a horn (*ceratias* type), was seen. Barrett (1).

470 B.C.; China. A broom star comet was seen. Ho (12).

467 B.C.; China, Greece. A broom star comet was seen. This event is often, but incorrectly, attributed to comet Halley. This is the comet that Plutarch noted appearing prior to the falling of the meteorite at Aegospotami, Greece. Ho (13), Barrett (2), P255.

433 B.C.; China. A broom star comet was observed. Ho (14), P258.

426 B.C. Winter; Greece. A comet appeared in the north around the time of the winter solstice. Barrett (4).

373–372 B.C. Winter; Greece. A comet was seen in the west at the time of the great earthquake and tidal wave at Achaea, Greece. From the Greek descriptions of the comet's motion, Pingré infers that its perihelion was located in Virgo or Libra and that its perihelion distance was quite small. Pingré considers this comet to be the one the Greek Ephorus reported to have split into two pieces. The accounts given by Aristotle and Seneca suggest the comet was seen in the winter of 373–372 B.C. while the account of Diodorus Siculus, an historian of the second half of the first century B.C., suggests the comet was seen in the following year. Barrett (5), P259.

361 B.C.; China. A broom star comet appeared in the west. Ho (15), P263.

345–344 B.C.; Italy. Comet seen in the west. Uncertain event. Barrett (6), P264.

341–340 B.C.; Greece. A comet appeared for only a few days in the equatorial zone near Leo. Barrett (7), P264.

305 B.C.; China. A broom star comet appeared. Ho (16).

303–302 B.C.; China, Greece. A broom star comet was seen. A Greek marble stele records a comet during the rule of Leostratus. This so-called Parian Marble records a series of events starting from Cecrops, legendary King of Athens, until 264–263 B.C. Ho (17), Barrett (8).

296 B.C.; China. A broom star comet was seen. Ho (18).

240 B.C. (P = May 25.1, d = 0.45 on June 4); China. Comet Halley was described as a broom star and it first appeared in the east, then at the north. During the month May 24 to June 23, it was seen in the west. Ho (19), P265.

238 B.C.; China. A broom star comet appeared in the west, then in the north moving southward toward Sagittarius. It lasted 80 days. Ho (20) notes that this account may refer to two comets.

234 B.C. January–February; China, Babylonia. A broom star comet was seen in the east. Ho (21). Fragmented Babylonian records note a comet seen in the east during the last part of the night sometime within the interval January 23 to February 20. Hunger (1989).

214 B.C.; China. A bright star or comet appeared in the west. Possibly a nova. Ho (22).

210 B.C. June–July; Babylonia. During the interval June 24 to July 22, a comet was seen with its tail directed toward the east. Hunger (1989).

204 B.C. August–September; China, Rome. A bushy star comet was observed near Arcturus for over 10 days before disappearing from sight. Ho (23), Barrett (10).

172 B.C.; China. A tailed star comet was observed in the east. Ho (24).

164 B.C. (P = November 12.6, d = 0.11 on September 29) Babylonia. According to Stephenson et al. (1985), analysis of Babylonian tablets in the British Museum suggests that Comet Halley was seen in the east before the lunar month beginning October 21 and in the west while in Sagittarius during the period October 21 to November 19. During this latter period, the comet passed 2.5 degrees west and 7.5 degrees north of the planet Jupiter. Its motion as computed by Yeomans and Kiang (1981) is consistent with these observations. No Chinese records of this apparition have been found.

163 B.C. September 5; Babylonia. During the first part of the night, a comet became visible above α Coronae with its tail pointing toward the south. Hunger (1989).

162 B.C. February 6; China. A thien-chhan comet, or celestial magnolia tree, appeared in the southwest. Ho (25).

157 B.C. October; China, Babylonia. A bushy star comet appeared in the west near Scorpius. Its tail pointed northeast, measured more than 15 degrees, and reached the Milky Way. It went out of sight after 16 days. Ho (26). Babylonian diaries suggest that a comet was noted from October 19 until November 15. Hunger (1989).

155 B.C. Winter; China. A broom star comet appeared from the southwest. Ho (27).

155 B.C. September; China. A broom star comet was seen at the northeast. Ho (28).

154 B.C. February; China. A tailed star comet was seen in the west. Ho (29).

148 B.C. May; China. Comet seen in the northwest. Ho (30).

147 B.C. May 13; China. A white broom star comet, with a 15-degree tail, appeared at night in the northwest in Orion. It moved away at dawn and became smaller and went out of sight after 15 days. Ho (31).

147 B.C. August 6 (P = June 28, d = 0.15 on August 4); China. A white tangle star comet appeared in the southwest below Scorpius. On August 8 it was located north of Scorpius and its tail reached about 90 degrees (?). It left on August 16. Ho (32). Seneca records a comet after the death of Demetrius, king of Syria, and a little before the Greek Achaean war in 146 B.C. It was described as large like the Sun, reddish like fire and bright enough to dissipate the darkness. Barrett (19).

147 B.C. October; China. There was a comet in the northwest. Ho (33).

138 B.C. April; China. A bushy star comet appeared in Hydra, and traveled north through the north polar region until reaching the Milky Way. Ho (34).

138 B.C. May; China, Babylonia. A bushy star comet appeared in Hercules and traveled as far as Vega. Ho (35). Babylonian diaries record a comet that had previously set in Libra reappeared in the west on May 28. Hunger (1989).

138 B.C. August; China. There was a bushy star comet in the northwest. Ho (36).

137 B.C. October; China, Greece. There was a comet seen in the northeast by the Chinese. Seneca notes that, at the beginning of the reign of the Greek Attalus III, a comet spread out into *unlimited* size. Ho (37), Barrett (20).

135 B.C. July; China. A bushy star comet was seen in the north. Ho (38).

135 B.C. September; China. A tailed star comet appeared in the east stretching across the heavens. It lasted 30 days before leaving. Ho (39).

134 B.C. September; China. A tailed star comet stretched across the heavens. Quite possibly this apparition is a confused transcription of the September 135 B.C. comet. Roman sources note that when the Greek Mithridates VI was born a comet shone for 70 days and the whole sky seemed to be ablaze. Ho (41), Barrett (21).

120 B.C. Spring; China, Babylonia. A bushy star comet was seen in the east. Ho (42). Babylonian diaries record a comet seen on May 18 in Aries; it became stationary to the east sometime later than May 20. The same comet was seen on June 16 with its tail directed toward the south and on July 13 during the beginning of the night. Hunger (1989).

119 B.C. May; China. A tailed star comet was again seen in the northwest. Ho (43), P271.

110 B.C. June; China. A bushy star comet was first seen in Gemini. After more than 10 days, it was seen in Ursa Major. Ho (44).

110 B.C. November 23; Babylonia. A comet was seen in the east with its tail directed toward the west. Hunger (1989).

108–107 B.C.; China. A bushy star comet was seen near the region of Canis Minor and Gemini. Ho (45).

102 B.C. (Approximately); China. A bushy star comet was seen among the stars of Boötes. Ho (46).

87 B.C. (P = August 6.5, d = 0.44 on July 27) China, Italy. Comet Halley. A medieval Chinese encyclopedia states that a bushy star comet appeared in the east during the month from August 10 to September 8. Kiang (1972) notes that Halley would have been seen in the west during that time and suggests that the month may have been incorrectly transcribed in the secondary Chinese source. The comet would have been seen in the east in the previous month. Since the motion of Comet Halley in 87 B.C. is quite well established from orbit extrapolations, it seems likely that it was indeed the one referred to in the Chinese medieval source. Stephenson et al. (1985) note that according to Babylonian records, a comet was visible "day beyond day" during the lunar month July 14 to August 11, and a reasonable interpretation of those records suggests that the comet was last seen on August 24. The motion of the comet as given by Yeomans and Kiang (1981) indicates that its solar elongation on August 24 was only 31 degrees and decreasing with time. The Babylonian account also records the first quantitative measurement of comet Halley's tail noting that it was observed to be 4 cubits long, or approximately 10 degrees. Ho (47), Barrett (27).

84 B.C. March; China. A bushy star comet was seen in the northwest. Ho (48).

83 B.C. (approximate year); China. A tangle star comet was seen in the west, east of Hercules. It passed near Altair and entered Pegasus. Ho (49).

69 B.C. February; China. A bushy star comet was seen in the west about 30 degrees from Venus. Ho (52).

69 B.C. July 23; China. A white guest star measuring 3 degrees and pointing southeast appeared and remained above Spica in Virgo. Possibly, this was a nova. Ho (53).

69 B.C. August 20; China. A guest star was seen at the northeast of Corona Borealis moving in a southerly direction. On August 27 it entered the region near southern Hercules with its white tail pointing southeast. Ho (54).

61 B.C. August; China. A bushy star comet was seen in the east. Ho (55).

49 B.C. April; China, Korea. A guest star stayed about 15 degrees north-east of Cassiopeia. It measured over 15 degrees and pointed toward the west. It left Cassiopeia and went to the northern enclosure. Ho (56), Barrett (32).

47 B.C. June–July; China. A bluish-white guest star, with rays less than 1 degree in length, appeared in southeastern Perseus. Ho (58).

44 B.C. May–June; China, Korea, Italy. A broom star comet was seen at the northwest. It was reddish-yellow and measured about 12 degrees. After a few days, it was located near Orion, measured over 15 degrees, and pointed toward the northeast. During the games that Octavian was holding in honor of the assassinated Julius Caesar, several Roman authors reported a comet was seen in the north for 3 to 7 days. Ho (59), Barrett (34).

32 B.C. February; China. A bushy star comet appeared in Pegasus. It (later?) measured about 90 to 105 degrees in length and about 1.5 degrees in width and was bluish-white in color. Ho (60).

12 B.C. August 26 (P = October 10.8, d = 0.16 on September 10) China, Italy. This apparition of comet Halley was extensively observed in China. It was first seen on August 26 as a bushy star comet near Canis Minor and last seen in Scorpius some 56 days later. This was the comet reported by the Romans as marking the death of the Roman General Agrippa. Ho (61), Barrett (40).

10 B.C.; China. A bushy star comet was seen in Boötes near Arcturus. Ho (62).

5 B.C. March; China. A broom star comet appeared near Capricornus for over 70 days. Ho (63).

4 B.C. February 23 (?); China, Korea. A bushy star comet was seen near Altair. Ho (64).

A.D. 13, December; China. A broom star comet was observed. Ho (65).

22 November–December; China. A bushy star comet appeared in Hydra. It moved toward the southeast and went out of sight after 5 days. Ho (66).

39 March 13; China. A broom star comet appeared near the Pleiades with a tail measuring over 45 degrees. It moved northwest into Pegasus and lasted 49 days. Ho (68).

46–47 between December 17 and January 15; Korea. A bushy star comet was seen in the south. It went out of sight after 20 days. Ho (69).

54 June 9; China, Korea. A broom star comet in Gemini developed a white vapor tail, 7 degrees long, pointing southeast. It moved toward the northeast and went out of sight after 31 days. Various Roman sources refer to a comet seen at the time of the emperor Claudius' death. Ho (70), Barrett (43), P284.

55 December 12; China. A guest star with rays measuring about 3 degrees moved in a southwest direction. After 113 days, it went out of sight in the northeast of Cancer. Ho (71).

59 July; Korea. A bushy star comet was seen in Perseus. Ho (72).

60 August 9; China, Italy. A broom star comet, with a tail of about 3 degrees, was seen in the north of Perseus. It moved slightly to the north and arrived at a point south of Virgo. After 135 days, it went out of sight. (Possibly the comet went over the north polar region and then south toward Virgo.) It is not impossible that this sighting and the comet recorded in Korea in July of the previous year are the same object. Seneca and various Roman sources mention a comet, during the reign of Nero, that was seen for several months moving from north to south. Ho (73), Barrett (44).

61 September 27; China. A guest star was seen northwest of Boötes pointing toward Corona Borealis. It went out of sight after 70 days. Ho (74).

64 May 3; China, Italy. A guest star with a white, 3-degree tail was seen to the south of η Virginis. It was seen for 75 days. A comet mentioned by the Romans Tacitus, Suetonius, and Pliny may well be this object. Ho (75), Barrett (45), P286.

65 July 29; China. A tailed star was seen extending 37 degrees in Hydra. It moved near Leo and into Perseus and its vapor reached ι and κ Ursae Majoris. It went out of sight after 56 days. Ho (76).

66 January 31 (P = January 26.0, d = 0.25 on March 20) China. Comet Halley was first sighted in the east on January 31, again on February 20, and finally about April 10 when it was 1.54 and 0.77 AU from the Sun and Earth respectively. Hence the comet was visible to the naked eye for some 74 days after perihelion. Ho (78).

71 March 6; China. A guest star was seen near the Pleiades. After 60 days it gradually went out of sight near Regulus. Ho (80).

75 July 14; China. A broom star comet appeared in Hydra, with a tail measuring about 4 degrees. It turned to the south of Coma Berenices and entered the region of Virgo and Leo. Ho (81).

76 October 7; China. A broom star comet, with a 3-degree tail, appeared in the region of Hercules-Ophiuchus-Aquila. It approached the western region of Capricornus and went out of sight after 40 days. Ho (82), Barrett (47).

77 January 23; China. A broom star comet with a tail of 12 degrees was seen in the western region of Aries. It moved slowly north to the region of Draco-Ursa Minor-Camelopardus and went out of sight after 106 days. Ho (83).

79 April; Korea, Italy. A broom star comet was first seen in the east and then in the north, disappearing after 20 days. This may be the comet that prompted the Roman emperor Vespasian to joke—"this hairy star is an omen for the king of the Persians"—since Persian kings wore their hair long and Vespasian was balding. Ho (84), Barrett (79), P290.

84 May 25; China. A guest star measuring 4 degrees appeared in the eastern, morning sky in Aries. It passed Cassiopeia into the north polar region where it remained for 40 days before going out of sight. Ho (85).

85 June 1; Korea. A guest star entered the north polar region. Ho (86).

101 January 12; China. A grayish vapor measuring 45 degrees rose from the north of Eridanus pointing at Canis Major. It was there for 10 days. Ho (87).

104 May 30; China. A white vapor, like loose cotton, developed in the north polar region. On June 10 it moved westward to the Pleiades and went out of sight on June 24. Ho (89).

110 January; China. A broom star comet was seen at the south of Eridanus pointing northeast. It was of grayish color and measured about 10 degrees. Ho (91).

110 July 27; China. A grayish guest star, as large as a pear, had 3-degree rays pointing southwest toward ι and κ Ursae Majoris. Ho (92).

117 January 9; China. A guest star was seen in the west. On January 20 it was in Aquarius. It went as far as the middle of Aries. Ho (93).

125–126 December–January; China. A guest star, possibly a nova, was seen in the region of Hercules-Serpens-Ophiuchus-Aquila. Ho (94).

126 March 23; China. A guest star entered the region of Coma Berenices-Virgo-Leo. Ho (95).

128 September–October; Korea. A tailed star stretched across the heavens. Ho (96).

132 January 29; China. A guest star was seen in Capricornus with rays more than 3 degrees pointing southwest. It was grayish in color and went out of sight in northern Aquarius. Ho (97, 98).

133 February 8; China. A guest star was seen southwest of Eridanus. It had a white vapor measuring about 3 degrees wide and about 75 degrees long. Ho (99).

141 March 27; (P = March 22.4, d = 0.17 on April 22) China. Comet Halley. On March 27 a broom star was seen in the east with a tail about 9 degrees long and pale blue in color; this was certainly the singly ionized carbon monoxide (CO^+) ion tail. The comet was observed until late April. Ho (100), Barrett (50).

149 October 19; China, Korea. A yellowish-white broom star comet with 7-degree rays pointing southeast appeared within the region of Hercules-Serpens-Ophiuchus-Aquila. It went out of sight on October 22. Ho (101).

153 November; Korea. A broom star comet was first seen in the east, then in the northeast. Ho (102).

154 January 31; Korea. A guest star trespassed against the Moon. Ho (103).

158 March–April; Korea. A bushy star comet was seen in Ursa Major. Ho (104).

161 June 14; China. A guest star appeared in Pegasus. When it came near Antares, it turned into a broom star comet developing a ray measuring about 7 degrees. Ho (105).

178 September; China. A broom star comet appeared north of Virgo with a tail several degrees long. It moved eastward, developed a red tail 70 to 90 degrees long and after 80 days went out of sight in Eridanus. Ho (106).

180 Winter; China. A broom star comet appeared near Sirius-Canis Major-Puppis moving eastward. It went out of sight after reaching Hydra. Ho (107).

182 August–September; China, Korea. A broom star comet was seen below Ursa Major moving eastward. It reached eastern Leo and went out of sight after 20 days. Ho (108).

186 November; Korea. A bushy star comet appeared in the northwest for 20 days. Ho (110).

188 March–April; China. A broom star appeared at the boundary of Andromeda and Pisces. After retrograding and entering the region of Draco-Ursa Minor-Camelopardus, it appeared three times and went out of sight after more than 60 days. Ho (111).

188 July 28; China. A guest star, as large as a 3-pint vessel, appeared at Corona Borealis. It moved southwest at first then southeast into Scorpius and went out of sight. Ho (112).

191 October; China, Korea. A white, banner comet with a tail measuring over 100 degrees appeared in southern Virgo. Ho (113).

193 November–December; China. A bushy star comet appeared near Spica and moved toward the northeast. After entering the region of Hercules-Aquila-Serpens, it went out of sight. Ho (114).

200 November 6; China. A bushy star comet was seen near the Pleiades. Ho (115).

204–205 December–January: China, Korea, Rome. A bushy star comet appeared in the region of Gemini and Cancer. It entered the region near Leo. Ho (116, 117).

206 February; China. A bushy star comet was seen with its head in Ursa Major but its tail penetrating the pole star. Ho (118).

207 November 10; China. A bushy star comet appeared in the region of Crater. Ho (119).

213 January–February; China. A bushy star comet appeared in Gemini. Ho (120).

217 November–December; China, Korea. A bushy star comet was seen in the northeast. Ho (121).

218 May–June; (P = May 17.7, d = 0.42 on May 30) China, Rome. Halley's comet was seen for approximately 40 days from early May to mid-June. For the first 20 days it was described as a bushy comet seen in the east. It moved from Gemini into Ursa Major and then into Ophiuchus. A comet caused panic in Rome during the revolt against the emperor Macrinus. Ho (122), Barrett (53).

222 November 4; China. A guest star appeared in Virgo. Ho (123).

225 December 9; China. A bushy star comet appeared in Leo Minor. Ho (124).

232 December 3; China. A bushy star comet appeared near Crater. Ho (125).

236 November 30; China. A bushy star comet appeared near Polaris measuring about 4 degrees. On December 15 it became a broom star comet in Hercules. Ho (126).

238 August 21; China. A broom star comet measuring 4 degrees appeared in Hydra. It retrograded, moved westward, and went out of sight after 41 days. Ho (127).

238 September 30; China. A guest star was observed retrograding in the north of Pegasus. On October 11 it trespassed against Hercules and went out of sight on October 16. Ho (128).

240 November 10; (P = November 10, d = 1.0 on November 30) China. A broom star comet measuring 30 degrees appeared in Scorpius, swept through Sagittarius, and trespassed against Venus. On December 19, it trespassed against Aquarius. Ho (129).

245 September 18; China. A white, broom star comet measuring 3 degrees appeared in Hydra moving southeast. After 23 days it disappeared. Ho (130).

247 January 16; China. A broom star comet measuring 1 degree appeared for 156 days near Corvus. Ho (131).

248 April–May; China. A bluish-white broom star with rays measuring 9 degrees pointing southwest was seen near the Pleiades. Ho (132).

248 August–September; China. A broom star comet appeared for 42 days near Crater measuring 3 degrees and moving toward Corvus. Ho (133).

251 December 21; China. A bushy star comet appeared for 90 days moving westward from Pegasus. Ho (134).

252 March 24; China. A white broom star comet measuring 75 to 90 degrees appeared in the west in Aries. Its rays pointed south and penetrated Orion. After 20 days it went out of sight. Ho (135).

253–254 December–January; China. A broom star comet measuring 75 degrees appeared in western Virgo pointing southwest. After 190 days it went out of sight. Ho (136).

254 December; China. A white vapor was seen coming from Sagittarius. Its width extended several tens of degrees and its length traversed across the heavens. Ho (137).

255 February; China. A broom star comet appeared extending approximately northwest from Capricornus to Corvus. Ho (138).

257–258 December–January; China. A white broom star comet was seen in Virgo. Ho (139).

259 November 23; China. A guest star was seen within the region of Coma Berenices-Virgo-Leo. After turning to the east, it moved southward and passed Corvus. It disappeared after 7 days. Ho (140).

260 July–August; Korea. A bushy star comet appeared in the east for 25 days. Ho (141).

262 December 2; China. A white broom star comet measuring less than a degree appeared in Virgo and changed its course toward the north. After 45 days it went out of sight. Ho (142).

265 June; China. A white broom star comet measuring more than 15 degrees was observed at Cassiopeia pointing toward the southeast. It went out of sight after 12 days. Ho (143).

268 February 18; China. A bluish-white broom star comet appeared in Corvus. It moved toward the northwest and then turned east. Ho (144).

269 October–November; China, Korea. A bushy star comet appeared in the region of Draco-Ursa Minor-Camelopardus. Ho (145).

275 January–February; China. A bushy star comet appeared within Corvus. Ho (146).

276 June 23; China. A bushy star comet was seen in Libra. During August the comet appeared near Arcturus, and during September it appeared in the region of Coma Berenices-Virgo-Leo, stretching as far as Crater and Ursa Major. Ho (147).

277 February–March; China. A bushy star comet appeared in the west. It appeared in Aries during the period April 20 to May 29, then near π Leonis (May 20–June 17). It appeared at the east (June 18–July 17) and then within the region of Draco-Ursa Minor-Camelopardus (August 16–September 14). Ho (148).

278 May–June; China. A bannerlike comet appeared in Gemini. Ho (149).

279 April; China. A bushy star comet appeared in northern Hydra. Within the period April 28 to May 27 it appeared near π Leonis, and during the period July 26–August 24 within the region of Draco-Ursa Minor-Camelopardus. Ho (150).

281 September; China. A bushy star comet appeared in western Hydra. Ho (151).

281 December; China. There was a bushy star comet in Leo. Ho (152).

283 April 22; China. A bushy star comet was seen in the southwest. Ho (153).

287 October–November; China. A bushy star comet measuring hundreds of degrees (?) appeared for 10 days in Sagittarius. Ho (154).

290 May; China. A guest star was seen in the region of Draco-Ursa Minor-Camelopardus. Ho (155).

295 May; (P = April 20.4, d = 0.32 on May 12) China. Comet Halley, observed during the month of May, began as a bushy star comet in northern Pisces and later became a broom star in the west. Apparently, the Chinese recognized that the comet seen in early May on the eastern horizon was the same object seen later that month on the western horizon. Ho (156).

299 October–November; Korea. A guest star trespassed against the Moon. Ho (157).

300 April–May; China. An ominous star was seen in the south, possibly a meteor or nova. Ho (158).

300–301 December–January; China, Korea. A broom star comet was seen west of Capricornus pointing approximately north. Ho (159).

301 May–June; China. A broom star comet appeared near Aquarius. Ho (160).

302 May–June; China, Korea. A broom star comet appeared in the day. Ho (161).

373

303 April; China. A broom star comet appeared in the east pointing toward Ursa Major. Ho (162).

305 September; China. A bushy star comet appeared in Taurus near the Pleiades. Ho (164).

305 November 21; China. A bushy star comet appeared in Ursa Major. Ho (165).

315 September–October; Korea. A bushy star comet was seen in the northeast. Ho (166).

329 August–September; China. A bushy star comet appeared in the northwest trespassing against Ursa Major. It went out of sight after 23 days. Ho (167).

336 February 16; China, Korea, Rome. A broom star comet in Andromeda appeared in the western evening sky. The death of the Roman emperor Constantine in May 337 was presaged by a hairy star of unusual size. Ho (168), Barrett (54), P301.

340 March 25; China. A bushy star comet was seen in the region of Coma Berenices-Virgo-Leo. Ho (169).

343 December 8; China. A white broom star comet measuring 10 degrees appeared in Virgo. Ho (170).

349 December 2; China. A white broom star comet measuring 15 degrees appeared in Virgo with its ray pointing toward the west. On January 29, 350, a broom star comet was again seen in Virgo. One Chinese source mentions a broom star comet seen on March 30, 350. Ho (171).

358 June 26; China. A broom star comet was seen extending from northern Perseus to northern Aries. Ho (172).

363 August–September; China. A bushy star comet was seen in Virgo. A contemporary Roman historian records a comet presaging the death of the Roman emperor Jovian who died in February 364. Ho (173), Barrett (55).

374 March 4; (P = February 16.3, d = 0.09 on April 2) China. Comet Halley appeared as a bushy star comet near western Aquarius. It traveled westward, becoming a broom star comet in Libra on April 2. The Chinese observational records are remarkably brief considering that the comet made a close Earth approach near opposition in early April. A bushy star comet was recorded as being seen on November 19 in the region of Hercules-Aquila-Serpens, but it could not have been comet Halley. Ho (175).

383 October–November; Korea. A bushy star comet appeared in the northwest. Ho (176).

389; Rome. Roman sources describe a strange and unusual star that was seen near Venus for 26 days. Pingré notes problems with the original reports and suggests the comet appeared near Jupiter, not Venus, in August. Barrett (57), P303.

390 August 7; (P = September 5, d = 0.10 on August 18) China, Korea, Rome. A bushy star comet appeared in Gemini. On September 8 it entered the region of northern Ursa Major. It was white in color and measured over 100 degrees. On September 17 it entered the north polar region and went out of sight. A Roman source reported a sign that appeared in the sky hanging like a column and blazing for 30 days. Ho (178), Barrett (58).

395 March; Korea. A bushy star comet appeared in the northwest for 20 days. Ho (180).

395 October; China. A tangle star comet, resembling loose cotton, moved toward the southeast and reached Aquarius. Ho (181).

396 July–August; China. A broom star comet was seen near the Pleiades. At first, a large yellow star appeared in Taurus for more than 50 days. In the period mid-December 396 to mid-January 397, the yellow star reappeared. The yellow star was possibly a supernova. Ho (182).

400 March 19; (P = February 25, d = 0.08 AU on March 31) China, Korea, Rome. A bushy star comet measuring 45 degrees appeared in the region between Andromeda and Pisces. Its upper portion reached Cassiopeia. It entered Ursa Major and between April 10 and May 9 it entered Leo. Roman historians noted a very large comet stretching from the sky to the ground. Ho (183), Barrett (61), P306.

400 September; Korea. A bushy star comet was seen in the east. Ho (184).

401 January 2; China. A bushy star comet was seen between the regions of Corona Borealis and Cygnus. Ho (185).

402 November–December; China. A white guest star, which resembled loose cotton, was seen at the west of the region Coma Berenices-Virgo-Leo. During the period January 9 to February 7, 403 the comet moved east into this region. Ho (186).

414 July 20; China. A bushy star comet appeared south of the Pleiades. Ho (187).

415 June 24; China, Korea. A broom star comet left the region of Hercules-Aquila-Ophiuchus-Serpens for the second time and swept a region in southern Hercules. It was then seen near the border between Libra and Ophiuchus. Ho (188).

416 October–November; China. A long broom star comet appeared in Ursa Major. On November 22 it entered the region of Coma Berenices-Virgo-Leo. It was seen for more than 80 days. Ho (189).

416–417 December–January; China. A broom star comet appeared from Cygnus and entered the region of Coma Berenices-Virgo-Leo. It then moved to Ursa Major and trespassed against Draco. After more than 80 days it reached the Milky Way and went out of sight. Ho (190).

418 June 24; China, Rome. A bushy star comet appeared within Ursa Major. On September 15 it appeared as a broom star comet with its handle arising from Leo and its rays extending to more than 100 degrees sweeping across northern Ursa Major. This account may refer to two different comets, one seen in June and the other in September. However, the Roman sources quoted by Pingré note that a cometlike phenomenon was seen during a solar eclipse on July 19, 418, and a comet was seen from mid-summer to the end of autumn. Ho (191), P309.

419 February 17; China, Korea. A bushy star comet appeared at the western wall of the region Coma Berenices-Virgo-Leo. Ho (192).

420 May; China. A tailed star comet extended across the heavens. Ho (193).

421 January–February; China. A guest star was seen in Crater. Ho (194).

422 March 26; China, Rome. A bushy star comet appeared in Aquarius. It moved toward the Milky Way and Cygnus and swept Altair. Roman sources note a comet, with a very long white ray, lasting 6 nights in March. Ho (195), P311.

422 December 18; China. A bushy star comet was seen in Pegasus. Ho (196).

423 February 13; China. A white bushy star comet measuring more than 30 degrees appeared in the eastern region of Pegasus and went out of sight after 20 days. A different Chinese source notes the comet in eastern Pisces in February with a 30-degree tail extending southeast to the Milky Way. Ho (197).

423 December 13; China. A bushy star comet appeared in northern Libra. Its tail measured 60 degrees, pointing northwest toward Boötes and facing Arcturus. It moved eastward and its length increased 9 to 10 degrees daily. After more than 10 days it went out of sight. During the interval from mid-December 423 to mid-January 424, a broom star comet was also recorded in Cetus. Ho (198).

435 June–July; China. A comet appeared in Leo. Hasegawa (298).

436 June 21; China. A bushy star comet was seen in Scorpius. Ho (199).

442 November 10; (P = December 15, d = 0.58 AU on December 7) China. A bushy star comet appeared in Ursa Major and entered Auriga. It passed through Taurus and reached Eridanus. More than 100 days later it disappeared in the west. Ho (201).

449 June–July; China. A broom star comet appeared north of the Pleiades. Ho (202).

449 November; China. A broom star comet entered the region of Coma Berenices-Virgo-Leo. Ho (203).

451 June 10; (P = June 28.2, d = 0.49 on June 30) China, Rome. Comet Halley was first seen June 10 near the Pleiades as a broom star. On July 13 it was recorded in southern Hercules. When the comet was last seen on August 15, it was 48 days past perihelion. This is the comet reported during the defeat of Attila the Hun at Chalons, France. Ho (204), P312.

453 February–March; China, Rome. A bushy star comet was seen in the west. This may be the comet reported just prior to the death of Attila in 453. Ho (205), P313.

454; Korea. A bushy star comet measuring 30 degrees was seen in the northwest. Ho (206).

460 November; China. A tailed star comet, over 15 degrees in length, appeared in Cetus. Ho (207).

461 April 20; China. A tailed star comet appeared in Cygnus. It was red in color and as long as a piece of cloth. Ho (208).

464–465 December–January; China. During the period from mid-December to mid-January a tailed star comet was seen near Vega. It was pure white in color. Ho (209).

467 February 6; China, Europe. A white vapor, called a long path, was seen stretching across the heavens from the southwest to the southeast. This type of comet may have two tails but the report does not rule out an auroral display. European sources report a large prodigy in this year. Ho (210), P314.

483 November–December; China. A guest star, the size of a peck measure and looking like a bushy star comet, appeared in Orion. Ho (211).

498 December; China. A broom star comet was seen in Leo. It passed through southern Cancer and reached the Milky Way. Ho (212).

501 February 13; China. A tailed star comet was seen stretching across the heavens. A comet recorded on April 14 may also refer to this object. Ho (213, 214).

507 August 15; China. A bushy star comet was seen in the northeast. Ho (215).

520 October 7; China. A broom star comet, as bright as a flame, was seen in the east. Ho (216).

530 August 29; (P = September 27.1, d = 0.28 on September 3) China, Rome. Comet Halley. On August 29 a broom star was seen in the northeast morning sky west(?) of Ursa Major. Its pure white tail was reported to be 9 degrees long. On September 4 it was seen as a northwest evening object with a 1-degree tail pointing to the southeast. It gradually turned to Libra and on September 23 it was barely visible. On September 27 it could not be seen (in the evening twilight?). Ho (217), P315.

533 March 1; China. A tailed star comet appeared. Ho (218).

535; China. A bushy star comet appeared in the region of Coma Berenices-Virgo-Leo. It passed Ursa Major and Pegasus before going out of sight. Ho (219).

537 February; China. A guest star was seen in the region of Draco-Ursa Minor-Camelopardus. Ho (220).

539 November 17; (P = November 6, d = 0.70 on November 27) China, Europe. A broom star comet appeared in Sagittarius measuring over 1 degree and pointing southeast. Its length gradually increased to over 15 degrees and it went out of sight on December 1 after reaching western Aries. Ho (221), P319.

541 February–March; China, France. A guest star appeared in the region of Draco-Ursa Minor-Camelopardus. Ho (222), P321.

560 October 9; China. A broom star comet was seen with its 6-degree rays pointing southwest. Ho (223).

561 September 26; China. A guest star appeared in Crater, possibly a nova. Ho (224).

565 April 21; China. A broom star comet was seen. Ho (225).

565 July 22; (P = July 15, d = 0.54 AU on September 13) China. A broom star comet appeared in Ursa Major. It entered southern Pegasus and its length gradually increased to over 15 degrees pointing northeast. After more than 100 days its length diminished to 4 degrees and it went out of sight in northern Aquarius. Ho (226).

568 July 20; China. A bushy star comet appeared in southern Gemini moving northward. After one month it reached Cancer and went out of sight. Ho (227).

568 July 28; (P = August 27, d = 0.09 AU on September 25) China. A guest star was seen in Libra. On September 3 a white guest star, resembling loose cotton, was seen near Antares. On September 27 it trespassed Delphinus and entered Pegasus. On October 16 it entered the region between Andromeda and Pisces and became smaller. On November 5 it was in western Aries. It is not clear whether this entire account refers to one comet, or to a nova seen on July 28 and a separate comet seen from September 3 to November 5. Ho (228).

574 April 4; (P = March 25, d = 0.89 AU on April 5) China. A bluish-white guest star, as large as a peach, appeared in Auriga. It gradually moved eastward while its length increased to 3 degrees. It traveled eastward through Ursa Major during the month of May and went out of sight after 93 days. Ho (229).

574 May 31; China. A bushy star comet was seen near the region of Draco-Ursa Minor-Camelopardus. It was the size of a fist and reddish-white in color. It pointed toward Leo and gradually moved southeast, while its length increased to 22 degrees. On June 9 it reached southern Ursa Major and went out of sight. Ho (230).

575 April 27; China. A bushy star comet appeared near Arcturus. Ho (231).

579 November; Korea. A tailed star comet stretched across the heavens and went out of sight after 20 days. Ho (232).

582 January 15–20; China, France. A broom star comet was seen in the southwest. Ho (233), P324.

583 February 20; China. A tangle star comet was seen. Ho (234).

588 November 22; China. A bushy star comet appeared in western Capricornus. Ho (235).

595 January 9; China, Korea, Europe. A broom star comet appeared in northern Aquarius. It reached northeastern Pisces. Korean records put the comet in Virgo. Ho (236), P324.

607 February 28; China. A tailed star comet that extended across the heavens was seen in eastern Pegasus and went out of sight after 20 days. Ho (237).

607 April 4; China. A tailed star comet was seen in the west extending across the heavens. It traveled through Pisces, Aries, and Virgo (?). The Chinese account states that this comet was again seen on October 21 in the south. It was seen in Virgo, Hercules, and nearly circled the sky before going out of sight in late January 608. This was certainly a separate comet altogether. Ho (237).

607 March–April; (P = March 15.5, d = 0.09 on April 19) China. Comet Halley. Three or four separate apparitions of cometlike objects result in a confusing set of observations in 607. Stephenson and Yau (1985) suggest that Halley was first recorded on April 18 in Gemini and Ursa Major. It passed Perseus, Auriga, and Gemini. It may also have been seen on March 30 in eastern Pegasus and followed for 20 days before being lost in the eastern morning twilight. Ho (237).

607 June 25; China. A bushy star comet was seen near the region of Ursa Major and Coma Berenices. Ho (238).

608 October 22; China. A broom star comet appeared in Auriga, swept Ursa Major, and went out of sight after reaching Scorpius. Ho (239).

615 July; China. A bushy star comet appeared southeast of Ursa Major. It measured one-half degree, looked black, and pointed and scintillated as it moved toward the northwest for several days until it reached Ursa Major. Ho (240).

617 July; China. A bushy star comet appeared in Leo. It was reddish-yellow, measured less than a half degree and went out of sight after a few days. Ho (241).

617 October; China. A broom star comet was seen in Pegasus. Ho (242).

626 March 26; China. A bushy star comet was seen near the Pleiades. On March 31, the comet was in Perseus. Ho (243).

634 September 20; China, Japan. A bushy star comet was seen in Aquarius for 11 days before going out of sight. The Japanese recorded a tailed star comet seen in the south during the month of September and (another?) broom star comet seen in the east during the month of February 635. Ho (244).

639 February; China, Japan, Korea. The Koreans recorded a bushy star comet seen in the northwest during the month of February. The Japanese recorded a tailed star comet seen on March 5, and the Chinese reported a bushy star comet appeared near the Pleiades on April 30. Ho (245).

641 August 1; China. A bushy star comet was seen in the region of Coma Berenices. It disappeared on August 26. Ho (246).

642 August 9; Japan. A guest star was seen. Ho (247).

647 September; Korea. A broom star comet appeared in the south and many stars drifted north. Ho (248).

662 Spring; Korea. A guest star was seen in the south. Ho (249).

663 September 29. A broom star comet measuring over 3 degrees appeared in Boötes. On October 1, it went out of sight. Ho (250).

666 February 15; China. A comet appeared in the region of Coma Berenices-Virgo-Leo. Hasegawa (390).

667 May 24; China. A broom star comet appeared at the northeast in the region of Auriga and Taurus. It was not visible on June 12. Ho (251).

668 May–June; China, Korea. A broom star comet appeared in northern Auriga. It went out of sight on June 7. Ho (252).

672 October; Korea. A broom star comet was seen on seven occasions in the north. Ho (253).

675 November 4; China. A broom star measuring 7 degrees in length appeared in southern Virgo. Ho (254), Hasegawa (395).

676 September 4; China, Japan, Korea, Europe. A broom star comet appeared in Gemini pointing toward the northeast. It measured over 4 degrees, increasing in size to 45 degrees. It entered Ursa Major, and after 58 days it went out of sight. Ho (255), P331.

681 October 17; China, Japan. A broom star comet, measuring 7 degrees, appeared in the west within the region of Hercules-Aquila-Serpens-Ophiuchus. As it moved eastward it became smaller, reached the region north of Altair, and went out of sight on November 2. Ho (256).

683 April 20; China, Korea. A broom star comet appeared in northern Auriga. After 25 days it went out of sight. Ho (257).

684 September 6; (P = October 2.8, d = 0.26 on September 7) China, Japan. Comet Halley. On September 6 or 7 a broom star was seen in the west with a tail more than 15 degrees long. It was observed for approximately 33 days. Ho (258).

684–685 December–January; Japan. A bushy star comet appeared at the zenith. It moved along with the stars of the Pleiades, and after approximately one month it went out of sight. Ho (258). An Italian source gives the interval of appearance as 684 December 25 through 685 January 6. Newton, R.R. (p. 679).

693 July 20; Korea. A broom star comet appeared in the east. Five days later, it appeared in the west. Ho (p. 214).

699 March; Korea. A white vapor spanned the heavens and a bushy star comet was seen in the east. Ho (259).

701 March–April; Korea. A broom star comet entered the Moon. Ho (260).

707 November 16; China. A broom star comet appeared in the west for 43 days before going out of sight. Ho (261).

708 July 28; China. A bushy star comet appeared in northern Aries. Ho (262).

709 September 16; China. A bushy star comet appeared in the region of Draco-Ursa Minor-Camelopardus. Ho (263).

712 July–August; China. A broom star comet moved from Leo to the region of Coma Berenices-Virgo-Leo. After reaching Arcturus, it went out of sight. Ho (264).

718 December 8; Japan. A broom star comet was observed. Ho (265).

725 February 11; Japan. A bushy star comet was seen. Ho (267).

729 January. Two comets were reported lasting 14 days, one in the east before sunrise and the other in the west after sunset. A single comet near the sun and with high declination would explain the reported sightings. P335, Chambers (250).

730 June 30; China. A broom star comet appeared in Auriga. On July 19 the bushy star comet was seen in Taurus. Ho (268).

738 April 1; China. A bushy star comet appeared in the region of Draco-Ursa Minor-Camelopardus. It passed the *box* of Ursa Major and after more than 10 days it was unobservable due to dark clouds. Ho (269).

744 Winter; Korea, Syria. An ominous star, as large as a five-peck measure, appeared in the central heavens. It went out of sight after 10 days. A great comet was seen in Syria. Ho (270), P336.

745 January 8; Japan. A bushy star comet was seen. Ho (271).

759 April; Korea. A broom star comet appeared. It went out of sight in the autumn. Ho (272).

760 May 17; (P = May 20.7, d = 0.41 on June 3) China. Comet Halley was first seen in the eastern sky before dawn with a white tail some 6 degrees long. First seen in northern Aries, it moved rapidly northeast. It passed through Taurus, Orion, Gemini, Cancer, and Leo. It was observed for approximately 50 days and was extinguished at a position 7 degrees west of β Virginis in western Virgo. Ho (273).

760 May 20; China. An ominous star measuring several tens of degrees was seen in the south (or possibly the west). During the month June 18 to July 16 it was extinguished. Ho (274).

761 May–June; Korea. A broom star comet was observed. Ho (275).

762. A comet like a beam was seen in the east. P337.

764 April–May; Korea. A bushy star comet appeared in the southeast. Ho (276).

767 January 21; China. A broom star comet appeared in Delphinus with its rays gradually invading northern Ophiuchus. It measured more than 1.5 degrees, and after 20 days it went out of sight. Ho (277).

768 Spring; Korea. A broom star comet was seen in the northeast. Ho (278).

770 May 26; (P = June 5, d = 0.30 on July 10) China, Japan, Korea. A white, broom star comet with a tail extending some 75 degrees was seen in the north. On June 19 it moved eastward approaching northern Auriga. On July 9 it was in northern Canes Venatici, and on July 25 the comet went out of sight. Ho (279).

773 January 15; China, Japan. A tailed star comet appeared east of Orion's belt. Ho (280).

776 January 11; China. A comet appeared in Delphinus and trespassed near α Herculis. It measured several degrees in length, and after 20 days it disappeared. Hasegawa (437).

813 August 4; Constantinople. A comet was seen that resembled two moons joined together. They separated, and having taken different forms, at length appeared like a man without a head. Despite this peculiar description, Pingré considers it to be a comet. This is a questionable appearance at best. P337.

814 April 17; China. A broom star comet appeared in the east. Ho (281).

815 April–May; China, Korea. A tailed star comet was seen in the region of Coma Berenices-Virgo-Leo. The Korean records note a comet appearing on September 7 in the region of Corvus and Crater with a 9-degree tail pointing west. Ho (282).

817 February 17; China, Europe. A broom star comet appeared in Taurus measuring more than 3 degrees and pointing southwest. After 3 days it came near Orion and went out of sight. Ho (283), P339, Newton, R.R. (p. 675).

821 February 27; China. A bushy star comet appeared in Crater. On March 7 the comet was above the planet Mercury near western Virgo. However, on March 7, Mercury was in Pisces, not Virgo. Ho (284).

821 July; China. A broom star comet appeared in northern Taurus measuring 15 degrees. It disappeared after 10 days. Ho (285), Hasegawa (447).

823 February 19; Japan. A bushy star comet appeared in the southwest for 3 days. Ho (286).

828 September 3; China. A broom star comet measuring 3 degrees appeared in Boötes. Ho (287).

834 October 9; China. A broom star comet appeared near Coma Berenices. Its tail measured over 15 degrees and pointed west. It moved northwest and went out of sight after 9 days. On October 31 a broom star comet again appeared in the east. Its rays measured 4 degrees and were very intense. Ho (289).

836 July–August; Korea. A bushy star comet was seen in the east. Ho (290).

837 March 22; (P = February 28.3, d = 0.04 on April 11) China, Japan, Europe. It is fortunate that this apparition, during which comet

Halley made its closest recorded approach to the Earth, is covered by the most detailed set of observations preserved in the Far Eastern records. On the night of March 22 a broom star appeared in the east with an 11-degree tail pointing west; it was then on the border between Pegasus and Aquarius. On the night of April 6 the comet was in Aquarius with a 15-degree tail pointing slightly south. On April 8 the comet was in western Aquarius with its tail increasing in both length and width. On April 9 the tail was split into two branches. On the evening of April 11 the comet was in Virgo and the 75-degree tail was undivided and pointing north. The next night the tail was pointing east, while the comet moved northwest. On April 13 the comet was in western Hydra and its tail reached its maximum length of over 90 degrees and pointed east. Thus, in a few weeks the Chinese reported the tail pointing in all four directions. On the night of April 28 the comet's tail was 4 degrees and pointing east. The comet went out of sight near Leo. Pingré records European observations through May 7. Ho (291), P340.

838 November 10; China, Japan, Korea. A broom star comet was seen measuring 30 degrees. On November 11 it was 37 degrees in length; on November 12 it was 45 degrees. On November 13 the comet was above Antares, with its 52-degree tail pointing toward Corvus. On the night of November 21 the comet appeared from the east, stretching across the heavens from east to west. One Chinese account has this comet going out of sight on December 28. Ho (292).

839 February 7; China, Japan, Europe. The comet appeared at the west, some 14 degrees from central Pegasus. Another Chinese source states the comet was in Aquarius on February 7. On March 12 it was seen at the north of Perseus. It went out of sight on April 13. Ho (292), P345.

840 March 20; China. A broom star comet appeared in eastern Pegasus and lasted 20 days. Ho (293).

840 December 3; China. A broom star comet appeared in the east. Ho (294).

841 July–August; China. A broom star comet appeared in northern Aquarius. Ho (295).

841 December 22; China, Japan, Europe. A broom star comet appeared in Piscis Austrinus. It was later found in Pegasus, then entered the region of Draco-Ursa Minor-Camelopardus. It went out of sight on February 9, 842. Ho (296), P346.

852 March–April; China, Japan. A broom star comet appeared in Orion. The Japanese note the comet was seen in the west with a tail larger than 50 degrees. Ho (297).

855 February 23; Japan, France. A tailed star comet was seen. Ho (298), P347.

857 September 22; China. A broom star comet measuring 4 degrees appeared west of Antares. Ho (299).

864 June 21; China, Japan, Europe. A broom star comet was seen in the northeast morning dawn. It was yellowish-white, measured 4 degrees, and was found in western Aries. The Japanese recorded a broom star comet on April 23, and French sources recorded a comet during the first 20 days of May. There is insufficient evidence to determine whether two separate comets appeared. Ho (300), P347, Newton, R.R. (p. 675).

868 January; Korea. A guest star trespassed against Venus. Ho (302).

868 January 21; Japan, Europe. A broom star comet was observed. The month is somewhat uncertain. In Europe a comet was recorded on January 29 in Ursa Minor. It advanced toward Triangulum and lasted 17 days. Ho (301), Hasegawa (471), P348.

868 February; China. A broom star comet appeared in Aries. This comet is possibly a later sighting of the comet seen in the previous month. Ho (303).

869 September–October; China, Europe. A broom star comet was seen in Perseus pointing northeast. Ho (304), P348.

875 March–April; Korea. A bushy star comet appeared in the east for 20 days. Ho (305).

875 June 5; Japan, France. A red broom star comet, with pointed rays, appeared in the northeast. On June 9 it measured over 15 degrees in northern Auriga. On June 24 a bushy star comet was seen. The death of the emperor Louis II was announced by a burning star that showed itself on June 7 in the north. Ho (306), P348.

877 February 11; Japan. A guest star appeared in Pegasus. Ho (307).

877 March; China, Europe. The Europeans record a comet in Libra seen in the west in March and lasting for 15 days. The Chinese record a comet in the period from June through July. P349.

882 January 18; France. A comet with a prodigiously long tail was seen. This is a questionable apparition. P350.

885; China. A broom star comet was seen near Castor, in Gemini. Ho (309).

886 June 13; China. A bushy star comet appeared in eastern Scorpius and passed Ursa Major and Boötes. Ho (310).

886 November 16; China. A comet known as the *long-path* type was seen coming from the west. It was white in color, 21 degrees in length, and bent at an angle. Eventually, it fell like a meteor. Ho (311). Possibly, this comet exhibited a tail directed toward the Sun as well as away from it.

890 May 23; Europe. A tailed comet was reported. Chambers (288).

891 May 12; China, Japan, Europe. A broom star comet appeared in Ursa Major. It moved eastward, swept Arcturus and entered the region of Hercules-Aquila-Serpens-Ophiuchus. It became over 100 degrees long and went out of sight on July 5. On May 11 the Japanese reported a guest star east of Antares. Ho (313), P350.

892 June; China, Europe. A white, cometlike banner was seen. It was 3 degrees long, shaped like hair and after several days it stretched from the midheaven to the horizon. European records put the comet in the tail of Scorpius, but note that it lasted 80 days and was followed by drought in April and May. If this is correct the European sighting would imply the comet was seen at the beginning of the year. Ho (312), P351.

892 December 28; China. A comet of the type called *celestial magnolia tree* was seen in the southwest. On December 31 it turned into a cloud and faded away. Hasegawa (486) places this comet in Sagittarius. Ho (314).

893 May 6; China. After many days of overcast sky, a broom star comet appeared in Ursa Major with a tail over 100 degrees. It moved eastward, swept Arcturus, and entered the region of Hercules-Aquila-Serpens-Ophiuchus. After more than 37 days, it became concealed by clouds. This entry has a peculiar similarity to the comet of 891. Ho (315).

894 February–March; China, Japan. A bushy star comet appeared in eastern Gemini. Ho (316).

894 August; China. An ominous, or evil, star appeared. Ho (317).

896 November–December; China. Three guest stars, one large and two small, appeared in northern Aquarius. Moving eastward together, they sometimes approached one another, then separated, giving the illusion that they were fighting among themselves. After 3 days the two smaller ones disappeared while the larger one faded away in northern Aquarius. Ho (318).

900 February; China. A guest star was seen in southern Hercules. It was as large as a peach and its rays concealed from view a small group of stars west of α Herculis. Ho (319).

900 August; Japan. A comet was seen. Hasegawa (493).

902 February–March; China. A guest star the size of a peach appeared in Cassiopeia. On March 2 a meteor arose from Ursa Major and reached the guest star, which was then remaining stationary. On March 4 the guest star was in northern Cassiopeia. In the following year it was still visible. (If this is a comet, this last statement is likely an error.) Ho (320).

904 July 15; China. A comet appeared in the east. It was many degrees in length and pointed to the southwest. Hasegawa (495).

904 November–December; Europe. A comet seen in the east lasted for 40 days. P352.

905 May 18; (P = April 26, d = 0.21 on May 25) China, Japan, Europe. A star resembling Venus appeared in the northwest evening sky. It emitted rays of 45 to 60 degrees and was blood-red in color. On May 19 its color resembled that of white silk. On May 22 the comet appeared in northern Gemini, but penetrated Ursa Major. Starting from Leo on June 12 it reached the western wall of the region of Hercules-Aquila-Serpens-Ophiuchus. Its brightness was very intense and it stretched across the heavens. On June 13 the sky was overcast; when it cleared on June 18, the comet had disappeared. Ho (321), P352.

907 April 7; Japan. A broom star comet was seen. Two days later it trespassed against Venus, then in Aries, and measured about 45 degrees. It went out of sight on April 15. Ho (322).

908 March; Korea. A bushy star comet was seen in the east. Ho (323).

911 June; China. A guest star trespassed against a group of stars in southern Hercules. Ho (324).

912 May 15; (P = July 18.7, d = 0.49 on July 16) China, Japan. Comet Halley. The Chinese observations of mid-May are discordant and were probably copied incorrectly, but contemporary Japanese observations during the second half of July are reasonably consistent with Halley's motion. The Japanese reported that a broom star comet was seen on the night of July 19 in the northwest. On July 21 and 22 it was the same. On July 24, Chinese records note that it was seen in the southeast although its appearance in the northwest seems more likely. On July 25 it was seen in the northwest direction, and on July 28 it was seen in the west direction. Ho (325).

918 November 7; Japan. A broom star comet appeared in the southwest and lasted three days. Ho (326).

923; China. A broom star comet was seen. Ho (327).

928 December 13; China. A bushy star comet appeared in the southwest measuring over 15 degrees. It was pointing southeast and was found in western Capricornus. Ho (328).

930 June–August; Japan. A guest star was seen in Aquarius. Ho (329), Hasegawa (508).

934 December 19; China. A broom star comet appeared in Aquarius and swept northern Capricornus. Ho (331).

936 September 21; China. A broom star comet appeared in Aquarius and swept northern Capricornus. It was small in magnitude and measured over 1 degree. This account has a peculiar similarity to that of the 934 December comet. Ho (332).

938 January 31; China. A bushy star comet was seen in the north. Ho (333).

939 February 24; Japan. A long star was seen in the sky. Hasegawa (512).

939 July(?); Italy. A comet of surprising grandeur was seen for eight nights. It threw out rays of extraordinary length. P354.

941 April; Japan. A star appeared in the west. It was bright and looked like a white rainbow. It had a small head but a large tail and it lasted two months before going out of sight. Ho (334).

941 August–September; China, Europe. A bushy star comet, several degrees long, was seen in the region of Hercules-Aquila-Serpens-Ophiuchus. It disappeared after 70 days. The Chinese accounts give various dates for the view period of this comet. Ho (335), P354.

942 October 18; France. A comet appeared for three weeks in the west and advanced gradually eastward to the meridian. P355.

943 November 5; China. A broom star comet near Spica appeared in the east, with its 15-degree tail pointing west. Ho (336).

947 February 20; Japan. A strange *lance star* appeared in the west. Ho (337).

947 September 12; China. A broom star comet appeared in the east, near the horizon. Its tail swept southern Leo Minor and on September 27, it went out of sight after having reached southern Leo. Ho (338).

948 March 2; Japan. A broom star comet was seen in the southwest. Ho (339), Hasegawa (521).

956 March 13; China. A bushy star comet appeared in southeastern Orion with its rays pointing southeast. Ho (340).

957 March 6; Japan. A white comet was seen 20 to 30 degrees in length. Hasegawa (523).

961 March 16; Japan. A broom star comet appeared in the southwest with its light resembling a wildfire. Ho (341), Hasegawa (526).

962 January 28; (P = December 28, 961, d = 0.35 on February 24) China. A guest star appeared in western Pegasus. It had a tail and emitted some faint rays. On February 19 it moved southwest and entered Libra. It went out of sight on April 2 when it reached a region near α Hydrae. Ho (342).

965 March 12; Japan. A guest star, possibly a meteor, was seen from southwest to northeast. Ho (343), Hasegawa (529).

967 January 8; Japan. A broom star comet was seen in the south. Ho (344), Hasegawa (530).

972 February 1; Japan. A broom star comet measuring over 6 degrees penetrated the Moon. Ho (345).

975 August 3; China, Japan, Europe. A broom star comet appeared in northwest Hydra measuring 60 degrees. In the morning, it was seen in the east pointing southwest. Moving westward, it passed through 11—of 28— lunar mansions before going out of sight after 83 days. Ho (346), P357.

977 March 16; Japan. At 8 P.M., two comets were seen, in the northeast and the southeast. Ho (347), Hasegawa (535, 536).

983 April 3; China. A guest star appeared in Virgo moving northward. Ho (348).

989 August 12; China. A guest star appeared west of Castor. It became dimmer and developed some rays and a tail that pointed southwest. Another Chinese account places this comet in the previous month. Ho (349).

989 August 12; (P = September 5.7, d = 0.39 on August 20) China, Japan, Korea. Comet Halley. On August 12 and 13, a bluish-white broom star appeared in northern Gemini. In the morning it was seen at the northeast for 10 days. Later, in the evening, it was seen at the northwest. It passed near Arcturus, and after 30 days reached a region east of Spica and disappeared. Ho (350).

990 February 2; China. A guest star appeared in Corvus. It retrograded and reached western Hydra. After traveling 40 degrees within 70 days, it went out of sight. Ho (351).

990 August–September (?); Europe. A star with a long tail appeared in the north. After some days, it was in the west with its tail pointing east. P359.

995 August 10; France. A comet was seen for 80 days. P359, Newton, R.R. (p. 676).

998 February 23; China, Japan. A broom star comet appeared in Pegasus with rays measuring over 1 degree. On March 8, it disappeared. Ho (352).

1000 December 14; France. A comet appeared for nine days. A meteor fell at about the same time and reports of the two events are confused. Some sources report the year as 999, rather than 1000. P360.

1003 February; Europe. A comet was seen in the west near sunset and then near sunrise in the east. P362, Newton (p. 674).

1003 December 21; China, Europe. A broom star comet trespassed against Cancer. On December 24, it trespassed against Cancer and Gemini. It was the size of a cup, bluish-white in color, and over 6 degrees long. After passing northern Gemini and Auriga, it entered eastern Orion and went out of sight. It lasted 30 days. Ho (354), P362.

1005 October 4; China, Europe. A guest star with rays like a bushy star comet appeared in Draco. Gradually it passed Cassiopeia and went out of sight after 11 days. Ho (355), P362.

1006; Korea. A broom star comet was observed. This record may refer to the nova observed in Lupus during the month of April 1006. Ho (357).

1009 May; Europe. A comet resembling a large beam was seen for four months in the southern parts of the sky. P365.

1011 February 8; China. A guest star appeared in Sagittarius. Ho (358).

1014 February 12; (P = April 6, d = 0.04 on February 25) China, Korea, Japan. A broom star comet appeared in the western evening sky. On February 27 it entered Auriga, and on March 7 it entered Perseus. Ho (359).

1017; Europe. A comet, like a large beam, was seen for four months. P366.

1018 August 3; (P = August 27, d = 0.38 on August 9) China, Japan, Korea, Europe. A broom star comet, measuring more than 30 degrees, appeared in the northwest. On August 13 the rays had become more intense. The Korean records mention the comet in the handle of Ursa Major on August 3 with the 60-degree tail pointing southwest. Ho (360), P366.

1019 February 6; Korea, Europe. A broom star comet appeared in Ophiuchus pointing west. Ho (361), P367.

1019 July 30(?); China. A broom star comet measuring about 4 degrees appeared in the handle of Ursa Major. Moving westward, it passed Leo and reached Hydra. It reached a length of 45 degrees and after 37 days, it went out of sight. There is some uncertainty in dating this record, and it may refer to the comet seen in August 1018. Ho (362).

1020 January 26; Korea. A broom star comet appeared in Ophiuchus. Ho (363).

1021 May 25; China, Korea. A guest star as large as a plum appeared in Leo. It moved rapidly past Regulus into western Virgo, concealing the star β Virginis. After 75 days it entered the horizon and went out of sight. Ho (364).

1023 France; Europe. A comet appeared in Leo. P368.

1032 July 15; China. A guest star appeared at the northeast above the horizon with rays shooting out like a comet. It went out of sight on July 27. Ho (366).

1033 March 5; China, Japan, Europe. A yellowish-white broom star comet with rays measuring about 3 degrees was seen in the northeast. In France, the comet was seen on March 9 in the morning sky. It lasted for three days. Ho (365, 367), Hasegawa (572), P369.

1034 September 20; China, Japan, Korea. A bushy star comet measuring 10 degrees long and one-half degree wide appeared in western Hydra. After 12 days, it went out of sight. Ho (368).

1035 January 15; China. A star with vaporous rays appeared in southeastern Pisces. Ho (369).

1037 March 19; Korea. Five comets, each measuring 7 to 9 degrees, were seen. Ho (370).

1041 September; Korea. A broom star comet about 45 degrees long appeared in the east and went out of sight after more than 20 days. Ho (371).

1041 November; Korea. A broom star comet about 45 degrees long appeared in the east and went out of sight after more than 10 days. Ho (372).

1049 March 10; China. A broom star comet appeared in Aquarius. In the morning, it was observed in the east pointing southwest. It reached western Aries before going out of sight after 114 days. Ho (373).

1053 February 25; Korea. A broom star comet, over 15 degrees long, appeared in Centaurus and then entered Crater. Ho (374).

1056 August–September; China, Japan. A broom star comet appeared in the region of Draco-Ursa Minor-Camelopardus and reached Hydra. It was white in color and measured more than 15 degrees. It went out of sight on September 25. Ho (376, 377), Hasegawa (585).

1056 December; Korea. A comet was seen in Corvus. Hasegawa (586).

1060 December 22; Japan. A broom star comet about 7 degrees long appeared in the south and disappeared after 5 days. Ho (378).

1063 June 8; Korea. A comet appeared near Arcturus. Hasegawa (590).

1063 August 22; Korea. A comet several degrees long was seen in Libra. Hasegawa (591).

1063 December (?); Korea. A comet appeared in the region of Hercules-Aquila-Serpens-Ophiuchus, then moved toward eastern Scorpius. The month of the comet's appearance is uncertain. Hasegawa (592).

1065 September 11; China, Korea. A guest star trespassed against Hydra. Ho (379).

1066 April 3; (P = March 20.9, d = 0.11 on April 24) China, Japan, Korea, Europe. Comet Halley. A broom star comet appeared in Pegasus. In the morning it was seen in the east with a length of about 10 degrees. It pointed southwest and reached Aquarius. It moved eastward and disappeared when near the Sun. On the evening of April 24 it was seen in the southwest. On April 25 it was in Auriga with its white vapor branched and stretched across the sky penetrating Gemini, Coma Berenices, and Leo, and reached Virgo and Libra. On April 26 the tail was 22 degrees long. The comet passed eastward from Pegasus all the way to Hydra and went out of sight after 67 days. Comprehensive Chinese observations allow the comet to be placed as a morning object in the east from April 3 through 22 and—after solar conjunction—as a western evening object from April 24 through June 6. The comet was last observed 77 days from perihelion, suggesting an unusually bright postperihelion apparition. According to an eleventh century manuscript in the archives of the cathedral in Viterbo, Italy the comet was observed in the eastern morning sky by a cleric for 15 days beginning on April 5. Reappearing in the western evening sky on April 24, "it looked like an eclipsed Moon, its tail rose like smoke halfway to the zenith, and it kept shining to about the beginning of June." Stein (1910), Ho (380).

1069 July 12; China. A guest star appeared at the longitudes of western Sagittarius. On July 23 it trespassed against Sagittarius and then went out of sight. Ho (381).

1070 December 25; China. A guest star appeared in Cetus. Ho (382).

1072 October 14; Korea. A firelike star appeared and trespassed Pegasus. Hasegawa (601).

1073 September 10; Korea. A guest star appeared in eastern Pegasus. Ho (383), Hasegawa (602).

1074 August 19; Korea. A guest star as large as a melon appeared in eastern Pegasus. Ho (384).

1075 November 17; China, Korea, Japan. A bluish-white star, looking like Saturn, appeared in the southeast at the longitudes of Corvus. On November 18 it grew a ray in the northwest measuring 4 degrees. On November 19 the ray measured 7 degrees, and the next night it measured 10 degrees pointing obliquely toward κ Corvi. On November 29 it entered the horizon and went out of sight. Ho (385).

1080 January 6; China, Korea. A broom star comet trespassed against Scorpius. A Korean record states that on February 5 a white vapor ran from the Pleiades to Crater and Corvus. Ho (386), Hasegawa (607).

1080 August 10; (P = September 10, d = 0.06 on August 5) China, Japan. A broom star comet appeared at the northwest in Coma Berenices.

Its white vapor measured 15 degrees and pointed obliquely to the southeast at the longitudes of Corvus. On August 13, it moved northwest and was found within the longitudes of Crater. On August 15 it measured 4 degrees and penetrated Coma Berenices obliquely. On August 20 it trespassed Leo and on August 24 it entered the horizon and went out of sight. On August 27, it reappeared in the morning at the longitudes of western Hydra and finally disappeared after a total of 36 days. Ho (387).

1090 March 31; Japan. Two strange stars were seen, one in the southeast and one in the southwest. Ho (388). Hasegawa (610) notes a Japanese record giving three comets.

1092 January 8; (P = February 22, d = 0.25 on January 7) China. A guest star appeared from the longitudes of eastern Orion, trespassing against and then concealing the stars at its sides. On January 9 it trespassed against the stars of eastern Eridanus, and on January 30 it entered northern Pisces. It went out of sight on May 7. Ho (389).

1097 October 2; (P = September 22, d = 0.52 on October 15) China, Korea, Japan. A broom star appeared in the west. On October 6 the comet appeared at the longitudes of Libra looking like Saturn. It was white in color and bright, with its tail measuring 4 degrees. On October 9 the rays measured 7 degrees. On October 16 it trespassed southern Hercules and the next night it moved slightly eastward and trespassed near α Ophiuchi. It went out of sight on October 25. Ho (390).

1106 February 9; China, Japan, Korea, Europe. On February 10 a broom star comet the size of the mouth of a cup appeared in the west. Its rays scattered in all directions as if broken into fragments. The comet was more than 60 degrees long, 4.5 degrees wide, and pointed obliquely toward the northeast. From northern Pisces, it traveled through Aries into Taurus before disappearing into twilight. The Japanese and Korean records note the comet was first seen in the southwest on February 9 and lasted over a month. According to Pingré, the comet was seen during the day and very close to the Sun as early as February 4 or 5. This comet is perhaps the progenitor of the Kreutz sungrazing family of comets. Curiously, a medieval European annalist stated that a meteor detached itself from the comet on February 16 and fell to Earth. Ho (391), P384, Newton, R.R. (p. 673), Marsden (1967, 1989).

1106 December 14; Korea. A broom star comet was observed. Ho (392).

1109 December; Europe. A comet with a tail pointing south was seen near the Milky Way. P389.

1110 May 29; (P = May 18, d = 0.49 on June 12) China, Japan, Korea. A broom star comet with rays measuring 9 degrees appeared in northern Pisces. It moved northward and entered the region of Draco-Ursa

Minor-Camelopardus. It then entered the horizon and went out of sight in the northwest. On June 9th the comet was seen in southern Cassiopeia. The next night it was in northern Cassiopeia, and on June 14 it had moved into Draco. A German annalist noted the comet from June 9 to June 30. Ho (393), Newton, R.R. (p. 678).

1113 August 15; Korea. A bushy star comet was seen in Pegasus. Ho (394).

1114 late May; Europe. A comet with a long tail was seen for several nights. P391.

1118 April 23; Japan. A comet was seen for only one night. Hasegawa (626).

1123 August 11; Korea. A bushy star comet was seen near the stars ε and δ in Ursa Major. Ho (395).

1126 July 19; China, Japan, Europe. A broom star comet measuring about 4.5 degrees was seen in the region of Draco-Ursa Minor-Camelopardus. Ho (396), P391.

1127 January 8; China. A white vapor arose at night in the region of Coma Berenices-Virgo-Leo and a broom star comet also appeared. Another Chinese source notes a comet stretching across the whole sky during the period between December 16, 1126, and January 14, 1127. Ho (397).

1130 December 30; Korea. A vapor measuring about 9 degrees and resembling a broom star comet appeared in northern Auriga. Ho (398).

1131 September–October; China. A broom star comet was seen. Ho (399).

1132 January 5; China. A broom star comet was seen. Ho (400).

1132 October 5; (P = August 30, d = 0.04 on October 7) China, Japan, Korea, Europe. The Japanese discovered the white broom star comet pointing west at the longitude of λ Orionis. On October 7 the Chinese noted a white broom star comet seen in northern Aries with intense rays, measuring over 45 degrees, and pointing northwest. On October 8 it moved south, reaching eastern Pisces, but its rays had become fainter and only 15 degrees in length. On the next night it moved further south and its rays measured only 3 to 4 degrees. According to the Japanese, it went out of sight on October 12 but the Chinese record its last appearance on October 27. Ho (401), P392.

1138 June–July; China. A guest star was guarding western Aries. Ho (402).

1138 September 3; China, Japan. A broom star comet was observed in the east. It went out of sight on September 29. On August 27 the Japanese reported a white broom star comet in the northwest measuring 8 to 9 degrees. It remained a few days before going out of sight. Ho (403).

1139 March 23; China. A guest star guarded eastern Virgo. Ho (404).

1145 April 15; (P = April 18.6, d = 0.27 on May 12) China, Japan, Korea, Europe. Like the previous return of Comet Halley, the observed period during this apparition is extraordinary. According to Pingré, Europeans first sighted the comet on April 15. In China it was apparently seen from April 26 until July 6, 78 days past perihelion. When first sighted by the Chinese on April 26, the comet was described as a broom star in the eastern, or morning, sky. On May 3 the comet, now described as a bushy star comet, was seen near northern Pisces. On May 9 the rays, which were some 20 degrees in length, pointed west. After entering solar conjunction, the comet emerged from the solar glare on May 16 in the west-northwest evening sky; its rays pointed east and were about 5 degrees in length, being partly covered with clouds. On May 17, after the skies had cleared, the broom star was seen with a 20-degree tail. On June 4 the broom star was close to the Moon and its rays were not seen. On June 8 its rays were 3 degrees in length, and on June 14 it was in western Hydra. On July 6 the broom star had dispersed. Ho (406), P393.

1146 December 29; China, Japan. A broom star comet appeared in the southwest in southern Pegasus. It measured about 15 degrees and gradually faded after more than 10 days. Ho (407).

1147 February 8; (P = January 28, d = 0.32 on December 29, 1146) China, Japan. A broom star comet over 15 degrees long was seen in the east. On February 13 it appeared in western Aquarius, with its rays measuring 15 degrees and approaching Delphinius. On February 14 it moved gradually northward, and on February 25 it faded away. Possibly, the comet seen in late December 1146 was this same object seen prior to its late January 1147 solar conjunction. Ho (408).

1155 May 5; Europe. A comet was seen. P394.

1156 July 23; China, Japan, Korea. A broom star comet appeared in the east. On July 25 it was seen in Gemini and was about 15 degrees in length. On July 30 it moved in a northeast direction, and on August 4 it trespassed against northern Canes Venatici. According to the Korean records, the comet was seen until August 25. Ho (409).

1161 July 22; China. A broom star comet appeared at the northeast of δ Ursae Majoris. Ho (410).

1163 August 10; Korea. A guest star trespassed against the Moon, which was then in Pisces. Ho (411).

1165 August; Europe. Two comets appeared before sunrise, one in the north and the other in the south. P394.

1166 April 23; China, Japan. A broom star comet appeared between the stars β Virginis and σ Leonis. It penetrated β Virginis, was white in color, and about 4 degrees long. After 20 days it went out of sight. Ho (412).

1175 August 10; China. A bushy star comet appeared in the northwest above Corona Borealis, in the northwestern region of Hercules. It was as small as Mars, but its rays radiated copiously in all directions. On August 15, it went out of sight. Ho (413).

1178 January 14; Japan. A broom star comet appeared at the southeast. On January 18 its rays became more intense. The Japanese also observed the comet on January 27th. Ho (414).

1181 August 6; China, Japan, Europe. A guest star appeared in southern Andromeda and trespassed against Cassiopeia. It went out of sight after lasting 185 days. Another Chinese source notes the guest star appearing on August 11 in Cassiopeia and lasting 156 days. Ho (415), P395.

1185 February 2; Japan. A banner comet measuring over 15 degrees was seen at the southeast. It was located in the eastern region of Hydra. Ho (416), Hasegawa (656).

1189 March 16; Japan. A reddish-white, broom star comet appeared in the east. It was over 15 degrees long and located in Coma Berenices. Ho (417).

1189 April 17; Japan. A white vapor appeared in the sky and pierced Ursa Major. Hasegawa (659).

1192 November 24; Japan. There were invocations for a comet. Hasegawa (660).

1198 November; England. A comet appeared for 15 days. P396.

1202 March; Japan. A broom star comet appeared. Ho (418).

1210 February–March; China. A guest star with rays spread out like a red dragon was seen in the region of Draco-Ursa Minor-Camelopardus. Ho (420).

1210 October 19; Japan. A broom star comet appeared in the west near the region of Hercules-Aquila-Serpens-Ophiuchus. It pointed east, measured over 15 degrees, and was seen throughout the night. On November 28 it was seen again. Ho (421).

1211 May–June; Korea, Europe. Pingré reports a comet seen for 18 days in Poland during the month of May. The Koreans recorded a white vapor appearing on June 18 and traveling through western Hydra, Crater, Corvus, the region of Coma Berenices-Virgo-Leo, and Ursa Major before disappearing. Hasegawa (668), P397.

1217 Autumn; Europe. A star was seen after sunset. It turned toward the south pointing a little westward. P398.

1220 January 25; Japan. A red broom star comet appeared in the northwest near the region of western Cassiopeia. Ho (422).

1220 February 6; Korea. A broom star comet appeared in Cepheus. Its tail measured about 4 degrees and pointed northwest. Ho (423).

1220 March 21; Korea. A bushy star comet appeared in Leo. Ho (423).

1221 January; Korea. A bushy star comet appeared near ε and δ Ursae Majoris. Ho (424).

1221 March 30; China. A comet was seen. Hasegawa (675).

1222 September 3; (P = September 28.8, d = 0.31 on September 6) China, Japan, Korea. Comet Halley. Korean observers first discovered the comet prior to solar conjunction on September 3 in southern Ursa Major. The observations describe it as a broom star, with a tail more than 5 degrees long pointing west. After the comet reached conjunction on September 5, Japanese observers described it as a broom star seen in the northwest. On September 8 it appeared in the northwest, its center as large as a half moon and white in color. However, the tail rays were red and more than 25 degrees long. Although the Korean observers record the comet as being visible during the day on September 9, this seems unlikely since it was 0.33 AU from the Earth at the time. However, Joseph N. Marcus and D.A.J. Seargent have suggested that this enhanced brightness might have been a result of forward scattering of sunlight from cometary dust particles since the Sun-comet-Earth angle was 136 degrees at the time of the reported daylight observation. On September 10 the broom star comet appeared in eastern Virgo with its tail pointing toward Arcturus. On September 25 it was in Libra, 4 days later it was seen in the southwest, and on October 8 was extinguished. Ho (425), Marcus and Seargent (1986).

1223 July; Europe. Early in July, a comet appeared in the western evening twilight. P400.

1224 July 11; China. A guest star guarded and trespassed against Scorpius. This record may refer to a nova rather than a comet. Ho (427).

1225 March 29; Japan. A comet was seen on March 29 and 31 and April 2. Hasegawa (680).

1230 December 4; (P = December 28, d = 0.13 on December 15) China, Japan. A guest star appeared in the west. The next night the comet was in the constellation Cygnus. On December 8 a star appeared from below Cygnus. It was as large as Saturn, but not as bright. On December 13 the comet entered the region of Hercules-Aquila-Serpens-Ophiuchus and departed December 26. On December 31 it passed a region west of Antares moving southeasterly. One Chinese source suggests the comet went out of sight as late as March 30, 1231. Not all accounts given for this comet are consistent. Ho (428).

1232 October 17; China, Japan. A white broom star comet appeared in the east. Its tail was more than 15 degrees long and bent like an elephant's tusk. It came from a region southwest of Spica and moved southward until October 27, when its tail measured 30 degrees. On October 31 it was not seen because of bright moonlight. Between 3:00 and 5:00 A.M. on November 11 it was seen in the southeast with a tail over 60 degrees. It lasted until December 14. Ho (429).

1239 May 27; Japan. An ominous star was seen in the northwest. Ho (431), Hasegawa (685).

1240 January 27; (P = January 21, d = 0.36 on February 2) China, Japan, Europe. A reddish-white broom star comet appeared in the southwest, measuring 4 degrees and pointing southeast. Two nights later it again appeared and was of the same size as Saturn, with its rays extending up to 6 degrees. On January 31 it was seen in Pegasus, and on February 1 it was seen at the side of Jupiter, the same size as Venus, with its rays measuring 7 degrees and pointing northeast. On February 2 it was facing Jupiter and was visible all night. On February 5 it was in eastern Pegasus, and on February 13 it was in southern Andromeda. On February 21 its rays were still faintly visible, and two nights later it was in Cassiopeia. On March 31 it went out of sight. Ho (432), P403.

1240 August 17; China. A guest star appeared in Scorpius. Ho (433).

1242 March 28; Japan. A comet was seen in the northwest. Hasegawa (688).

1245 February 24; (P = April 1, d = 0.11 on February 22) Japan. A guest star appeared in the southeast. On February 25 it appeared in the region of Ophiuchus and Aquila, and the next night it was seen in western Capricorn. On February 28 it resembled loose cotton. On March 30 a broom star comet was seen in eastern Pegasus. It measured 3 degrees and went out of sight on April 4. Ho (435).

1264 July 21; (P = July 20, d = 0.18 on July 29) China, Japan, Korea, Europe. A comet appeared in the northwest. On July 26 the broom star comet was in northern Hydra, with a tail reaching 100 degrees and illuminating the heavens. On July 31 it was in Cancer, on August 2 it was in Gemini, and on August 17 it was at a longitude corresponding to the eastern region of Orion. Toward the third week in September its rays slightly decreased, and on October 10 it finally went out of sight. The Japanese records note that its tail was only about 4 degrees on July 28, but its rays extended across the heavens on July 31. Pingré notes that it was seen in France as early as July 14 or 17. Aegidius of Lessines observations are summarized by Thorndike (1950). Ho (436), P406.

1265 Autumn; Europe. A comet appeared at the beginning of autumn and lasted until the end of that season. It was visible from midnight. A sixteenth century European annalist noted a star of unusual brilliance that poured out smoke like a furnace. P411, Newton, R.R. (p. 674).

1266 January 17; Japan. A broom star comet appeared in the east. It lasted until sometime within the interval February 7 to March 8. Hasegawa (699) has the comet first observed on January 18 in the southwest. Ho (437).

1266 August; Europe. Before morning twilight a comet was seen near the constellation Taurus. P413.

1268 August 27; Japan. A broom star comet was seen. Hasegawa (701) gives August 13 as the first date of observation and notes it was seen in the north. Ho (438).

1269 August; Scotland. A very fine comet was observed in the east during August and September. P415.

1273 February 5; Japan, Korea. A broom star comet was seen by the Japanese on February 5 and by the Koreans on February 17. Ho (439).

1273 April 9; China. A bluish-white guest star, with the appearance of loose cotton, appeared in Auriga. It moved from a region near ε and δ Ursae Majoris to a region in Boötes south of Arcturus. It lasted 21 days. The Japanese recorded a guest star on April 12 seen in the northwest. Ho (439), Hasegawa (705).

1273 October 17; Japan. A comet was seen in the west between 7:00 and 9:00 P.M. Ho (439).

1277 March 8; China, Japan, Korea. A broom star comet appeared in the northeast measuring over 4 degrees. Korean sources mention a banner comet seen on April 1. Ho (440).

1293 November 7; (P = October 28, d = 0.17 on November 22) China, Japan, Korea. A broom star comet measuring over 1 degree in length entered the region of Draco-Ursa Minor-Camelopardus. On November 25 it reached the region between η and γ Virginis; later, on December 23, it reached the box of Ursa Major. Ho (441).

1297 March 12; Japan, Korea. The Koreans noted a broom star comet appearing in Gemini. It lasted six days, but on March 25 it reappeared in Gemini. The Japanese noted that a comet measuring 9 degrees was seen in the west between 7:00 and 9:00 P.M. on March 13. Ho (442).

1297 September 14; China, Japan. An ominous star seen by the Japanese in the east between 7:00 and 9:00 P.M. appeared in southern Andromeda. Four days later, it was seen again in approximately the same location. Ho (443).

1299 January 6; (P = March 31, d = 0.75 on February 2) China, Japan, Korea. A comet approximately 2 degrees long appeared in the southern morning sky. On January 24 it was seen in the constellation Columba. Peter of Limoges noted that the comet had a long tail and was dark blue in color. On February 24 it was in the 14th degree of Taurus with a latitude of 5 degrees south. It remained visible until March 5. Ho (444), Thorndike (1950).

1299 October 23; Japan. A comet measuring 30 degrees in length appeared in the southeast between 3:00 and 5:00 A.M. Ho (445).

1301 September 1; (P = October 25.6, d = 0.18 on September 23) China, Japan, Korea, Europe. Comet Halley. Pingré notes that Europeans first discovered the comet on September 1. If so, this would imply that the comet was already unusually bright. Korean observers first sighted the broom star on September 14. Chinese observers first noted the white broom star comet, some 7 degrees long, on September 16 in Gemini. Later it passed into Ursa Major, Canes Venatici, and Corona Borealis and measured 15 degrees in length. Then it shortened to slightly more than 1 degree as it went to the south of δ and ε Ophiuchi. It went out of sight on October 31. Thorndike (1950) notes that the French cleric Peter of Limoges, Petrus Lacepiera, made observations of the comet's position with respect to the stars during the interval September 30 to October 6, when the comet moved southeast from Scorpius into Sagittarius. On September 15, between 3:00 and 5:00 A.M., Japanese observers noted a broom star comet in the east measuring more than 4 degrees; on September 23, between 7:00 and 9:00 P.M., they noted that it reappeared, this time in the northwest with its tail longer than 15 degrees. Ho (446), P420, Jervis (1985).

1301 November–December; Europe. Before Christmas, a comet was seen in the west after sunset. It set before midnight and lasted 15 days. On December 1 it was in Aquarius or Pisces. P423.

1303 July 27; Japan. A white comet measuring over 1 degree in length was seen in the northeast. Hasegawa (722).

1304 February 3; China, Japan, Korea. A white broom star comet was seen in Pegasus. Until April 18 it measured abut 1 degree in length and pointed southeast, then its length gradually increased. It pointed northwest, swept northern Lacerta, and went out of sight after 74 days. Korean observers placed the comet near the border of Andromeda and Pisces on February 7. Ho (447), Hasegawa (723).

1304 December 24; (P = January 19, 1305, d = 0.14 on December 22) Japan, Korea. A broom star comet appeared in northern Aquarius, and on December 30 it had entered southern Pegasus. Japanese observers noted that on December 26 it was seen in the west with a white tail, more than 1 degree, pointing eastward. Ho (448), Hasegawa (724).

1305 April; Europe. From April 15 through April 21, a large comet with a long tail was seen. P424.

1305 June 19; Japan. At 4:00 A.M., a comet was seen in the northeast. It was not seen after July 1. Hasegawa (726).

1307 March 20; Japan. There were invocations for an evil star. The star disappeared on the third day of the praying. Hasegawa (727).

1307 August 24; Korea. A broom star comet appeared in eastern Scorpius. Ho (449).

1313 April 13; China, Japan, Korea. A broom star comet appeared in Gemini. It remained visible for two weeks. Ho (450), P425.

1315 October 29; China, Japan, Europe. A guest star appeared in the region of Coma Berenices-Virgo-Leo. On November 28 it turned into a broom star comet and trespassed against the region of Draco-Ursa Minor-Camelopardus. It passed Corvus and reached eastern Pegasus after traversing 15 lunar mansions. It went out of sight on March 11, 1316. Ho (451), P426. According to Thorndike (1950), the French physician and astrologer, Geoffrey of Meaux, observed this comet from mid-December through February 12 and because of its height above the horizon and brightness, it was visible day and night without setting.

1330 April 7; Japan. A comet was seen in the northwest on April 7 and 12. Hasegawa (736).

1337 May 4; China, Korea. A bushy star comet appeared in Cassiopeia. It went out of sight on July 31 in Corona Borealis. Ho (454), Hasegawa (738).

1337 June 26; (P = June 14, d = 0.39 on July 21) China, Japan, Korea, Europe. A broom star comet appeared in the northeast and was located near the Pleiades moving toward the region of northern Perseus. It was a large white comet with a linear dimension of over 1 degree. On June 27 it traveled southwest and increased its speed daily until June 30, when its tail was about 4 degrees. On July 6 it swept Cassiopeia, on July 14 it swept Ursa

Minor, and on the next night it passed Polaris. On July 27 it trespassed Corona Borealis, and on July 29 it swept γ Herculis. On August 4 it swept Serpens Cauda, and on August 7 its rays were barely seen under the brightness of the Moon. On August 19 its rays became much weaker, but still could be seen northwest of Antares. It went out of sight on August 28 after passing a total of 15 lunar mansions. Ho (455, 452), Hasegawa (739), P429.

1338 April 15; Japan, Europe. Japanese invocations for a guest star were held. The comet was moving eastward in Gemini and set about midnight. On April 17 it was some 24 degrees within Gemini, with a latitude of 17 to 18 degrees. It lasted for about 2 weeks. Hasegawa (741), P433.

1340 March 24; (P = May 13, d = 0.35 on March 29) China, Japan, Korea. A white broom star comet, resembling loose cotton, appeared in northwestern Scorpius. It was about one-half degree long pointing southwest. It gradually moved northwest, and on March 25 was no longer visible, disappearing in Leo after having appeared for a total of 32 days. Ho (456, 453), Hasegawa (742), P434.

1345 July 31; (P = August 23, d = 0.05 on July 31) Japan, Korea, Europe. A broom star comet appeared in the region of Draco-Ursa Minor-Camelopardus. On August 3 it appeared near Castor in Gemini. On August 2 the Japanese noted that a white broom star comet, about 6 degrees long, was seen in the northeast. European records note that the comet first appeared in Ursa Major and went out of sight after reaching Leo, where the Sun was located. Ho (457), P435.

1347 August; Europe. A comet was seen in Taurus and in the head of Medusa. It was seen in Italy for 15 days and elsewhere for 2 months. P435.

1348 September 7; Japan. A comet appeared in the evening. Hasegawa (745).

1349 January–February; Japan. A guest star was seen in the interval between January 19 and February 17. Ho (458).

1351 November 24; (P = November 19, d = 0.05 on November 29) China, Europe. A bushy star comet was seen in northern Pisces. On November 26 it was seen in western Aries and the next night in northern Aries. On November 29 it appeared near the Pleiades and the next night was near Aldebaran in Taurus, but barely visible. Ho (459), P437.

1356 May 3; Korea. A guest star trespassed against the Moon, then in Gemini. This may have been a nova. Ho (460).

1356 September 21; China. A bluish-white broom star comet, about 1 degree long pointing southwest, was seen in the east. It appeared in western Hydra. It went out of sight on November 4 and had been seen moving northwesterly for over 40 days. Ho (461).

1360 March 18; China, Europe. A broom star comet appeared in the east. Ho (462), P438.

1362 March 5; (P = February 25, d = 0.44 on March 25) China, Japan. A bluish-white broom star comet measuring over 1 degree in length was seen in southern Pegasus. On March 17 it trespassed against the stars in northern Pegasus, its rays measuring over 30 degrees. On March 28 it wasn't seen, but it left a bent, white vaporous structure stretching across the heavens toward the west and sweeping Arcturus. On April 1 it passed the Sun and the rays were not visible. The comet was near the Pleiades and as large as a wine glass with a dull and faint color. On April 7 it went out of sight. Ho (463).

1362 April 25; China. A tailed star comet several tens of degrees long appeared in northern Aquarius. It disappeared after more than 40 days. Ho (464).

1362 June 29; China, Japan, Korea. A white broom star comet over 1 degree long appeared in the region of Ursa Minor and Draco. It moved in a southeast direction and pointed southwest. On July 6 its rays swept ι Draconis. The Koreans observed it on July 29 in northern Cassiopeia with a tail more than 1 degree in length. On August 2 it went out of sight. Ho (465).

1363 March 16; China. A broom star comet appeared in the east. Ho (466).

1364 March 30; Korea. A broom star comet was seen south of the region of Coma Berenices-Virgo-Leo, one was seen by the side of Arcturus, one was seen in Ursa Major, and one was seen in northern Libra with a red color and measuring more than one degree. A literal interpretation of the Korean record would suggest four different comets. Ho (467).

1366 October 25; (P = October 18, d = 0.03 on October 26) China, Japan, Korea. Periodic comet Tempel-Tuttle. A bushy star comet, the color of loose cotton and the size of a peck measure, appeared near the star δ Ursae Majoris. It moved southeast, passed and trespassed against Draco. On October 26, the comet was in southern Pegasus and the next night it was in western Aquarius. During its southeastern journey it trespassed Draco and Lyra, and went out of sight in western Aquarius. Ho (468), Hasegawa (758).

1367 February 20; Korea. A comet was seen on the horizon. Hasegawa (759).

1368 February 7; China. A broom star comet appeared in northern Taurus. Ho (469).

1368 April 8; (P = May 5, d = 0.14 on April 2) China, Japan, Korea. A broom star comet measuring over 12 degrees and pointing northeast was seen in Perseus. On April 26 it went out of sight northeast of Capella. Ho (470), Hasegawa (760, 761).

1370 January 31; Korea. A broom star comet appeared in the northeast. Ho (471).

1371 January 15; Europe. A very great comet was seen in the north with its tail directed toward the south. P442.

1373 April–May; China. From April 23 to May 22 a broom star comet entered the region of Draco-Ursa Minor-Camelopardus three times. Ho (472).

1374 March 11; Japan, Korea. A broom star comet measuring over 15 degrees was observed in the east. It lasted 45 days before going out of sight. Ho (473), Hasegawa (765).

1375 November; China. A comet appeared in Sagittarius. Hasegawa (766).

1376 June 22; (P = July 31, d = 0.18 on June 27) China, Japan, Korea. A white comet stopped in Cetus. It passed into eastern Pisces and Perseus, entered the region of Draco-Ursa Minor-Camelopardus, swept Ursa Major, and pointed toward Draco. It then entered Hydra and on August 8 went out of sight. The Japanese records note that a broom star comet appeared in the northeast on July 10. On July 25 it was seen at the northwest measuring over 15 degrees, and in the morning it appeared again at the northeast. Ho (474).

1378 September 26; (P = November 10.7, d = 0.12 on October 3) China, Japan, Korea, Europe. Comet Halley. The Chinese discovered the comet, measuring over 15 degrees, on September 26 near Capella. A week later, Korean observers noted that its position was less than 10 degrees from the north celestial pole. It swept Ursa Minor, then η Draconis on October 4. It was last seen on October 11 in Scorpius; it was cloudy after October 11 until November 9. Japanese observers were still offering prayers against the comet on November 15, so it may have been followed until then. However, Chinese observers recorded it only through October 11. Ho (475), P442.

1379 April; Japan. A broom star comet appeared in the east. Ho (476), Hasegawa (769).

1379 September 23; Japan. A comet appeared. Hasegawa (770).

1381 November 7; Japan, Korea. A broom star comet appeared in Libra. It measured over 15 degrees and went out of sight after 15 days. Ho (477).

1382 March 11; Korea, Europe. A bushy star comet was seen at the north. Ho (478), P443.

1382 September 19; Korea, Europe. A broom star comet measuring over

15 degrees in length appeared at the eastern wall of the region of Coma Berenices-Virgo-Leo. Though perhaps not trustworthy, a European account notes a comet appearing in the month of December. Ho (480), P443.

1384 September–October; China. A comet swept the constellation Crater at night. Hasegawa (775).

1385 October 23; (P = October 24, d = 0.74 on November 9) China, Japan. A star appeared in the region of Coma Berenices-Virgo-Leo. On October 30 it entered Crater measuring over 15 degrees. On November 4 it trespassed against Hydra. Ho (481), Hasegawa (776).

1388 March 29; China. A star appeared in eastern Pegasus. Ho (482).

1389 June–July; China. A white comet more than 15 degrees in length was seen in ecliptic longitudes similar to that of western Capricornus. Hasegawa (779).

1391 May 23; China, Korea. Two broom star comets were seen. One entered the region of Draco-Ursa Minor-Camelopardus while the other trespassed against northern Camelopardus and swept Cepheus. Korean records note a broom star appearing on May 11 for more than 10 days and a guest star appearing on May 22 in the region of Draco-Ursa Minor-Camelopardus. Ho (483).

1392 March 18; Korea. A broom star comet stretched across the heavens. Ho (484).

1397 December 25; Japan. A guest star was seen in the northwest. Ho (485), Hasegawa (785).

1399 October 7; Japan, France. A guest star was seen in the southern sky. Pingré notes that in November a star of extraordinary brilliance was seen with its tail turned toward the west. It lasted only a week. Ho (486), Hasegawa (787), P445.

1402 February 20; (P = March 21, d = 0.71 on February 19) Japan, Korea, Europe. A broom star comet measuring about 9 degrees appeared northeast of Pisces with its rays pointing east. On February 22 it appeared in the east with its rays radiating in all directions. On March 8 the rays continued to be of the same magnitude, and on March 19 it went out of sight. Pingré notes that this comet was first seen about February 8 in Europe; afterwards, during the fourth week in March, it became so bright as to appear in the daytime for eight days. Europeans followed the comet until the middle of April. Ho (487), P446, Hind (1877).

1402 June; Europe. From June to September an immense comet was seen in the west. It perhaps reached its greatest brilliancy at the end of August and was noted to be visible during the daytime. P449.

1403 December 18; Korea. A broom star comet appeared in the northeast. Ho (488).

1404 March 1; Korea. A bushy star comet was seen in the east. Ho (489).

1407 December 15; China. A broom star comet was seen. Ho (491).

1408 July 14; Japan. A guest star was seen. Ho (492).

1410 December 23; Korea. A guest star appeared. Hasegawa (799), P452.

1413 August 18; Korea. A comet appeared. Hasegawa (800).

1414 April 8; Japan. An evil star was seen. Ho (493).

1415 September; China. A broom star comet was seen in Sagittarius. Ho (494).

1416 July 29; Japan. An ominous star was seen. Ho (495).

1417 March 18; Korea. A comet was seen at the east. Hasegawa (804).

1419 June 12; Japan. An object like a tailed star comet was seen at the northeast after 11:00 P.M. Ho (496).

1421 January 9; Japan. A broom star comet measuring 7 degrees appeared at the northwest between 7:00 and 9:00 P.M. Ho (497).

1421 December 27; Japan. A guest star was seen. Ho (498).

1430 November 14; China. A tangle star comet appeared in southern Pisces moving southeasterly. It passed through Cetus and into Fornax and went out of sight after eight days. Ho (501).

1431 January 4; China. A bright, yellowish-white star appeared in eastern Eridanus. It disappeared after 15 days but appeared again on April 29. This sighting might have been a nova. Ho (502).

1431 May 15; China. A bushy star comet measuring over 7 degrees appeared in Gemini. Ho (503).

1432 February 3; China. A broom star comet appeared in the east measuring over 15 degrees. Its tail swept southern Cygnus and its course was toward the southeast. It went out of sight, but on October 26 reappeared in the west. It then went out of sight after 17 days. Hasegawa (1980) considers the comet reappeared on February 29 rather than October 26. One Chinese source has the comet reappearing on March 29 and lasting 17 days. Possibly there were 2 comets in 1432. Ho (504).

1433 September 15; (P = November 8, d = 1.23 on October 9) China, Japan, Korea, Europe. A broom star comet measuring over 15 degrees appeared in Boötes. On October 2 it entered Corona Borealis. On October 12 it swept the star κ Ophiuchi. On November 2 it was in the western heavens and was very small; two days later it went out of sight. The Italian Paolo

Toscanelli first observed the comet on October 4 as it moved southward into Corona Borealis. He followed it until October 31. Ho (505), P453, Jervis (1985).

1434 April 14; Japan. A broom star comet was seen in the east. Ho (506).

1434 September 11; Japan. A broom star comet was seen in the east. Ho (507).

1437 March 11; Korea. A guest star, possibly a nova, appeared in Scorpius and went out of sight after 14 days. Ho (508).

1438 March 16; Japan. A guest star was seen. Ho (509).

1439 March 25; (P = May 9, d = 0.31 on March 30) China, Japan, Korea, Europe. A broom star comet as large as a pellet appeared in eastern Hydra. On April 2 it was more than 75 degrees, moved toward the west, and swept western Leo. It extended northward and trespassed Cancer. Korean records put the comet in Gemini on April 12. Ho (510), P454.

1439 July 12; China. A broom star comet measuring over 15 degrees appeared near the star Aldebaran in Taurus. It pointed southwest and went out of sight after a total of 55 days. Ho (511).

1443 May 3; Japan. An evil star was observed. Ho (512).

1444 August 6; China, Japan, Korea, Europe. A broom star comet appeared at the eastern wall of the region of Coma Berenices-Virgo-Leo. It measured over 15 degrees, and its length increased daily. On August 15, it entered a region near the star Spica, in Virgo, and went out of sight. Ho (513), P454.

1449 December 20; (P = December 9, d = 0.50 on January 26, 1450) China, Korea. A broom star comet appeared at the region of Hercules-Aquila-Serpens-Ophiuchus. It measured over 3 degrees, passed the longitudes of eastern Scorpius, and went out of sight on January 12. Korean records note it was visible through the end of January 1450. Observing from Florence, Paolo Toscanelli followed the comet from December 26, 1449 through February 13, 1450 using a ruler to determine its position with respect to neighboring stars. Ho (514), Jervis (1985).

1452 March 21; China. A bushy star comet appeared in Taurus. Ho (515).

1453 January 4; China. A star appeared in Cancer moving slowly towards the west. Ho (516).

1456 May 27; (P = June 9.6, d = 0.45 on June 19) China, Japan, Korea, Europe. Comet Halley. The Chinese first observed the broom star comet in northern Aries on May 27 with a 3-degree tail pointing southwest. By June 7 the tail length had increased to over 15 degrees, and on June 22 the comet was seen in northern Hydra with a tail more than 13 degrees long sweeping northern Leo. On June 28 it appeared in eastern Hydra, measuring over 10 degrees and moving southwest. On July 6, it measured over one degree and entered the region of Coma Berenices-Virgo-Leo. The Italian Toscanelli observed the comet from June 8 to July 8. When the comet was near solar conjunction on June 17, he made note of it both in the morning and evening sky. Peurbach, in Vienna, tried to measure the comet's parallax—the first attempt of its kind. Ho (517), P459, Hellman, Jervis (1985).

1457 January 14; China, Japan. A broom star comet again appeared in Taurus measuring one-half degree. It moved toward the southeast, gradually increased in length, and went out of sight on January 23. Paolo Toscanelli recorded its position from January 23 to 27, 1457. Ho (518), Jervis (1985).

1457 June 15; (P = August 8, d = 0.27 on July 2) China, Japan, Korea, Europe. A broom star comet with a half-degree tail pointing southwest was seen in southern Pegasus and appeared to be vibrating. On June 22 it measured over 15 degrees and concealed a region near Pegasus. It moved from the southwest to the northeast in Pegasus, then into Andromeda, Perseus, and Gemini. Korean observers recorded the comet through July 16, while the Japanese noted the comet was a morning object on July 30, with a tail 7 degrees long, and was still visible on August 19. Paolo Toscanelli, in Florence, noted the comet's positions from July 6 through August 29, 1457. Ho (519), P464, Jervis (1985).

1457 October 26; China, Korea. A broom star comet with a half-degree tail pointing north appeared near Spica in Virgo. Ho (520).

1458 December 24; (P = November 7, d = 0.21 on December 23) China, Korea. A white star, pointing west, was seen in Hydra. On December 27 its body became smaller and appeared like loose cotton in northern Leo. On December 31 it developed a ray measuring one-half degree, and it went out of sight on January 12, 1459, in Gemini. Korean observers followed the comet through January 13, 1459, and record what may be another comet beginning again on February 1. On February 13 it was in northwestern Aries, and observations continued on several days until April 8 when its light was recorded as faint. Ho (521).

1459 October–November; China. A broom star comet appeared. Ho Peng Yoke and Ang Tian-Se (30).

1461 July 30; China. A star as white as powder appeared in northern Ophiuchus. On August 2 it turned into white vapor and went out of sight. Ho (522).

1461 August 5; China. A broom star comet appeared in the east and pointed southwest. It entered the longitude of Gemini, and on September 2 it went out of sight. Ho (523).

1462 June 29; (P = August 4, d = 0.29 on July 2) China. A darkish white star appeared in Cassiopeia and trespassed against Ursa Major. On July 16, while still in Ursa Major, it gradually became smaller. Ho (524).

1465 March; China. A broom star comet measuring over 45 degrees was seen in the northwest. It first appeared in the interval between February 25 and March 26 and was seen for 3 months. Ho (525).

1468 September 18; (P = October 7, d = 0.67 on October 2) China, Japan, Korea, Europe. A star appeared in Hydra moving northeast. After 5 days, its rays exceeded 45 degrees and pointed southwest. Henceforth, it appeared in the eastern morning sky and at dusk in Pegasus. It trespassed against northern Canes Venatici and Ursa Major, then turned and entered the region of Hercules-Aquila-Serpens-Ophiuchus. After leaving this region, it gradually diminished in size, trespassed against Scutum, and went out of sight on December 8. One Chinese source noted a bushy star comet in Ursa Major during the interval between July 19 and August 17. Ho (526), P467.

1469 March–April; Japan. A guest star was seen (possibly a nova). Ho (527).

1469 September 3; Japan, Korea. A broom star comet was seen in the east. Korean observers followed the comet until October 10. Ho (528).

1471 March–April; China. A bushy star comet appeared in Virgo. Ho Peng Yoke and Ang Tian-Se (36).

1471 May 20; Japan. A guest star was seen. Ho (529).

1471 Autumn; Poland. A very great comet was seen before sunrise. It lasted a month in Virgo and Libra. This reference may well belong to the comet first seen on December 25, 1471. P469.

1471 December 25; (P = March 1, 1472, d = 0.07 on January 23, 1472) China, Korea, Japan, Europe. The Chinese noted a broom star comet appearing in Virgo pointing west on January 16. It moved northward trespassing against a region near Arcturus and sweeping eastern Coma Berenices. It reached Leo with its tail pointing west. On January 24 its rays grew and stretched across the heavens from east to west. It moved northward for over 28 degrees, trespassing against many stars in Ursa Major and even appearing at midday. On January 27 it moved south and trespassed against stars in western Aries and eastern Pisces, where it was observed on February 17. Gradually it diminished, but took a long time to disappear altogether. Korean observers noted the comet in Virgo, with faint rays, as early as January 7, then reported its tail as being some 30 degrees in length in the third week of January, decreasing to 15 degrees in the last week and steadily there-after, with the last Korean report of the comet, only faintly visible, on February 21. Pingré notes European observations from December 25, 1471. Paolo Toscanelli, then 75 years old, recorded its position from January 8 through 28, 1472. On January 8, it was in Libra, and on January 22 it was moving very quickly only 15 degrees from the north pole. On the evening of January 23 it was moving more than one degree per hour, and on January 26 it was seen in Aries. Its extreme angular motion on the sky January 22 and 23 was due to its close Earth approach. Ho (530), P471, Jervis (1985).

1476 February; Japan. A guest star was seen between January 27 and February 24. Ho (531).

1476 December; Europe. A small pale blue comet was seen from the last of December 1476 until January 5, 1477. P477.

1480 March 17; Japan. A broom star was seen in the southeast. Ho (532).

1482 June–July; Japan. A large star, perhaps a guest star, was seen at the east. Ho (533).

1489 November 6; China. A broom star comet was seen. Within the month from November 23 to December 21 a guest star was seen in the region of Hercules-Aquila-Serpens-Ophiuchus. Ho Peng Yoke and Ang Tian-Se (38).

1490 December 31; (P = January 8, 1491, d = 0.52 on January 13, 1491) China, Japan, Korea. A broom star comet appeared in Cygnus with its tail pointing northeast. It trespassed against western Pegasus and on January 10, 1491, it was in the middle region of Pegasus. On January 22 it trespassed Cetus. Korean observers followed it until February 14, 1491. Ho (534, 535). Hasegawa (1979, 1980) considers there were three naked-eye comets seen in January through February 1491 and provides orbits for two of them. However, his orbit for comet 1491 I satisfies all the observations reported here fairly well.

1495 January 7; China. A slow-moving star appeared in Ophiuchus, approaching Sagittarius. On March 20 it entered southwestern Pegasus. Ho (536).

1499 August 16; (P = September 9, d = 0.06 on August 17) China, Korea. A star appeared in southeastern Hercules, passed into Draco, Ursa Minor, and Ursa Major, and went out of sight on September 6. Ho (537).

1500 May 8; (P = April 30, d = 0.83 on June 2) China, Japan, Korea. A broom star comet appeared near the border between Capricornus and Aquarius. It moved into Pegasus and increased its length to over 5 degrees. It then became smaller, approached Draco, and trespassed against Ursa Major. On July 10 it went out of sight. Pingré notes European records of a comet in Sagittarius and Capricornus during the month of April. Ho (538), P479.

1502 November 28; China. A star appeared in western Hydra, moved eastward into Crater, then back to western Hydra. It disappeared on December 8. Ho (539).

1506 July 31; China, Japan. A star like a pellet, with a darkish-white color and faint rays, appeared near northeastern Orion. Its length gradually increased to 3 degrees like a broom, then reached Ursa Major in the northwest. On August 11, the comet showed a bright ray extending 4 degrees to the southeast. After 3 days, it measured 7 degrees and swept southern Ursa Major. Finally it entered the region of Coma Berenices-Virgo-Leo. The Japanese reported a white comet that appeared in the northwest between 7:00 and 9:00 P.M. on August 7. It was seen in Ursa Major with a 15-degree tail. Ho (540).

1520 January–February; China. During the interval January 20 to February 18, a broom star comet was observed. Ho (541).

1521 February 7; China. A star like a fire appeared in the southeast. It turned white, measured 9 to 10 degrees and stretched from east to west. It then bent like a hook and disappeared after some time. This was possibly an auroral display. Ho (542).

1523 July–August; China. During the interval of July 13 to August 10, a bushy star comet appeared in the region of Hercules-Aquila-Serpens-Ophiuchus. Ho (543).

1529 February 9; China. A tailed star stretched across the heavens. Ho (544).

1529 August; China, Korea. A tailed star again appeared during the interval of August 4 to September 2. Korean sources noted a white broom star comet appearing in the west on September 1. It measured 6 to 7 degrees and on September 18 it shifted to the east. Ho (545).

1531 February 5; Japan. A broom star comet was seen. After a month and a half, it went out of sight. Ho (546).

1531 August 5; (P = August 26.2, d = 0.44 on August 14) China, Japan, Korea, Europe. Comet Halley. A broom star comet appeared in Gemini measuring over 1 degree. Its rays increased in length, and when it reached Crater it measured over 10 degrees and swept central Gemini at the northeast. It swept across Coma Berenices and, to the southeast, brushed a region near Spica in Virgo. It gradually diminished in size and went out of sight after 34 days. Japanese sources note the comet appeared in the northeast between 5:00 and 7:00 A.M. on August 9. It measured 7 degrees and was bluish-white. Later it was seen in the northwest. Korean sources note the

411

comet on August 10 as being white in color and 15 degrees in length. Halley (1752) notes the comet was observed from August 13 to 23 by Peter Apian at Ingolstadt. Pingré suggests the first European sightings were in late July or early August. Ho (547), P487, Hellman (1971).

1532 March 9; China. A darkish-white star with pointed rays appeared at the southeast. After 19 days it went out of sight. Ho (548).

1532 June 21; Japan. A broom star comet was seen at the northeast. After a few weeks it went out of sight. Ho (549).

1532 September 2; (P = October 18, d = 0.67 on September 21) China, Japan, Korea, Europe. A broom star comet measuring about 1 degree appeared in southern Gemini. (Chinese sources have it moving northeast into Cygnus but that seems unlikely.) It then gradually increased to over 15 degrees and swept a region near the star Spica in Virgo. It appeared until December 21. Korean sources first record the comet on September 14 near the border between Cancer and Hydra. On October 2 it shifted to an object on the eastern horizon and appeared white in color with a 15-degree tail. It appeared until December 30, when its light became faint. Japanese sources record the comet in the east on September 15. It was observed by Peter Apian, Girolamo Fracastoro, and Johannes Vögelin. From their observations, its motion can be tracked from southern Gemini southeastward into Leo, Virgo, and Libra. Ho (550), P491, Jervis (1980), Hellman (1971).

1533 June 27; (P = June 15, d = 0.42 on August 2) China, Japan, Korea, Europe. On July 1 a broom star comet measuring over 7 degrees appeared in Auriga with its tail sweeping into western Perseus. Its length gradually increased to over 15 degrees, then it trespassed against western Cassiopeia. On September 16 it went out of sight. A Korean account places the white comet in northern Auriga on June 27 with a tail measuring 10 to 12 degrees. It entered Perseus, went west through Cassiopeia, and reached Cygnus. On August 26 its size gradually diminished. The Japanese recorded the broom star comet in the north on July 15. European observers included Copernicus, Peter Apian, Gemma Frisius, and Girolamo Fracastoro. Ho (551), P496, Jervis (1985), Hellman (1971).

1534 June 12; China. A star appeared in southwestern Cassiopeia. It moved eastward and went out of sight after 24 days. Ho (552).

1536 March 24; China. A star appeared in Draco. It moved eastward and entered the Milky Way. On April 27 it went out of sight. Ho (553).

1537 March 8; Japan. A broom star appeared at the northwest. Ho (554).

1538 January 9; (P = December 30, 1537, d = 0.94 on December 17, 1537) Japan, Korea, Europe. The Japanese recorded a broom star comet on January 9. Korean sources note a broom star comet appeared in the west on January 21. It was white and measured about 45 degrees. On January 29 it developed a faint vapor. On January 17 Peter Apian saw the comet, with a 30-degree tail, some 5 degrees into Pisces with a latitude of 17 degrees north. Five days later, Gemma Frisius put it 9 degrees in Pisces with a latitude of 11 degrees north. Ho (555), P498.

1539 April 30; (P = May 12, d = 0.2 on May 31) China, Korea, Japan, Europe. A broom star comet was seen in Leo. Its rays pointed southeast and measured about 4 degrees. After 10 days it went out of sight. Korean records note a broom star comet in Leo on April 20. Its tail was white and 7 to 9 degrees long. During the interval of May 18 to June 15 it passed into Hydra. In Europe it was observed by Gemma Frisius and Peter Apian. Ho (556), P500.

1545 December 26; China. A star appeared in Draco. It entered western Sagittarius and turned to a northeast course. In the following month it went out of sight. Ho (557).

1549 March 7; Korea. A broom star comet was seen in the northeast. From March 29 to April 26 it shifted to the east. Ho (558).

1554 June 23; China, Korea. A broom star comet appeared beside δ Ursae Majoris. It then moved near the horizon and disappeared after 27 days. The Korean records note that a broom star comet with a white tail measuring about 2 degrees appeared in Ursa Major on June 26 and went out of sight on July 11. Ho (559).

1555 October–November; China. A comet appeared in the Pleiades. It moved north to Ursa Major. Hasegawa (913).

1556 February 27; (P = April 22, d = 0.08 on March 13) China, Japan, Korea, Europe. A broom star comet was seen in Corvus. Moving northeast into Virgo, it entered the region of Draco-Ursa Minor-Camelopardus. Its white tail measured about 7 degrees. On April 20 it reached eastern Pegasus, but both its size and brightness had gradually diminished. On May 10 it went out of sight. In Europe it was first noted on February 27 and followed by Paul Fabricius, Joachim Heller, Erasmus Flock, and others. Heller's observations on February 27 put it a few degrees southwest of Spica moving northeast. In mid-March, it moved rapidly northeast through Boötes and Draco. By April 1 it had passed through Cassiopeia into Andromeda. Ho (560), P502, Hellman (1971), Crommelin (1917).

1556 August; China. A comet appeared. It was some degrees in length and disappeared at the end of the month. Hasegawa (915).

1557 October 10; (P = September 22, d = 0.30 on October 25) China, Korea, Europe. A broom star comet appeared in Serpens pointing northeast. It disappeared on November 13. Korean records put the comet in Sagittarius when it faded from view. Ho (561). Pingré also mentions a comet in Sagittarius during October. P507.

1558 August 8; (P = September 14, d = 0.14 on August 11) Japan, Korea. Korean records note that a broom star comet with a white tail measuring 6 to 7 degrees appeared in the region of Coma Berenices-Virgo-Leo. From August 14 to September 11 it diminished in size and shifted its position to the region of Hercules-Aquila-Serpens-Ophiuchus. Japanese records note that a broom star comet was seen on August 9. European records place the comet under Coma Berenices on August 6, and on the 15th under the tail of the great bear, Ursa Major. On August 20 it was located 28.5 degrees from the end of the tail of Ursa Major and 30.5 degrees from Arcturus. P508, Ho (562), Hellman (1971).

1567 January 10; Korea. A tailed star comet one-half degree broad and 7 to 8 degrees long appeared in the south. It was wide at the top but narrow at the base. Ho (563).

1569 November 9; China. A broom star comet, pointing northeast, appeared in the region of Hercules-Aquila-Serpens-Ophiuchus. It went out of sight on November 28. Ho (564). European observations put the comet in Ophiuchus, Sagittarius, and Capricornus during November. P509.

1573 July; China, Europe. A comet was seen in Pisces. P511, Hasegawa (928).

1576 July–August; China. During the interval July 26 to August 23 a broom star comet was seen. Ho (566). Hasegawa (929) notes a Korean record citing the comet on September 2.

1577 July–August; Korea. During the interval July 15 to August 13, a broom star comet was seen. Ho (567).

1577 October 19; (P = November 8, d = 0.12 on November 3) Berlin, Germany. Leonard Thurneysser observed the comet in Capricorn on October 19 and followed it for 10 days. On October 29 it was in Sagittarius with its tail pointing toward Aquila. Landgraf (1977).

1577 November 8; (P = October 27, d = 0.63 on November 10) China, Japan, Korea, Europe. The Japanese noted the broom star comet in the southwest evening sky on November 8. It had a nucleus as bright as the Moon and a white tail over 60 degrees long. The Chinese first note the comet in the southwest on November 14. It appeared darkish-white and measured several tens of degrees. The vapor formed a white rainbow stretching from eastern Scorpius across Sagittarius to western Aquarius. It went out of sight after one month. European observers followed it until January 26, 1578. Tycho Brahe's long series of observations from November 13 through

January 26 indicate the comet moved northeastward from Aquila into western Pegasus. On November 13 Brahe estimated its coma diameter and tail length to be 7 arc minutes and 22 degrees respectively. Ho (568), P511, Hellman (1971).

1578 February 22; China, Korea. Chinese records note a large star like the Sun came out from the west, encircled by a number of stars at the west. This may be an account of a fireball. Korean records note a tailed star comet stretching across the heavens like a white chain. It went out of sight after a few days. Ho (569).

1578 November–December; China. A comet was seen in the east with a white tail stretching across the sky some 75 to 90 degrees in length. Hasegawa (932).

1580 October 1; (P = November 28, d = 0.23 on October 12) China, Japan, Korea, Europe. A broom star comet appeared at the southeast. It increased in length every night and stretched across the Milky Way. After more than 70 days, it went out of sight. The Japanese recorded the comet first on October 9. Mästlin discovered the comet on October 2, and eight days later both Tycho Brahe and Hagacius first observed it. Tycho made detailed observations through December 12. Its motion was generally westward from beneath the box of Pegasus through Aquila and Ophiuchus. It was last seen on January 14. Ho (570), P521.

1581 September; China. A comet was seen in the west and its light shone the Earth. After more than 30 days it went out of sight. Hasegawa (935).

1582 May 12; (P = May 6, d = 0.83 on May 9) China, Japan. Tycho Brahe discovered the comet on May 12, less than 11 degrees from the bright star Capella, and observed it until the 17th. On May 20 the Chinese noted a broom star comet appearing in the northwest like a chain with its tail pointing toward Auriga. After more than 20 days it went out of sight. On May 13 the Japanese noted a broom star comet in the northwest evening sky measuring over 100 degrees like a white cloud or rainbow. Ho (571), P544.

1582 September–October; China. A comet was seen stretching across the heavens. It was blue in color and pointed to the northwest. After about 49 days it disappeared. Hasegawa (938).

Transition from Julian Calendar to Gregorian Calendar

1584 July 11; China. A star appeared northwest of Antares. Ho (572).

1585 October 13; (P = October 8, d = 0.14 on October 18) China, Japan, Korea, Europe. A broom star comet about 1 degree long appeared in southern Aquarius. Every evening it was found moving eastward and diminishing in size. It was last seen on November 27. Tycho Brahe was prevented by bad weather from seeing it before October 28, but the Landgrave of Hesse's astronomer, Christoph Rothmann, recorded its motion from October 18 through November 18 as it moved northeast from eastern Aquarius through Pisces. Ho (573), P550.

1587 August 30; Japan. An object like a guest star was seen throughout the day. Perhaps it was a nova. Ho (574).

1587 October; Korea. During the interval October 2 to 30 a broom star comet was seen at the west. Its tail was bent and measured 45 to 60 degrees. Its rays illuminated the ground and after three months, it went out of sight. Ho (575).

1590 March 5; (P = February 8, d = 0.25 on March 3) China, Europe. A comet was seen in the southeast. After more than 10 days it disappeared. On March 5 it was discovered in Pisces by Tycho Brahe and described as having a tail 7 degrees long. Brahe followed its eastward motion until October 16. Hasegawa (944), P554.

1591 April 13; China. A star appeared in the northwest. It looked like a broom star comet and measured over 1 degree. It passed eastward through Pegasus and then measured about 3 degrees. On April 23 it entered western Aries. Ho (576).

1592 December 2; China. From December 2 to 4 a guest star moved westward in Cassiopeia. It remained visible until March 1593. Ho (577).

1593 July 30; (P = July 19, d = 0.45 on August 27) China, Japan, Korea. A broom star comet appeared in Gemini. On August 19 it retrograded and trespassed against Cassiopeia before entering the region of Draco-Ursa Minor-Camelopardus. On August 3 the Japanese noted a comet at midnight in the northwest measuring 15 degrees; they record it last seen on August 16 but the Koreans record it last on September 19. One of Tycho Brahe's assistants observed it from August 4 through September 3. Its motion was generally northwest from Gemini, winding up in the Milky Way between Cygnus and Cepheus on September 3. Ho (578), P557.

1596 July 19; (P = July 25, d = 0.56 on July 15) China, Japan, Korea, Europe. The Koreans recorded a broom star comet the same size, or brightness, as Capella appearing in Gemini. The Chinese recorded a broom star comet on August 5 appearing in the northwest. It was also moving toward the northwest and measured over 1 degree when it entered Crater. It went out of sight on August 22. These two accounts are difficult to reconcile since Crater is southeast—not northwest—of Gemini. European observations clearly suggest the comet was first seen in Auriga in mid-July. It then moved eastward through Gemini and Leo. Tycho Brahe first observed the comet on July 24 and followed it until August 3. Ho (579), P560.

1600 September 2; Korea. A broom star comet was seen in Ursa Major. By September 27 its rays had diminished when it shifted inside the region of Coma Berenices-Virgo-Leo. Ho (580).

1601 December 20; Korea. A comet with a long tail was seen in the northeast. Hasegawa (951).

1602 August 26; Korea. A comet with a tail about 11 degrees long was seen for a long time moving southeast from southwestern Ursa Major into Coma Berenices. Hasegawa (952).

1607 September 21; (P = October 27.5, d = 0.24 on September 29) China, Japan, Korea, Europe. Comet Halley. When first discovered by the Chinese on September 21 it was reported in Gemini, moving slowly toward the northwest with its tail pointing toward the southwest. (It was actually moving southeast on September 21.) On October 12, it was moving eastward in Scorpius and was last seen on October 26 as it moved into solar conjunction. Johannes Kepler, at Prague, made observations from September 26 to October 26, while Longomontanus, at Malmo, Sweden and Copenhagen, Denmark observed from October 1 to 26. William Lower and Thomas Harriot observed the comet from September 22 to October 6 and from September 21 to October 22 respectively. Harriot's observations were the most accurate position measurements recorded during this apparition. Ho-Peng-Yoke and Ang Tian-Se (75), P3, Halley (1752), Bessel (1804), Yeomans (1977), Roche (1985).

1607 November–December; China, Korea. Korean records report a comet in southwestern Ursa Major, while Chinese records note a comet was seen in the west with its red tail pointing east. Hasegawa (956).

1609: A large star appeared in the southwest casting its pointed rays in all directions. Ho-Peng-Yoke and Ang Tian-Se (76).

1615 August–September; China. A comet, shaped like a broom, was seen in the morning. Hasegawa (959).

1618 August 25; (P = August 17.6, d = 0.52 on August 20) Europe, China, Korea. The comet was discovered in Hungary on August 25 and two days later by Kepler near Linz, Austria, where it rose in the morning sky with its tail pointing west. The Koreans noted the comet on August 28 below Ursa Major with a bluish-white tail more than 15 degrees long. On September 1, Kepler observed the comet 10 degrees within Leo with a north latitude of 23.5 degrees. Five days later, the comet scarcely had a tail when viewed with the naked eye. As seen through Kepler's telescope, it was rather large and resembled a cloud. (This was probably the first use of a telescope for observing a comet.) Kepler last observed it on September 25 in Cancer. P4, Hasegawa (960).

**1618 November 16; (P = November 8.9, d = 0.36 on December 6)
Europe, China, Korea.** This comet's tail was first seen rising above the
morning horizon by several observers. Its head, although below the horizon,
was in Libra. For northern hemisphere observers it was best seen only in December and late November, when it had moved away from the Sun's glare.
On November 18, Jesuit observers in Rome noted the tail to be 40 degrees in
length. The Swiss Jesuit Johann Baptist Cysat followed the comet from December 1, 1618, through January 22, 1619, and the English astronomer John
Bainbridge followed it from November 28 through December 26, 1618. As
pointed out in Chapter 3, both Cysat and Bainbridge used telescopes for
some of their observations. On November 26 Chinese observers noted the
broom star comet in Libra measuring over 15 degrees and pointing southeast. The tail gradually pointed to the northwest and swept the stars of Ursa
Major. On November 30, Korean observers noted a comet in the east with a
tail several degrees long. P6, Ho-Peng-Yoke and Ang Tian-Se (77).

**1618 November 10; (P = October 27.9, d = 0.17 on November 19)
Europe, China, Korea.** This comet was seen between November 10 and
December 9, 1618 moving westward from a region near Libra into western
Hydra. On the morning of November 20, Kepler noted its tail and observed
its head near the head of the Centaur. P5, Landgraf (1985).

1619 February–March; China. Between February 14 and March 15, a
broom star comet appeared in the southeast measuring 100 degrees. It cast
its rays downward and its tail was bent and sharp-pointed. In one moment it
was seen in the northeast and at another it was seen in the west. Ho-Peng-
Yoke and Ang Tian-Se (78).

1621 January 22; Korea. A comet was seen. Hasegawa (966).

1621 February 8; Korea. A comet appeared. This comet is likely to be
identical with the previous entry. Hasegawa (967).

1623 November 24; Japan. A comet was seen in the west. Hasegawa
(969).

1624 July–August; China, Japan. The Chinese recorded that a large star
entered the Moon, while the Japanese noted a comet appeared in the south.
Hasegawa (970).

1625 January 26; Europe. A comet was observed in Eridanus and Cetus
by Wilhelm Schickard observing near Tübingen, Germany. It was seen from
January 26, 1625, until February 12. Olbers (1824).

1627 Summer; Korea. A comet appeared with the Moon. Hasegawa
(972).

1628 August 13; Japan. A comet was seen in the southeast. Hasegawa
(973).

1629 November 13; Korea. A comet appeared. Hasegawa (974).

1630 January–February; Japan. A comet was seen in the northwest. Hasegawa (975).

1633 November; China. During the period November 2 to 30, a broom star comet appeared. Ho-Peng-Yoke and Ang Tian-Se (79).

1638 June–July; Korea. A reddish comet appeared and lasted for one month. Hasegawa (977).

1639 October 26; (P = November 29.4, d = 0.08 on October 26) Korea, China. Korean observers reported a guest star that appeared in eastern Orion, and the next day it moved to southern Orion and covered an unknown star. The Chinese reported a comet seen in Orion sometime during the period between July 30 and October 25. On October 27 Placidus de Titis noted a comet with a small tail in Canis Major. Hasegawa (978), Ho-Peng-Yoke and Ang Tian-Se (80), Olbers (1831).

1640 December 21; A broom star comet appeared. Possibly it appeared on October 22, but the December date seems more likely in light of John Evelyn's notation in his diary that a comet was seen about the time of the Earl of Strafford's trial in 1640. Strafford was impeached on November 11, 1640. Ho-Peng-Yoke and Ang Tian-Se (81), Chambers (506).

1647 September 29. A comet some 12 degrees long was seen soon after sunset in Coma Berenices. It moved eastward through Boötes into Corona Borealis and lasted one week. P9.

1652 December 17; (P = November 13.2, d = 0.13 on December 20) China, Europe, Africa. The Chinese reported the comet on December 22 as an extraordinary star with a pale vapor. It moved toward the northwest and entered Taurus. European observers noted the comet first on December 18 and followed it until the first few days of January 1653. Hevelius, at Danzig, followed the comet from December 20 until January 8 and described it as being a pale and livid color and as large as the Moon. At the Cape of Good Hope, the comet was reported first on December 17 in the east-southeast evening sky. Its position was southward from the head of Orion with the tail pointing northward. On December 24, its head was 1 degree from the Pleiades, and the tail, pointing east-southeast, was described as reaching its greatest brightness. Ho-Peng-Yoke and Ang Tian-Se (82), P9, McIntyre (1949), Hind (1879), Knobel (1897).

1653 August 21; Korea. A comet appeared and became small near Castor and Pollux in Gemini. It then moved northwest to eastern Camelopardus. Its blue-white tail, one-half degree in length, pointed to the southwest. It disappeared on September 15. Hasegawa (982).

1656 April 1. Peter Mundy, an Englishman sailing off the coast of India, recorded a comet that he observed in the morning sky. Its tail pointed upwards. After the weather cleared on April 6, he again noted it with a tail 6 to 7 degrees in length spreading upward to the southeast. It was located about 12 degrees southwest of Venus. Maunder (1934).

1656 September 2; Korea. A comet was seen in Gemini resembling the great star in Auriga, Capella. Its blue-white tail was less than one-half degree in length. It disappeared on September 17. Hasegawa (983).

1661 February 3; (P = January 27.4, d = 0.61 on January 29) Europe. Hevelius observed the comet from February 3, when it was in eastern Aquila, throughout the remainder of the month and into March. He made physical observations through his telescope noting multiple structures in its head. It moved slowly westward through Aquila and was last observed on March 28 in western Aquila. This comet was mistakenly thought by many, including Edmond Halley, to be a return of the comet of 1532. P10, Halley (1705).

1661 December 16; Korea. A guest star was seen in western Aquarius moving northeast. Another Korean record noted the star disappeared on January 1, 1662. Hasegawa (985).

1664 November 17; (P = December 5.0, d = 0.17 on December 29) China, Korea, Japan, Europe. The great comet of 1664 was seen from November 17, 1664, through March 20, 1665, rising in the morning hours in November and December, then switching to an evening object for the first few months of 1665. The Chinese first noted the broom star comet on November 18 in Corvus. On November 26 its grayish tail was about a degree in length and pointing toward the southwest. On December 16 it was in Crater with its tail over 5 degrees and pointing to the northwest. The next night it was in western Hydra with its tail pointing north. On December 29 it was in Gemini, and on January 1, 1665 it moved northwest and reached western Taurus. Two nights later, with its tail pointing northeast, it reached northern Aries. On January 8 it reached northwestern Aries with its blue tail pointing east. On January 20, it was in northern Pisces with a tail that measured approximately 3 degrees. In Europe, the comet was extensively observed by Christiaan Huygens in Leiden, Johannes Hevelius at Danzig, Adrien Auzout and Pierre Petit in France, and several others. With the aid of a telescope, the comet was followed until March 20 when its increasing distance from the Sun and Earth, as well as its approaching solar conjunction, made additional observations impossible. P10, Ho-Peng-Yoke and Ang Tian-Se (83), Petit (1665), Maunder (1934).

1665 March 27; (P = April 24.7, d = 0.57 on April 4) China, Korea, Japan, Europe. The comet of 1665 was observed extensively throughout

420

Europe from March 27 through April 20, when it came too close to the Sun for further observations. Hevelius observed it from April 6 through 20. Chinese observers first noted the comet on March 28 in northwestern Aquarius. On April 13 it was in eastern Pegasus with its tail measuring more than 7 degrees. On April 17 it entered northern Pisces. P22, Ho-Peng-Yoke and Ang Tian-Se (84).

1666 February; Korea. A comet was seen in the winter by Korean observers and an observer in Ceylon, now Sri Lanka, made note of a comet in February. P23, Hasegawa (988).

1668 March 3; (P = February 28.1, d = 0.80 on March 5) China, Korea, Japan, India, Africa, Europe. The comet was first seen at the Cape of Good Hope on March 3 and two days later in Brazil and in Lisbon, Portugal. It was most easily visible to observers in southern latitudes. On March 10 J.D. Cassini, in Bologna, Italy noted it extending from Cetus to the middle of Eridanus, some 30 degrees in length. In Goa, India positions of its head were mapped from March 9 through March 21, and at the Cape of Good Hope from March 3 through 23. Chinese observers noted the comet from March 7 through March 30, when it went out of sight. On March 7 it was described as a stretch of white light in the southwest, measuring over 9 degrees and pointing toward the southeast. On March 18 it extended over 40 degrees in Eridanus. This comet is probably a member of the Kreutz group of sungrazing comets. P22, Ho-Peng-Yoke and Ang Tian-Se (85), Henderson (1843), Marsden (1967, 1989).

1672 March 2; (P = March 1.9, d = 1.02 on March 15) Europe. The comet was discovered by Hevelius in Pegasus on March 2 and followed by him until April 22, when it was in northern Orion. Cassini followed it from March 26 to April 7. P23, Berberich (1888).

1673 March 10; China. A strange star appeared in western Aries. It was white, like a peach in size, and with a short tail pointing to the east. It was also seen the next night. Ho-Peng-Yoke and Ang Tian-Se (86), Hasegawa (991).

1676 February 14; China, Europe. A comet was seen by a French Jesuit on February 14 in Eridanus. It was last seen on March 9. On February 18, the Chinese reported an extraordinary white star in northern Eridanus. P23, Ho-Peng-Yoke and Ang Tian-Se (87).

1677 April 27; (P = May 6.5, d = 0.54 on April 17) Europe. The comet was discovered by Hevelius at Danzig on April 27 and followed by him until May 8, when it entered into morning twilight. On May 3 John Flamsteed reported a tail length of 6 degrees. P24.

1678 September 11; (P = August 27.1, d = 0.26 on August 28) Europe. Philippe de La Hire first detected this comet in Aquarius. It was seen for the last time on October 7 in Pisces. P24.

1680 November 14; (P = December 18.5, d = 0.42 on November 30) Europe, China, Japan, Korea, North America. This first telescopic discovery of a comet was made by the German astronomer Gottfried Kirch on the morning of November 14. It remained a morning object until the first few days of December, when it entered into solar conjunction. On December 18 it was seen at noon in the Philippines, less than 2 degrees from the Sun. John Flamsteed, at Greenwich, first detected its tail in the evening sky on December 20, with the head being observable two days later as it exited the solar glare. Extensive observations were made by many European observers. In the British colony of Maryland in North America, Arthur Storer noted its tail was 15 to 20 degrees in length on November 29. The Chinese reported the comet first on November 23 in Crater. It had a white tail more than 1 degree long pointing toward the west. On December 21, the Chinese observed it coming out of solar conjunction with a darkish-white color and a tail over 60 degrees long pointing toward the northeast. The last observations were made by Isaac Newton on March 19, 1681. Its motion from discovery to last observation was generally eastward, passing nearly around the entire sky from western Leo into Auriga. P25, Ho-Peng-Yoke and Ang Tian-Se (88), Newton (1687), Broughton (1988), Selga (1930).

1682 August 15; (P = September 15.3, d = 0.42 on August 31) Europe, China, Africa. Comet Halley. The comet was first seen in England on August 15; by the Jesuits of Orleans, France on the night of August 23; by Arthur Storer in the British colony of Maryland from August 24 through September 22; by Cassini, Abbé Picard, and Philippe de La Hire in Paris, August 25 through September 21; by Samuel Dörffel in Plauen, Germany August 25 through September 20; by Hevelius in Danzig, Poland August 26 through September 13; and by John Flamsteed at Greenwich, England August 30 through September 19. Halley himself observed the comet that was to bear his name from September 5 to 19. When discovered, the comet was above Gemini. It then moved in a southeast direction going out of sight in eastern Virgo. The Chinese noted its tail as more than 3 degrees on August 26, growing to 9 degrees three days later. Although Flamsteed's astrometric positions are the most accurate, having a mean error of approximately 13 arc minutes, the observations by Storer were surprisingly good, achieving an accuracy of 38 and 13 arc minutes in longitude and latitude. Though his observing equipment was crude, his accuracy exceeded that of Hevelius and Halley himself. At the Cape of Good Hope, the comet was discovered on September 8 and followed until September 24, when it was last observed.

P28, Halley (1752), Ho-Peng-Yoke and Ang Tian-Se (89), Broughton (1988), McIntyre (1949).

1683 July 23; (P = July 13.6, d = 0.32 on September 4) Europe, China. John Flamsteed first observed the comet on July 23 and followed it until September 5. Hevelius observed it from July 30 to September 4. The Chinese discovered the comet on August 2 in Auriga moving slowly toward the southwest. P28, Ho-Peng-Yoke and Ang Tian-Se (90).

1684 July 1; (P = June 8.8, d = 0.17 on June 27) China, Europe. At Rome, Francesco Bianchini observed this comet from July 1, in Virgo, through July 17 when it was in Boötes. On July 1 the Chinese noted the comet as a bright white star in Corvus. P28, Ho-Peng-Yoke and Ang Tian-Se (91).

1686 August 12; (P = September 16, d = 0.32 on August 16) China, Korea, Africa, South America. First seen at the Cape of Good Hope, the comet's position was put at the left shoulder of the Hare, Lepus, with its 35-degree tail stretching into Gemini. Two days later, in Brazil, the comet was noted just below the belt of Orion as bright as a first-magnitude star with a tail 18 degrees in length. On August 16 and 17, Jesuits in Siam (Thailand) reported the tail length as 15 degrees. The comet was observed until September 22. Chinese observers reported that on September 3 it moved 10 degrees and when it reached southern Cancer showed traces of a tail. Korean observations put the comet in northwestern Hydra on September 14. P28, Ho-Peng-Yoke and Ang Tian-Se (92), Hasegawa (999), McIntyre (1949).

1688 November 2; China. Chinese observers noted that an extraordinary white star appeared in southern Andromeda for three days. Ho-Peng-Yoke and Ang Tian-Se (93).

1689 November 24; (P = November 30.7, d = 0.74 on December 14) Africa, India. At the Cape of Good Hope a tailed star was seen in the southeast morning sky on November 24 and 25. It was again seen in the morning sky on December 9 with a tail more than 4 degrees long. On December 8 the comet's tail in the arms of Centaurus was noted by Father Richaud in Pondicherry, India. The comet moved in a southern direction with its tail reaching a length of 60 degrees before going out of sight in the first few days of January 1690. This comet may be a member of the Kreutz sungrazing comet group. P29, McIntyre (1949), Marsden (1967).

1690 September–October; China. Chinese observers noted a new yellow star in Sagittarius that lasted two nights. This is probably a nova rather than a comet. Hasegawa (1002) gives a date of October 23 while Ho-Peng-Yoke and Ang Tian-Se (94) prefer a date of September 29.

1695 October 28; South America. While in Brazil, the French Jesuit P.

Jacob discovered this comet's tail on October 28 an hour before sunrise. Two days later the tail was described as reaching Virgo with the comet's head in Libra. The tail reached a length of 30 to 40 degrees and the comet was last reported on November 19, just before full moon. If this comet is a member of the Kreutz sungrazing comet group, its perihelion passage time would be October 23.1. P33, Marsden (1967).

1698 September 2; (P = October 17.5, d = 0.21 on September 7) Europe. Philippe de La Hire and Jean-Dominique Cassini observed this comet first between β and κ Cassiopeia on September 2 and followed it as it first moved west then rapidly south. In mid-September, the Moon's brightness interfered with observations. When last observed on September 28 it was in Scorpius. P36, Hind (1876).

1699 February 17; (P = January 13.9, d = 0.18 on February 18) China, Europe. This comet was first observed in the northern heavens by the Jesuit missionary De Fontenay in Peking, China. He followed its southeastern motion until February 26 when it was in northern Orion. The comet was also followed by Cassini and Maraldi I at Paris from February 20 until March 2. P36, Hind (1879).

1699 October 26; (P = October 10.9, d = 0.07 on October 27) Germany. Gottfried Kirch observed a faint, but still naked-eye nebulosity in Argo moving in a generally southward direction. He was unable to locate the object the next morning. This single observation has been identified as a return of the short period comet Tempel-Tuttle. P36, Schubart (1965), Yeomans (1981).

Bibliography

Chapter 1

Aristotle's *Meteorologica*

Barnes, Jonathan, ed. 1984. Meteorology. *In* Aristotle's Complete Works. The revised Oxford translation. Volume 1. Princeton, N.J.: Princeton University Press. (Bollingen series, v. 71:2) pp. 555–625.

West, M.L. 1960. Anaxagoras and the meteorite of 467 B.C. British Astronomical Association Journal 70(Oct.):368–369.

Seneca's *Natural Questions*

Clarke, John. 1910. Physical Science in the Time of Nero; being a translation of the Quaestiones naturales of Seneca. London: Macmillan. liv, 368 p.

Corcoran, Thomas H., trans. 1972. Naturales quaestiones of Lucius Annaeus Seneca, Volume 2. Cambridge, Mass.: Harvard University Press (The Loeb Classical Library). 312 p.

The *Natural History* of Pliny the Elder

Rackham, H., trans. 1938. Pliny Natural History, Volume 1. Cambridge, Mass.: Harvard University Press (The Loeb Classical Library). 378 p.

The *Tetrabiblos* of Ptolemy and the *Centiloquy*

(Ptolemy) 1822. Tetrabiblos, or Quadripartite. Being four books of the influence of the stars ... Containing extracts from the Almagest ... and the whole of his Centiloquy. London: Davis and Dickson. xxvii, 240 p.

General Works

Burke, John G. 1986. Cosmic Debris: Meteorites in History. Berkeley: University of California Press. 445 p.

Haney, Herbert L. 1965. Comets: A Chapter in Science and Superstition in Three Golden Ages—the Aristotelian, the Newtonian, and the Thermonuclear. Ph.D. Thesis, University of Alabama. 410 pp.

Hellman, Clarice Doris. 1971. The Comet of 1577: Its Place in the History of Astronomy. Reprint of the 1944 edition with addenda, errata, and a supplement to the appendix. New York: AMS Press. 488 p.

Pingré, Alexandre Guy. 1783–1784. Cométographie, ou traité historique et théorique des comètes. Paris, 2 volumes.

Ruffner, James A. 1966. The Background and Early Development of Newton's Theory of Comets. Ph.D. Thesis, Indiana University, Bloomington. 363 p.

Chapter 2

The Period to about 1200: Theology and Superstition

Barker, Peter and Goldstein, Bernard R. 1988. The role of comets in the Copernican revolution. Studies in History and Philosophy of Science 19:299–319.

Celichius, Andreas. 1578. Theologische erinnerung von dem neuen cometen. Magdeburg, Germany.

Dudith, Andreas. 1579. De cometarum significatione commentariolius. Basel, Switzerland.

Hellman, Clarice Doris. 1971. The Comet of 1577: Its Place in the History of Astronomy. Reprint of the 1944 edition with addenda, errata, and a supplement to the appendix. New York: AMS Press. 488 p.

Pedersen, O. and Pihl, M. 1974. Early Physics and Astronomy. New York: American Elsevier. 413 p.

Pingré, Alexandre Guy. 1783–1784. Cométographie, ou traité historique et théorique des comètes. 2 volumes. Paris.

Thorndike, Lynn. 1950. Latin Treatises on Comets between 1238 and 1368 A.D. Chicago: University of Chicago Press. 274 p.

Vigenère, Blaise de. 1578. Traicté des cometes, ou estoilles chevelues, apparoissantes extraordinairement au ciel. Paris. 171 p.

White, Andrew D. 1910. A History of the Warfare of Science with Theology in Christendom. New York: D. Appleton and Co. 2 volumes.

1200–1577—The Rebirth of Scientific Observations

Apian, Peter. 1531. Practica auff dz. 1532 jar. . . . Landshut, Germany.

Apian, Peter. 1540. Astronomicum Caesareum. Ingolstadt, Germany.

Ashbrook, Joseph. 1965. Tycho Brahe's nose. Sky and Telescope 29 (June):353 and 358.

Beaver, Donald deB. 1970. Bernard Walther: Innovator in astronomical observation. Journal for the History of Astronomy 1:39–43.

Cardano, Girolamo. 1663. Opera Omnia. Lyons, France.

Celoria, Giovanni. 1884. Sull' Apparizione della cometa di Halley avvenata nell anno 1456. Astronomische nachrichten 111:65–72.

Celoria, Giovanni. 1921. Sulle osservazioni di comete Fatte da Paolo dal Tosconelli . . . In Pubblicazioni de . . . osservatorio di Brera, No. 55, Milan, Italy.

References

Copernicus, Nicholas. 1543. De revolutionibus orbium coelestium. Nuremberg, Germany.

Curtze, M. 1878. Inedita Copernicana. Leipzig, Germany.

Digges, Leonard. 1576. A Prognostication Euerlasting. London.

Fracastoro, Girolamo. 1538. Homocentricorum, sive de stellis. Venice, Italy.

Gingerich, Owen. 1971. Apianus's Astronomicum Caesareum and Its Leipzig Facsimile. Journal for the History of Astronomy 2:168–177.

Hagecius, Thaddaeus. 1574. Dialexis de novae et prius incognitae stellae inusitatae magnitudinis & splendidissimi luminis apparitione. Frankfurt. Facsimile reprint, Prague, 1967. pp. 150–167.

Halley, Edmond. 1752. Astronomical Tables. With the precepts both in English and Latin. London.

Hellman, Clarice Doris. 1971. The Comet of 1577: Its Place in the History of Astronomy. Reprint of the 1944 edition with addenda, errata, and a supplement to the appendix. New York: AMS Press. 488 p.

Jervis, Jane L. 1980. Vögelin on the comet of 1532: Error analysis in the 16th century. Centaurus 23(3):216–229.

Jervis, Jane L. 1985. Cometary Theory in Fifteenth-century Europe. Dordrecht, Netherlands: D. Reidel Publishing Co. 209 p.

Mizauld, Antoine. 1549. Cometographia crinitarum stellarum. Paris.

Pingré, Alexandre Guy. 1783–1784. Cométographie, ou traité historique et théorique des comètes. 2 volumes. Paris.

Pogo, Alexander. 1934. Earliest diagrams showing the axis of a comet tail coinciding with the radius vector. Isis 20:443.

Regiomontanus (Johannes Müller). 1531. Ioannis de Monteregio Germani. Viri undecunque doctissimi, de comete magnitudine, longitudineque ac de loco eius vero, problemata XVI. Nürnberg, Germany.

Ruffner, James A. 1966. The Background and Early Development of Newton's Theory of Comets. Ph.D. Thesis, Indiana University, Bloomington. 363 p.

Thorndike, Lynn. 1945. Peter of Limoges on the Comet of 1299. Isis 36:3–6.

Thorndike, Lynn. 1950. Latin Treatises on Comets between 1238 and 1368 A.D. Chicago: University of Chicago Press. 274 p.

Vögelin, Johannes. 1533. Significatio cometae qui anno 1532 apparuit. Vienna.

Zinner, Ernst. 1968. Leben und wirken des Johannes Müller von Königsberg, genannt Regiomontanus. 2nd ed. Osnabrück, Germany.

The Comet of 1577

Brahe, Tycho. 1588. Tychonis Brahe Dani de mundi: aetherei recentioribus phaenomenis liber secundus qui est de illustri stella caudata ab elapso fere triente Nouembris anni 1577, usg: in finem ianuarij sequentis conspecta. Uraniborg.

Christianson, J.R. 1979. Tycho Brahe's German treatise on the comet of 1577: A study in science and politics. Isis 70:110–140.

Dreyer, Johann Louis Emil, Raeder, J., and Nystrom, E. 1922. Tychonis Brahe Dani opera omnia, Volume 4. Copenhagen. pp. 379–396.

Dreyer, Johann Louis Emil. 1890. Tycho Brahe: A Picture of Scientific Life and Work in the Sixteenth Century. Edinburgh, Scotland: Adam and Charles Black.

Gemma, Cornelius. 1578. De prodigiosa specie, naturaq, cometae, qui nobis effulsit altior lunae sedibus, insolita prorsus figura, ac magnitudine, anno 1577. Antwerp, Belgium.

Gingerich, Owen. 1977. Tycho Brahe and the great comet of 1577. Sky and Telescope 54(Dec.):452–458.

Hagecius, Thaddaeus. 1578. Descriptio cometae, qui apparuit Anno Domini M.D. LXXVII. Prague, Czechoslovakia.

Hagecius, Thaddaeus. 1580. Thaddaei Hagecii ab hayck epistola ad Martinum Mylium. Görlitz, Germany.

Hellman, Clarice Doris. 1971. The Comet of 1577: Its Place in the History of Astronomy. Reprint of the 1944 edition with addenda, errata, and a supplement to the appendix. New York: AMS Press. 488 p.

Mästlin, Michael. 1578. Observatio et demonstratio cometae aetherei, qui anno 1577. Tübingen, Germany.

Pedersen, O. 1976. Some early European observatories. Vistas of Astronomy 20:17–28.

Roeslin, Helisaeus. 1578. Theoria nova coelestium. Strasbourg, France.

Ruffner, James A. 1966. The background and early development of Newton's theory of comets. Ph.D. Thesis Indiana University, Bloomington. 363 p.

Westman, Robert S. 1972. The comet and the cosmos: Kepler, Mästlin and the Copernican hypothesis. Studia Copernicana, volume 5. Wroclaw, Poland. pp. 7–30.

Westman, Robert S. 1975. Michael Mästlin's adoption of the Copernican theory. Studia Copernicana 13:53–63. Wroclaw.

White, Andrew D. 1910. A history of the warfare of science with theology in Christendom. 2 volumes. New York: D. Appleton and Co.

Early Chinese Contributions to the History of Comets

Ho Peng Yoke. 1964. Ancient and medieval observations of comets and novae in Chinese sources. Vistas in Astronomy 5:127–225.

Kiang, Tao. 1971. The past orbit of Halley's comet. Royal Astronomical Society of London, Memoirs. 76:27–66.

Needham, Joseph. 1959. Science and Civilisation in China. Volume 3. New York: Cambridge University Press.

Needham, Joseph, Beer, Arthur, and Ho Ping Yü. 1957. Spiked comets in ancient China. Observatory 77:137–138.

Xi Ze-zong. 1984. The cometary atlas in the silk book of the Han tomb at Mawangdui. Chinese Astronomy and Astrophysics 8:1–7.

General Works

Johnson, Francis R. 1968. Astronomical Thought in Renaissance England. Reprint of 1937 edition. New York: Octagon Books.

Thorndike, Lynn. 1923–1958. A History of Magic and Experimental Science. 8 volumes. New York: Columbia University Press.

Chapter 3

The Ephemeral Comets of Johannes Kepler

Armitage, Angus. 1966. John Kepler. London: Farber and Farber. 194 p.

Caspar, Max. 1959. Kepler. Translated from the 1948 German edition by C. Doris Hellman. London and New York: Abelard-Schuman. 401 p.

Hellman, Clarice Doris. 1975. Kepler and comets. Vistas of Astronomy 18: 789–796.

Kepler, Johannes. 1604. Ad vitellionem paralipomena. Quibus astronomiae pars optica traditur. . . . Frankfurt, Germany.

Kepler, Johannes. 1608. Aussführlicher bericht von dem newlichen in monat Septembri und Octobri diss 1607. Halle, Germany.

Kepler, Johannes. 1619. De cometis libelli tres. Augsburg, Germany.

Ruffner, James A. 1971. The curved and the straight: Cometary theory from Kepler to Hevelius. Journal for the History of Astronomy 2:178–194.

Shagrin, Morton L. 1974. Early observations and calculations on light pressure. American Journal of Physics 42:927–940.

Westman, Robert S. 1972. The comet and the cosmos: Kepler, Mästlin and the Copernican hypothesis. Studia Copernicana 5:7–30. Wroclaw, Poland.

Westman, Robert S. 1975. Michael Mästlin's adoption of the Copernican theory. Studia Copernicana 13:53–63. Wroclaw, Poland.

The Devil's Advocate, Galileo Galilei

Drake, Stillman and O'Malley, C.D. 1960. The controversy on the comets of 1618. Philadelphia: University of Pennsylvania Press. 380 p.

Galilei, Galileo. 1623. Il saggiatore nel quale con bilancia esquisita e giusta si ponderano le cose contenute nella libra astronomica e filosofica di Lothario Sarsi. Rome.

Grassi, Horatio. 1619. De tribus cometis annus MDC XVIII. Rome.

Grassi, Horatio. 1619. Libra astronomica ac philosophica qua Galilaei Galilaei opiniones de cometis a Mario Guiducio. . . . Perugia, Italy.

Grassi, Horatio. 1626. Ratio ponderum librae et simbellae. Paris.

Guiducci, Mario. 1619. Discorso delle comete. Florence, Italy.

Guiducci, Mario. 1620. Lettera al M.R.P.T. Galluzzi . . . nella quale si giustifica dell imputazioni dateqli da L. Sarsi Sigensano nella libra astronomica, e filosofica. Florence, Italy.

Kepler, Johannes. 1625. Tychonis Brahei Dani hyperaspistes. Frankfurt, Germany.

Ruffner, James A. 1971. The curved and the straight: Cometary theory from Kepler to Hevelius. Journal for the History of Astronomy 2:178–194.

429

Early Seventeenth Century Views on Comets

Bainbridge, John. 1619. An astronomicall description of the late comet from the 18. of Novemb. 1618 to the 16. of December following. London. 42 p.

Cysat, Johannes B. 1619. Mathemata astronomica de loco, motu, magnitudine et causis cometae qui sub finem anni 1618 et initium anni 1619 in coelo fulsit. Ingolstadt, Germany.

Descartes, René du Perron. 1824. Oeuvres, Volume 3. Paris.

Gassendi, Pierre. 1658. Syntagma philosophicum. 8 volumes. Lyon, France.

Rigaud, Stephen P. 1972. Supplement to Dr. Bradley's miscellaneous works. Reprint of 1833 Oxford edition. Sources of Science Series, No. 97. New York: Johnson Reprint Corporation.

Ruffner, James A. 1966. The Background and Early Development of Newton's Theory of Comets. Ph.D. Thesis, Indiana University, Bloomington. 363 p.

Shirley, John W. 1983. Thomas Harriot: A Biography. Oxford: Clarendon Press.

Snel, Willebrord. 1619. Descriptio cometae. Qui anno 1618 mense Novembri primum effulsit. Huc accessit Christophori Rhotmanni descriptio accurata cometae anni 1585. . . . Leiden, Netherlands.

Ward, Seth. 1653. Idea trigonometriae demonstratae . . . item praelectio de cometis, et inquisitio in Bullialdi astronomiae philolaicae fundamenta. Oxford, England.

General Works

Armitage, Angus. 1951. Master Georg Dörffel and the rise of cometary astronomy. Annals of Science 7(4):303–315.

Guillemin, Amédée. 1877. The World of Comets. Translated and edited by James Glaisher. London. 548 p.

Johnson, Francis R. 1968. Astronomical Thought in Renaissance England. Reprint of 1937 edition. New York: Octagon Books. 357 p.

Pingré, Alexandre Guy. 1783–1784. Cométographie, ou traité historique et théorique des comètes. 2 volumes. Paris.

Van Helden, Albert. 1977. The invention of the telescope. American Philosophical Society Transactions 67: part 4.

Chapter 4

Closed Orbits: The Comet of 1664 as a Permanent Object

Armitage, Angus. 1950. Borell's hypothesis and the rise of celestial mechanics. Annals of Science 6:268–282.

Auzout, Adrien. Letter dated March 3, 1664/1665 and letter to Pierre Petit dated June 17, 1665. *In* The Correspondence of Henry Oldenburg, edited by A.R. Hall and M.B. Hall. Volume 1. Cambridge.

Borrelli, Giovanni Alfonso. 1665. Del movimento della cometa apparasa il mese di Dicembre 1664. Written under the pseudonym Pier Maria Mutoli. Pisa, Italy.

Cassini, Jean-Dominique. 1966. Hypothesis motus cometae novissimi, received at the Royal Society February 1665. *In* The Correspondence of Henry Oldenburg, Volume 1, edited by A.R. Hall and M.B. Hall. Document 369b. Cambridge.

Comiers, Claude. 1665. La nature et presages des comètes. Lyon, France.

Danforth, Samuel. 1665. Astronomical Description of the Late Comet or Blazing Star. Cambridge, Massachusetts.

Delambre, Jean-Baptiste J. 1821. Histoire de l'astronomie moderne, Volume 2, pp. 686–804. Paris.

Guillemin, Amédée. 1877. The World of Comets. Translated and edited by James Glaisher. London: 548 p.

Petit, Pierre. 1665. Dissertation sur la nature de comètes. Paris.

Pingré, Alexandre Guy. 1783–1784. Cométographie, ou traité historique et théorique des comètes. 2 volumes. Paris.

Ruffner, James A. 1966. The Background and Early Development of Newton's Theory of Comets. Ph.D. Thesis, Indiana University, Bloomington. 363 p.

Yeomans, Donald K. 1977. The origin of North American astronomy—seventeenth century. Isis 68:414–425.

Zach, Baron von. 1817. Litterarischer fund von Herrn Oberhofmeister Freiherrn von Zach. Zeitschrift für astronomie 3:347–348.

Rectilinear Orbits: The Comet of 1664 as a Transitory Object

Anonymous. 1665. An account of Hevelius, his Prodromus Cometicus together with some animadversions made upon it by a French philosopher. Philosophical Transactions of the Royal Society of London 1(6):150–151.

Bennett, J.A. 1975. Hooke and Wren and the system of the world. British Journal for the History of Science 8:32–61.

Boulliau, Ismael. 1665. Astronomica philolaica. Paris.

Glorioso, Joannes C. 1624. De cometis dissertatio astronomico—physica publica habita in gymnasio patavino Anno Domini MDCXIX. Venice, Italy.

Hevelius, Johannes. 1647. Selenographia: sive, lunae descriptio. Danzig, Poland.

Hevelius, Johannes. 1665. Prodromus cometicus quo historia cometae anno 1664 exorti . . . nec non dissertatio de cometarum omnium motu. Danzig, Poland.

Hevelius, Johannes. 1666. Descriptio cometae anno aerae Christianae 1665 exorti . . . cui addita est: mantissa prodromi cometici. Observationes omnes prioris cometae 1664. . . . Danzig, Poland.

Hevelius, Johannes. 1668. Cometographia, totam naturam cometarum . . . hypotheseos exhibens. Danzig, Poland.

Hooke, Robert. 1678. Cometa or remarks about comets. A 1678 paper published in the Cutler Lectures of Robert Hooke. *In* Early Science in Oxford, by R.T. Guther. Volume 8. pp. 217–271.

Huygens, Christiaan. Oeuvres complète de Christiaan Huygens. Publiées par la société Hollandaise des sciences. 22 volumes, 1888–1950. The Hague, Netherlands. For material on comets, see especially 1664–1665 correspondence in Volume 5.

Pingré, Alexandre Guy. 1783–1784. Cométographie, ou traité historique et théorique des comètes. 2 volumes. Paris.

Ruffner, James A. The Background and Early Development of Newton's Theory of Comets. Ph.D. Thesis, Indiana University, Bloomington. 363 p.

Von Hohenlohe, A., Volkoff, I., Franzgrote, E., and Larsen, A.D. 1971. Johannes Hevelius and His Catalog of Stars. Provo, Utah: Brigham Young University Press.

Chapter 5

Brewster, David. 1835. Memoirs of the Life, Writings, and Discoveries of Sir Isaac Newton. Volume 1. Edinburgh, p. 11.

Kirch and Dörffel

Armitage, Angus. 1951. Master Georg Dörffel and the rise of cometary astronomy. Annals of Science 7:303–315.

Dörffel, Georg Samuel. 1672. Bericht von dem neulichsten in mertzen dieses 1672 jahres erschienenem cometen. . . . Plauen, Germany.

Dörffel, Georg Samuel. 1672. Wahrhafftiger bericht von dem cometen. . . . Plauen, Germany.

Dörffel, Georg Samuel. 1680. Neuer comet-stern welcher im November des 1680-sten jahres erschienen. . . . Plauen, Germany.

Dörffel, Georg Samuel. 1681. Astronomische betrachtung des grossen cometen. . . . Plauen, Germany.

Kirch, Gottfried. 1681. Neue himmels-zeitung darinnen sonderlich und ausführlich von den zweyen neuen grossen im 1680. . . . Nürnberg, Germany.

Newton and Flamsteed

Baily, Francis, ed. 1835. An Account of the Revd. John Flamsteed, the First Astronomer Royal. London: Reprinted by Dawsons of Pall Mall. 1966.

Flamsteed, John. 1680–1681. Letter, Flamsteed to Crompton, 12 February. *In* The Correspondence of Isaac Newton, Volume 2, edited by H.W. Turnbull. Cambridge: Cambridge University Press, 1960. p. 336.

Flamsteed, John. 1680–1681. Letter, Flamsteed to Halley, 17 February. *In* The Correspondence of Isaac Newton, Volume 2, edited by H.W. Turnbull. Cambridge: Cambridge University Press, 1960. pp. 336–339.

Flamsteed, John. 1680–1681. Letter, Flamsteed to Newton, 7 March. *In* The Correspondence of Isaac Newton, Volume 2, edited by H.W. Turnbull. Cambridge: Cambridge University Press, 1960. pp. 348–356.

Halley, Edmond. 1687. Letter, Halley to Newton, 5 April. *In* The Correspondence of Isaac Newton, Volume 2, edited by H.W. Turnbull. Cambridge: Cambridge University Press, 1960. p. 474.

Kriloff, A.N. 1925. On Sir Isaac Newton's method of determining the parabolic orbit of a comet. Monthly Notices of the Royal Astronomical Society of London 85:640–656.

Newton, Isaac. 1680–1681. Letter, Newton to Flamsteed, 28 February. *In* The Correspondence of Isaac Newton, Volume 2, edited by H.W. Turnbull. Cambridge: Cambridge University Press, 1960. pp. 340–347.

Newton, Isaac. 1681. Letter, Newton to Crompton, April. *In* The correspondence of Isaac Newton, Volume 2, edited by H.W. Turnbull. Cambridge: Cambridge University Press, 1960. pp. 358–362.

Newton, Isaac. 1681. Letter, Newton to Flamsteed, 16 April. *In* The correspondence of Isaac Newton, Volume 2, edited by H.W. Turnbull. Cambridge: Cambridge University Press, 1960. pp. 363–367.

Newton, Isaac. 1687. Philosophiae naturalis principia mathematica, London.

Summary

Newton, Isaac. 1676. Letter, Newton to Hooke, 5 February. *In* The correspondence of Isaac Newton, Volume 2, edited by H.W. Turnbull. Cambridge: Cambridge University Press, 1960.

General Works

Robinson, James H. 1916. The Great Comet of 1680: A Study in the History of Rationalism. Northfield, Minnesota: Northfield News Press. 126 p.

Ruffner, James A. 1966. The Background and Early Development of Newton's Theory of Comets. Ph.D. Thesis, Indiana University, Bloomington. 363 p.

Westfall, Richard S. 1980. Never at Rest: A Biography of Isaac Newton. Cambridge: Cambridge University Press. 908 p.

Chapter 6

The Predicted Return of Comet Halley

Baily, Francis, ed. 1835. An Account of the Revd. John Flamsteed, the First Astronomer Royal. London: Reprinted by Dawsons of Pall Mall. 1966.

DeMorgan, Augustus. 1847. Cabinet Portrait Gallery of British Worthies. Volume 12. London: p. 10.

Flamsteed, John. 1702. Letter, Flamsteed to Mr. A. Sharp, 30 May. *Abstract published in* An Account of the Revd. John Flamsteed, the First Astronomer Royal, edited by Francis Baily. London: Reprinted by Dawsons of Pall Mall. 1966. p. 203.

Gregory, David. 1715. The Elements of Astronomy, Physical and Geometrical. London.

Halley, Edmond. 1695. Letter, Halley to Newton, 7 September. *In* Correspondence and Papers of Edmond Halley, edited by Eugene F. MacPike. London: Taylor and Francis. 1937. p. 91.

Halley, Edmond. 1695. Letter, Halley to Newton, 28 September. *In* Correspondence and Papers of Edmond Halley, edited by Eugene F. MacPike. London: Taylor and Francis. 1937. pp. 91–92.

Halley, Edmond. 1695. Letter, Halley to Newton, early October. *In* Correspondence and Papers of Edmond Halley, edited by Eugene F. MacPike. London: Taylor and Francis. 1937. pp. 92–94.

Halley, Edmond. 1695. Letter, Halley to Newton, 15 October. *In* Correspondence and Papers of Edmond Halley, edited by Eugene F. MacPike. London: Taylor and Francis. 1937. p. 95.

Halley, Edmond. 1695. Letter, Halley to Newton, 21 October. *In* Correspondence and Papers of Edmond Halley, edited by Eugene F. MacPike. London: Taylor and Francis. 1937. pp. 95–96.

Halley, Edmond. Letter, Halley to Newton, 1696(?). *In* Correspondence and Papers of Edmond Halley, edited by Eugene F. MacPike. London: Taylor and Francis. 1937. pp. 96–97.

Halley, Edmond. Entry in Journal Book of Royal Society for 3 June. In Correspondence and Papers of Edmond Halley, edited by Eugene F. MacPike. London: Taylor and Francis. 1937. p. 238.

Halley, Edmond. 1696. Entry in Journal Book of Royal Society for 1 July. *In* Correspondence and papers of Edmond Halley, edited by Eugene F. MacPike. London: Taylor and Francis. 1937. p. 238.

Halley, Edmond. 1705. Astronomiae cometicae synopsis. Oxford.

Halley, Edmond. 1705. A Synopsis of the Astronomy of Comets. London.

Halley, Edmond. 1749. Tabulae astronomicae. London.

Halley, Edmond. 1752. Astronomical Tables. London.

Hughes, David W. 1984. Edmond Halley's observations of Halley's comet. Journal for the History of Astronomy. 15:189–197.

Hughes, David W. The accuracy of Halley's cometary orbits. Vistas of Astronomy 28:585–593.

Hughes, David W. 1987. The history of Halley's comet. Philosophical Transactions of the Royal Society of London 323:349–367.

Thrower, Norman J.W., ed. 1981. The Three Voyages of Edmond Halley in the *Paramore* 1698–1701. Second series, No. 156, The Hakluyt Society, London.

The Race for Recovery

Barker, Thomas. 1755. Extracts of a letter of Thomas Barker, Esq., to the Reverend James Bradley, D.D. Astronomer Royal, and F.R.S. concerning the return of the comet expected in 1757, or 1758. Philosophical Transactions of the Royal Society of London 49(pt. 1):347–350.

Cheseaux, Jean Philippe Loys de. 1744. Traité de la comète qui a paru en Décembre 1743 & Janvier, Fevrier & Mars 1744. Lausanne and Geneva.

Clairaut, Alexis C. 1759. Mémoire sur la comète de 1682. . . . *In* Journal des sçavans. January. Paris. pp. 38–45.

Clairaut, Alexis C. 1760. Théorie du mouvement des comètes. . . . Paris.

Clairaut, Alexis C. 1762. Recherches sur la comète des années 1531, 1607, 1682 et 1759. . . . St. Petersburg.

Euler, Jean-Albert. 1762. Meditationes de perturbatione motus cometarum ab attractione planetarum orta. St. Petersburg.

Euler, Leonhard. 1746. Opuscula varii argumenti. Berlin. p. 276.

Halley, Edmond. 1759. Tables Astronomiques de M. Halley, pour les planetes et les comètes . . . et l'histoire de la comète de 1759. Edited by J.-J. Lalande. Paris.

Jamard, T.J. 1757. Mémoire sur la comète qui a été observée en 1531, 1607, 1682, & que l'on attend en 1757, ou plutard en 1758. Paris.

Kiang, Tao. 1971. The past orbit of Halley's comet. *In* Royal Astronomical Society Memoirs, London. 76:27–66.

Lalande, Joseph-Jérôme de. 1757. Lettre au subject de la comète dont on attend le retour. *In* Mémoires pour l'histoire des sciences et beaux-arts. November, Paris. pp. 2850–2863.

Lalande, Joseph-Jérôme de. 1771. Astronomie. Volume 3. Paris. p. 380.

Taton, René. 1979. Clairaut et le retour de la comète de Halley. *In* Arithmos-Arrythmos skizzen aus der Wissenschaftsgeschichte; festschrift für Joachim Otto Fleckenstein zum 65 geburtstag, edited by Karin Figala and Ernst H. Berninger. Munich: Minerva Publikation.

Waff, Craig B. 1990. The first International Halley Watch: Guiding the worldwide search for comet Halley. *In* Standing on the Shoulders of Giants: A Longer View of Newton and Halley, edited by Norman J.W. Thrower. Los Angeles: University of California Press. pp. 373–411.

Waff, Craig B. and Skinner, Stephen. 1987. Tales from the first International Halley Watch (1755–1759): 2. Thomas Stevenson of Barbados and his two-comet theory. *In* International Halley Watch Newsletter No. 10, edited by Stephen J. Edberg. Jet Propulsion Laboratory publication 410-20-10. Pasadena. pp. 3–9.

Recovery at Last

Anonymous. 1759. Anzeige dass der im jahr 1682. Erschienene und von Halley nach der Newtonianischen theorie auf gegenwärtige zeit vorherverhundigte comet wirklich sichtbar sey; und was derselbe in der folge der zeit für erscheinungen haben werde, von einem liebhaber der sternwissenschaft. Leipzig, published by Bernhard Christoph Breitkopf, January 24.

Delisle, Joseph-Nicolas. 1759. Letter dated 16 June. *In* Journal des Savans, August 1759 volume. Paris.

Delisle, Joseph-Nicolas. 1759. Lettre de M. de l'Isle, de l'Académie Royale des Sciences, &c., a l'auteur du Mercure, sur le retour de la comète de 1682. . . . *In* Mercure de France. Volume 1, July, Paris. pp.148–156.

Delisle, Joseph-Nicolas. 1766. Sur la comète de 1759, ou le retour de celle de 1682. *In* Mémoires de mathématique et de physique, tirés des registres de l'Académie Royale des Sciences, de l'année 1760. Paris. pp. 380–465.

Heinsius, Gottfried. 1759. Letter to Leonhard Euler dated April 21. Abstract of letter published in Die Berliner und die Petersburger Akademie der Wissenschaften im Briefwechsel Leonard Eulers, edited by A.P. Juskevic and E. Winter. Teil 3. Berlin: Akademie-Verlag. 1976. pp. 120–121.

Jones, Kenneth Glyn. 1969. Messier's nebulae and star clusters. New York: American Elsevier.

Pontécoulant, Philippe Gustave Le Doulcet, comte de. 1835. A history of Halley's comet, with an account of its return in 1835, and a chart showing its situation in the heavens. English translation by Charles Gold. London. 44 p.

General Works

Anonymous. 1765. Sur le retour de la comète de 1682. *In* Histoire de L'Académie Royale des Sciences (for 1759). Paris. pp. 119–164.

Broughton, Peter. 1985. The first predicted return of comet Halley. The Journal for the History of Astronomy. 16:123–133.

Lalande, Joseph-Jérôme de. 1765. Mémoire sur le retour de la comète de 1682. *In* Histoire de L'Académie Royale des Sciences (for 1759), Mémoires. Paris. pp. 1–40.

Lalande, Joseph-Jérôme de. 1803. Bibliographie astronomique; avec l'histoire de l'astronomie depuis 1781 jusqu'a 1802. Paris.

Ronan, Colin A. 1970. Edmond Halley, Genius in Eclipse. London: MacDonald & Co. 251 p.

Yeomans, Donald K., Rahe, Jürgen, and Freitag, Ruth S. 1986. The history of comet Halley. Royal Astronomical Society of Canada Journal 80(2):62–86.

Chapter 7

Theoretical Work

Barker, Thomas. 1757. An account of the discoveries concerning comets, with the way to find their orbits. . . . London.

Bošković, Rudjer J. 1746. De cometis. Rome.

Bošković, Rudjer J. 1749. De determinanda orbita planetae ope catoptricae ex datis vi, celeritate et directione motus in dato puncto. Rome.

Bošković, Rudjer J. 1774. De orbitus cometarum determinandis. . . . Mémoires de mathématique et de physique (Savants etranger). 6:198–215. Paris.

Bouguer, Pierre. 1735. De la détermination de l'orbite des comètes. Histoire de l'Académie Royale des Sciences, mémoires, année 1733, pp. 331–350. Paris.

Cunningham, Clifford J. 1988. The Baron and his celestial police. Sky and Telescope 75(March):271–272.

Dubiago, Alexander D. 1961. The Determination of Orbits. New York: Macmillan Company. Translated from the 1949 Russian edition.

Du Séjour, Achille-Pierre-Dionis. 1775. Essai sur les comètes en général; et particulierement sur celles qui peuvent approcher de l'orbite de la terre. Paris. 364 p.

Du Séjour, Achille-Pierre-Dionis. 1782. A la détermination des orbites des comètes. Histoire de l'Académie Royale des Sciences, mémoires, année 1779. Paris. pp. 51–168.

Du Séjour, Achille-Pierre-Dionis. 1786 and 1789. Traité analytique des mouvemens apparens des corps célestes. 2 volumes. Paris. 738 p. and 680 p.

Euler, Leonhard. 1743. Determinatio orbitae cometae anno 1742. Miscellanea Berolinensia, ad incrementum scientiarum ex scriptis societati regiae scientiarum. Volume 7. Berlin. pp.1–90.

Euler, Leonhard. 1744. Theoria motuum planetarum et cometarum.... Berlin. 186 p.

Fabritius, W. Du Séjour und Olbers. 1883. Ein beitrag zur geschichte des cometenproblems. Astronomische nachrichten 106:87–94.

Gauss, Carl Friedrich. 1809. Theoria motus corporum coelestium. Hamburg.

Gibbs, Josiah Willard. 1889. On the determination of elliptic orbits from three observations. Memoirs of the National Academy of Sciences. Washington, D.C. 4(part 2):81–104.

Lagrange, Joseph Louis. 1785. Théorie des variations périodique des mouvements des planetes. Nouveaux mémoires de l'Académie Royale des Sciences et Belleslettres, année 1783. Berlin. pp. 161–190.

Lambert, Johann Heinrich. 1761. Insigniores orbitae cometarum proprietates. Augsburg, Germany. 128 p.

Laplace, Pierre-Simon Marquis de. 1784. Mémoire sur la détermination des orbites des comètes. Histoire de l'Académie Royale des Sciences, mémoires, année 1780. Paris. pp. 13–72.

Leuschner, Armin Otto. 1913. A short method of determining orbits from three observations. Publications of the Lick Observatory. 7(part 1):1–20.

Marsden, Brian G. 1977. Carl Friedrich Gauss, astronomer. Royal Astronomical Society of Canada Journal 71:309–323.

Newton, Isaac. 1687. Philosophiae naturalis principia mathematica, London.

Olbers, Wilhelm Matthias. 1797. Abhandlung über die leichteste und bequemste methode die bahn eines cometen aus einigen beobachtungen zu berechnen. Weimar, Germany. 106 + 80 p.

Olbers, Wilhelm Matthias. 1802. Monatliche correspondenz zur beförderung der erd und himmels kunde, herausgegeben von Fr. von Zach. Volume 5. Gotha, Germany. p. 173.

Zach, Franz Xaver von. 1802. Über den zwischen Mars und Jupiter richtig vermutheten nun wirklich entdeckten neuen haupt-planeten unseres sonnensystems Ceres Ferdinandea. *In* Monatliche correspondenz zur beförderung der erd und himmels kunde, herausgegeben von Fr. von Zach. Volume 5. Gotha, Germany. p. 172.

The Orbit Computers

Ashbrook, Joseph. 1955. The comets of the Chevalier d'Angos. Sky and Telescope, 14:501.

Bessel, Wilhelm. 1807. Elemente de elliptischen bahn des kometen von 1769. *In* Astronomisches Jahrbuch für das Jahr 1810. Berlin. pp. 121–124.

Bradley, James. 1726. Observations upon the comet, that appear'd in the months of October, November, and December, 1723. Royal Society of London, Philosophical Transactions for the years 1724, 1725. 33:41–49.

Bradley, James. 1741. Observations upon the comet that appear'd in the months of

January, February, and March 1737, made at Oxford. Royal Society of London, Philosophical Transactions for the years 1737, 1738. 40:111–118.

Cunningham, Leland E. 1939. Harvard announcement card No. 498, dated August 3. Cambridge, Mass.

Delambre, Jean-Baptiste Joseph. 1814. Astronomie; théorique et practique. Volume 3. Paris.

Delambre, Jean-Baptiste Joseph. 1827. Histoire de l'astronomie au dix-huitiéme siècle. Paris. 796 p.

Gauss, Carl F. 1806. Comet vom jahr 1805. Monatliche correspondenz zur beförderung der erd und himmels kunde, herausgegeben vom Fr. von Zach. Volume 14. Gotha, pp. 75–86.

Halley, Edmond. 1705. A Synopsis of the Astronomy of Comets. London.

Halley, Edmond. 1749. Tabulae astronomicae. London.

Herschel, William. 1781. Account of a comet. Philosophical Transactions of the Royal Society of London 71:492–501.

Lacaille, Nicolas-Louis de. 1751. Sur les observations et la théorie des comètes qui ont paru depuis le commencement de ce siècle. Histoire de l'Académie Royale des Sciences, mémoires, année 1746. Paris. pp. 403–446.

Lemonnier, Pierre Charles. 1743. La théorie des comètes. . . . Paris, 192 p.

Leverrier, Jean-Joseph Urbain. 1857. Théorie de la comète périodique de 1770. Annales de l'Observatoire Impérial de Paris, memoires, Volume 3. Paris. pp. 203–270.

Lexell, Anders J. 1770. Recherches et calculs sur la vraie orbite elliptique de la comète de l'an 1769 et son temps périodique. St. Pétersbourg. 159 p.

Lexell, Anders J. 1772. Réflexions sur le temps périodique des comètes en général et principalement sur celui de la comète observée en 1770. St. Pétersbourg.

Lexell, Anders, J. 1779. Recherches sur la période de la comète, observée en 1770, d'après les observations de M. Messier. *In* Histoire de l'Académie Royale des Sciences, mémoires, année 1776. Paris. pp. 638–651.

Lexell, Anders J. 1783. Recherches sur la nouvelle planète découvert par Mr. Herschel. St. Pétersbourg.

Méchain, Pierre-François-André. 1782. Recherches sur les comètes de 1532 & de 1661. Paris. 64 p.

Newton, Isaac. 1687. Philosophiae naturalis principia mathematica, London.

Pingré, Alexandre-Guy. 1783–1784. Cométographie ou traité historique et théorique des comètes. 2 volumes. Paris.

Saron, Jean-Baptiste-Gaspard de. 1782. Note given in Lalande's article entitled "Mémoire sur la planète de Herschel." Histoire de l'Académie Royale des Sciences, mémoires, année 1779. Paris. p. 529.

Saron, Jean-Baptiste-Gaspard de. 1793. Comètes de 1793. Connaissance des temps de 1795. Paris. p. 286.

Struyck, Nicholaas. 1740. Inleiding tot de algemeene geographie, benevens eenige sterrekundige en andere verhandelingen. Amsterdam.

Struyck, Nicholaas. 1752. Viae cometarum, secundum hypothesin quae statuit illos curso suo parabolam circa solem describere. Royal Society of London, Philosophical Transactions for the Years 1749, 1750. Volume 46. pp. 89–92.

Struyck, Nicholaas. 1753. Vervolg van de beschryving der staartsterren. . . . Amsterdam. 1753

Wirtz, Carl W. 1915. Der kurzperiodische komet 1766 II. Astronomische nachrichten 201:65–82.

Yeomans, Donald K. 1985. Comets in 1979. Royal Astronomical Society, Quarterly Journal. London. 26:114.

Zanotti, Eustachio. 1744. The parabolic orbit for the comet of 1739 observed by Signor Eustachio Zanotti at Bologna. Philosophical Transactions of the Royal Astronomical Society of London. 41:809.

Cometary Thought in the Eighteenth Century

Briggs, J. Morton. 1967. Aurora and enlightenment. Eighteenth-century explanations of the aurora borealis. Isis 58:491–503.

Buffon, Georges-Louis Leclerc, comte de. 1749. Histoire naturelle, générale et particulière, avec la description du cabinet du Roi. Volume 1. Paris. pp. 131–132.

Buffon, Georges-Louis Leclerc, compte de. 1778. Histoire naturelle, générale et particulière, contenant les époques de la nature. Volume 12. Paris. p. 65.

Bernoulli, Jakob. 1682. Conamen novi systematis cometarum, pro motu eorum sub calculum revocando & apparitionibus praedicendis adornatum. Amsterdam.

Cheseaux, Jean Philippe Loys de. 1744. Traité de la comète qui a paru en Décembre 1743 & en Janvier, Fevrier & Mars 1744. Lausanne and Geneva. 296 p.

Dreyer, J.L.E. 1884. On the multiple tail of the great comet of 1744. Copernicus. 3:104–111. Dublin.

Euler, Leonhard. 1748. Recherches physiques sur la cause de queues des comètes, de la lumiere boreale, et de la lumiere zodiacle. Histoire de l'Académie Royale des Sciences et des Belle Lettres, année 1746. 2:117–140. Berlin.

Fontenelle, Bernard le Bovier de. 1686. Entretens sur la pluralité des mondes. Paris.

Halley, Edmond. 1726. Some considerations about the cause of the universal deluge, laid before the Royal Society, on the 12th of December 1694. Philosophical Transactions of the Royal Society of London 33:118–119.

Jaki, Stanley L. 1978. Planets and Planetarians. New York: John Wiley and Sons.

Lambert, Johann H. 1761. Cosmologische briefe über die einrichtung des weltbaues. Augsburg, Germany.

Lalande, Joseph-Jérôme de. 1771. Astronomie. Volume 3. Paris. p. 370.

Laplace, Pierre S. 1805. Traité de mecanique céleste. Volume 4. Paris. p. 230.

Mairan, Jean-Jacques Dortous de. 1752. Eclaircissements sur le traité physique et historique de l'aurore boreale, qui fait la suite de l'Académie Royale de Sciences, année 1731. Histoire de l'Académie Royale des Sciences, mémoires, année 1747:363–435.

Milne, David. 1828. Essay on Comets. Edinburgh.

Oliver, Andrew. 1772. Essay on Comets. Salem, Massachusetts.

Roemer, Elizabeth. 1960. Jean-Louis Pons, discoverer of comets. *In* Astronomical Society of the Pacific, Leaflet No. 371, May. 8 p.

Schagrin, Morton L. 1974. Early observations and calculations on light pressure. American Journal of Physics 42:927–940.

Sekanina, Zdenek and Yeomans, Donald K. 1984. Close encounters and collisions with the Earth. Astronomical Journal 89:154–161.

Whiston, William. 1696. A new theory of the Earth. London.

General Works

Alexander, A.F. O'D. 1965. The Planet Uranus, a History of Observations, Theory and Discovery. London: Faber and Faber. 316 p.

Bell, E.T. 1965. Men of Mathematics. New York: Simon and Schuster.

Gillispie, Charles C., ed. 1970–1980. Dictionary of Scientific Biography. 16 volumes. New York: Charles Scribner's.

Grant, Robert. 1966. History of Physical Astronomy. Reprinted from London, 1852 edition. New York: Johnson Reprint Corporation. 637 p.

Jones, Kenneth Glyn. 1969. Messier's Nebulae and Star Clusters. New York: American Elsevier.

Marsden, Brian G. 1974. Cometary motions. Celestial Mechanics 9:303–314.

Tammann, Gustav A. 1966. Jean-Philippe Loys de Cheseaux and his discovery of the so-called Olbers's paradox. Scientia, Revue internationale de synthèse scientifique. Bologna, January–February. pp. 1–11.

Chapter 8

Comet Encke and the Resisting Medium

Asten, Friedrich Emil von. 1872. Untersuchungen über die theorie des Encke'schen cometen. I. Berechnung eines wichtigen theiles der absoluten Jupiterstörungen des Encke'schen cometen. Mémoires de l'Académie Impériale des Sciences de Saint-Pétersbourg, series 7. 18(10):81 p.

Asten, Friedrich Emil von. 1878. Untersuchungen über die theorie des Encke'schen cometen. II. resultate aus den erscheinungen 1819–1875. Mémoires de l'Académie Impériale des Sciences de Saint-Pétersbourg, series 7. 26(2).125 p.

Backlund, Jöns Oskar. 1884. Untersuchungen über die bewegung des Encke'schen cometen 1871–1881. Mémoires de l'Académie Impériale des Sciences de Saint-Pétersbourg, series 7. 32(3).50 p.

Backlund, Jöns Oskar. 1886. Comet Encke 1865–1885. Mémoires de l'Académie Impériale des Sciences de Saint-Pétersbourg, series 7. 34(8).43 p.

Backlund, Jöns Oskar. 1910. Encke's comet, 1895–1908. Royal Astronomical Society, Monthly Notices. 70:429–442.

Bessel, Friedrich Wilhelm. 1804. Berechnung der Harriot'schen und Torporley'schen beobachtungen des cometen von 1607. Monatliche correspondenz. 10:425–440.

Bessel, Friedrich Wilhelm. 1836. Beobachtungen über die physische beschaffenheit des Halley'schen kometen und dadurch veranlasste bemurkungen. Astronomische nachrichten 13:185–232.

Bessel, Friedrich Wilhelm. 1836. Bemerkungen über mögliche unzulänglichkeit der die anziehungen allein berücksichtigenden theorie der kometen. Astronomische nachrichten 13:345–350.

Bond, George P. 1849. On some applications of the method of mechanical quadratures. American Academy of Arts and Sciences, Memoirs, 4(N.S.): 189–208.

Encke, Johann Franz. 1819. Ueber einen merkwürdigen kometen, der wahrscheinlich bei dreijähriger umlaufszeit schon zum viertenmale bei seiner rückkehr zur sonne beobachtet ist. *In* Berliner Astronomische jahrbuch für das jahr 1822:180–202. Berlin.

Encke, Johann Franz. 1820. Ueber die bahn des Pons'schen kometen, nebst berechnung seines laufs bei seiner nächsten wiederkehr im jahr 1822. *In* Berliner Astronomische jahrbuch für das jahr 1823:211–223. Berlin.

Encke, Johann Franz. 1823. Fortgesetzte nachricht über den Pons'schen kometen. *In* Berliner Astronomische jahrbuch für das jahr 1826:124–140. Berlin.

Encke, Johann Franz. 1852. Ueber eine neue methode der berechnung der planetenstörungen. Astronomische nachrichten 33:377–398.

Encke, Johann Franz. 1852. Neue methode zur berechnung der speciellen störungen in Nr. 791 und 792 der Astron. Nachrichten. Astronomiche nachrichten. 34:349–360.

Encke, Johann Franz. 1859. Uber die existenz eines widerstehenden mittels im weltraume. Royal Astronomical Society, Monthly Notices. 19:70–75.

Faye, Hervé. 1858. Sur les comètes et sur l'hypothèse d'un milieu résistant. Comptes rendus, 47:836–850.

Haerdtl, Eduard F. von. 1889. Die bahn des periodischen kometen Winnecke in den jahren 1858–1886. Denkschriften der Kaiserlichen akademie der wissenschaften. 56:151–185. Vienna.

Kamienski, Michael. 1933. Über die bewegung des kometen Wolf I in dem zeitraume 1884–1919. Acta astronomica, series A. 3:1–56.

Marsden, Brian G. and Sekanina, Zdenek. 1971. Comets and nongravitational forces. IV. Astronomical Journal 76:1135–1151.

Möller, Didrik Magnus Axel. 1861. Fortgesetzte untersuchungen über die bahn des Faye'schen cometen. Astronomische nachrichten 54:353–362.

Möller, Didrik Magnus Axel. 1861. Über die hypothese von Valz über die dichtigkeit des äthers. Astronomische nachrichten 55:321–334.

Möller, Didrik Magnus Axel. 1865. Elemente und ephemeride des Faye'schen cometen. Astronomische nachrichten 64:145–158.

Möller, Didrik Magnus Axel. 1872. Beiträge zu der neuen bearbeitung periodischen cometen. Vierteljahrsschrift der astronomischen gesellschaft 7:85–97.

Olbers, Heinrich Wilhelm M. 1819. Eine merkwürdige astronomische entdeckung und beobachtungen des kometen vom Jul. 1819. *In* Berliner Astronomische jahrbuch für das jahr 1822:175–180. Berlin.

Oppolzer, Theodor Ritter von. 1880. Ueber den periodischen cometen Winnecke (comet III 1819) und das widerstand leistende medium. Astronomische nachrichten 97:149–154.

Recht, A.W. 1940. An investigation of the motion of periodic comet d'Arrest (1851 II). Astronomical Journal 48:65–80.

Rümker, Karl. 1826. Auszug aus zweien briefen des Herrn Rümker au Herrn Doctor und Ritter Olbers in Bremen. Astronomische nachrichten 4:103–112.

Struve, Otto. 1952. Comet theories. Sky and Telescope 11:269–271, 273.

Whipple, Fred L. 1950. A comet model. I. The acceleration of comet Encke. Astrophysical Journal 111:375–394.

Zach, Franz Xavier von. 1793. Etwas aus den, vom Herrn von Zach im jahr 1784 in England auf gefundenen Harriotischen manuscripten, vornemlich original-beobachtungen der beyden kometen vom 1607 und 1618. *In* Sammlung astronomischer abhandlungen (supplemental volume to Berliner astronomischen jahrbuch). pp. 1–41. Berlin.

The Short Life of Comet Biela

Anonymous. 1833. Ueber Biela's comet. Astronomische nachrichten 10:219–222.

Bessel, Friedrich Wilhelm. 1806. Beobachtungen der bey den im jahr 1805 erschienenen kometen, vom Hrn. Dr. Olbers in Bremen und berechnung der elemente ihrer bahnen, vom Hrn. Bessel daselbst. *In* Berliner Astronomische jahrbuch für das jahr 1809, pp. 134–136. Berlin.

Bessel, Friedrich Wilhelm. 1806. II comet vom jahr 1805. Monatliche correspondenz 14:71–74.

Biela, Wilhelm von. 1826. Schreiben des Herrn Hauptmanns v. Biela an den Herausgeber. Astronomische nachrichten 4:433–436, 469–472.

Challis, James. 1846. Biela's (or Gambart's) comet. Royal Astronomical Society of London, Monthly Notices 7:73–77.

Clausen, Thomas. 1826. Ueber den von Herrn v. Biela am 27 sten Februar entdeckten cometen. Astronomische nachrichten 4:465–468.

Damoiseau, Marie Charles Théodore. 1830. Sur la comète périodique de 6.7 ans. *In* Connaissance des temps, additions. pp. 52–55.

Gambart, Jean F.A. 1826. Schreiben des Herrn Gambart. Astronomische nachrichten 4:435, 469–470.

Gauss, Carl Friedrich. 1806. Aus einem schreiben des Dr. Gauss. Monatliche correspondenz 14:75–86.

Gauss, Carl Friedrich. 1806. Beobachtungen der Ceres, Pallas und Juno und des zweyten kometen von 1805, nebst elemente der bahn des letztern. *In* Berliner Astronomische jahrbuch für das jahr 1809. Berlin. pp. 137–140.

Herrick, Edward C. and Bradley, Francis. 1846. Biela's comet. American Journal of Science and Arts 1(second series):293–294, 446–447.

Hujer, Karel. 1983. On the history of Wilhelm von Biela and his comet. Royal Astronomical Society of Canada Journal 77:305–309.

Marsden, Brian G. and Sekanina, Zdenek. 1971. Comets and nongravitational forces. IV. Astronomical Journal 76:1135–1151.

Maury, Matthew Fonatine. 1846. Biela's comet, 1846. Washington Observations 2:422–423.

Maury, Matthew Fonatine. 1846. Duplicity of Biela's comet. Royal Astronomical Society of London, Monthly Notices 7:90–92.

Newton, Hubert A. 1886. The story of Biela's comet. *In* American Journal of Science 31(3rd series):81–94.

Santini, Giovanni. 1844. Elemente und ephemeride des Bielaschen cometen. Astronomische nachrichten 21:171–174.

Secchi, Angelo. 1852. Beobachtungen des Biela'schen cometen und bemerkungen über denselben und seinen ausgefundenen begleiter. Astronomische nachrichten 35:191–192.

Secchi, Angelo. 1852. Entdeckung und beobachtungen eines kleinen cometen. Astronomische nachrichten 35:89–90.

Struve, Otto W. 1857. Beobachtungen den Biela'schen cometen im jahr 1852 angestellt am grossen refractor der Pulkowaer sternwarte. Mémoires de l'Académie Impériale des Sciences de Saint-Pétersbourg, series 6. 6:133–156.

Wichmann, Moritz Ludwig Georg. 1846. Biela's (or Gambart's) comet. Royal Astronomical Society of London, Monthly Notices 7:74–77.

Comets and Meteor Showers

Adams, John Couch. 1866–1867. On the orbit of the November meteors. Royal Astronomical Society of London, Monthly Notices 27:247–252.

Benzenberg, Johann F. and Brandes, Heinrich W. 1800. Versuch die entfernung, die geschwindigkeit und die bahn der sternschnuppen zu bestimmen. Annalen der physik 6:224–232.

Bertholon, Pierre. 1791. Observations d'un globe de feu. Journal des sciences utiles 4:224–228.

Bruhns, Karl C. 1875. Ueber den Pogson'schen cometen im jahre 1872. Astronomische nachrichten 86:219–224.

Burke, John G. 1986. Cosmic Debris: Meteorites in History. Berkeley: University of California Press.

Chladni, Ernst F.F. 1794. Ueber den ursprung der von Pallas gefundnen und andern ähnlichen eisenmassen. Riga, J.F. Hartknoch.

Chladni, Ernst F.F. 1819. Über feuer-meteore und über die mit denselben herabgefallenen massen. Vienna.

Clarke, W.B. 1834. Loudon's Magazine of Natural History 7:385–390.

d'Arrest, Heinrich L. 1867. Ueber einige merkwürdige meteorfälle beim durchgange der erde durch die bahn des Biela'schen cometen. Astronomische nachrichten 69:7–10.

Ellicot, Andrew. 1804. Account of an extraordinary flight of meteors (commonly called shooting stars). American Philosophical Society, Transactions. 6:28–29.

Galle, Johann G. 1867. Ueber den muthmasslichen zusammenhang der periodischen sternschnuppen des 20 April mit dem ersten cometen des jahres 1861. Astronomische nachrichten 69:33–36.

Halley, Edmond. 1714. An account of several extraordinary meteors or lights in the sky. Royal Society of London, Philosophical Transactions. 29:159–164.

Herrick, Edward C. 1838. On the shooting stars of August 9th and 10th 1837, and on the probability of the annual occurrence of a meteoric shower in August. American Journal of Science 33:176–180.

Hughes, David W. 1982. The history of meteors and meteor showers. Vistas of Astronomy 26:325–345.

Humboldt, Friedrich W.H.A. von. 1819. Fireballs. Philosophical Magazine and Journal of Science 53:312–314.

Humboldt, Friedrich W.H.A. von. 1872–1873. Cosmos; a sketch of a physical description of the universe, Volume 1. Translated by E.C. Otté. New York: pp. 519–527.

Kirkwood, Daniel. 1861. Danville Quarterly Review, Dec.

Kirkwood, Daniel. 1867. Meteoric Astronomy: A treatise on Shooting Stars, Fireballs, and Aerolites. Appendix A. Philadelphia: J.B. Lippincott & Co.

Klinkerfues, W. 1873. Schreiben des Herrn Prof. Klinkerfues an den herausgeber. Astronomische nachrichten 80:349–350.

Le Verrier, Urbain J.J. 1867. Sur les étoiles filantes de 13 Novembre et du 10 Aout. Comptes rendus 64:94–99.

Locke, John. 1834. Cincinnati Daily Gazette, August 11 and 12.

Newton, Hubert A. 1863. Evidence of the cosmical origin of shooting stars derived from the dates of early star showers. American Journal of Science and Arts, series 2. 36:145–149.

Newton, Hubert A. 1864. The original accounts of the displays in former times of the November star-shower, together with a determination of the length of its cycle, its annual period, and the probable orbit of the group of bodies around the sun. American Journal of Science and Arts, series 2. 37:377–389 and 38:53–61.

Newton, Hubert A. 1865. Abstract of a memoir on shooting stars. American Journal of Science and Arts, series 2. 39:193–207.

Olbers, Heinrich Wilhelm M. 1837. Die sternschnuppen. *In* Jahrbuch für 1837, edited by H.C. Schumacher. pp. 278–282.

Olivier, Charles P. 1925. Meteors. Baltimore: Williams and Wilkins. 276 p.

Olmsted, Denison. 1834. Observations on the meteors of November 13, 1833. American Journal of Science 25:411 and 26:132–174.

Olmsted, Denison. 1836. On the causes of the meteors of November 13, 1833. American Journal of Science 29:376–383.

Oppolzer, Theodor von. 1862. Ueber die bahn des cometen II 1862. Astronomische nachrichten 58:249–250.

Oppolzer, Theodor von. 1863. Bahn-bestimmung des cometen II 1862. Astronomische nachrichten 59:49–58.

Oppolzer, Theodor von. 1866–1867. Bahnbestimmung des cometen I, 1866. Astronomische nachrichten 68:241–250.

Oppolzer, Theodor von. 1867. Schreiben des Hern Th. Oppolzer an den herausgeber. Astronomische nachrichten 68:333–334.

Peters, Christian F.W. 1866–1867. Bemerkung über den sternschnuppenfall von 13 November und 10 August 1866. Astronomische nachrichten 68:287–288.

Pogson, Norman R. 1873. Extract of a letter from N.R. Pogson . . . dated Dec. 5, 1872. Royal Astronomical Society of London, Monthly Notices 33:116.

Pringle, John. 1759. Some remarks upon the several accounts of the fiery meteor (which appeared on Sunday the 26th of November, 1758) and upon other such bodies. Royal Society of London, Philosophical Transactions 51(part 1):259–274.

Quetelet, Lambert A.J. 1839. Catalogue des principales apparitions d'étoiles filantes. Académie royale de Bruxelles, mémoires. Brussels. 12:1–56.

Rittenhouse, David. 1780. Letter to Benjamin Franklin dated December 31, 1780. *In* The Scientific Writings of David Rittenhouse, edited by Brooke Hindle. New York: Arno Press, 1980.

Rittenhouse, David. 1786. Observations on the account of a meteor. American Philosophical Society, Transactions 2:173–176.

Schiaparelli, Giovanni V. 1866 and 1867. Intorno al corso ed all'origine probabile delle stelle meteoriche. Five letters to Angelo Secchi published in Dal bullettino meteorologico dell'osservatorio del Collegio Romano. 5(8, 10, 11, 12) and 6(2).

Schiaparelli, Giovanni V. 1867. Sur la relation qui existe entre les comètes et les étoiles filantes. Astronomische nachrichten 68:331–332.

Stoney, G. Johnstone and Downing, A.M.W. 1899. Perturbations of the Leonids. *In* Royal Society of London, Proceedings 64:403–409.

Twining, Alexander C. Investigations respecting the meteors of Nov. 13th, 1833. American Journal of Science 26:320–352.

Walker, Sears C. 1843. Researches concerning the periodical meteors of August and November. American Philosophical Society, Transactions. 8:87–140.

Weiss, Edmund. 1866–1867. Bemurkungen über den zusammenhang zwischen cometen und sternschnuppen. Astronomische nachrichten 68:381–384.

Weiss, Edmund. 1868. Beiträge zur kenntniss der sternschnuppen. Astronomische nachrichten 72:81–102.

General Works

Arago, Dominique F. 1861. A Popular Treatise on Comets. London: Longman, Green, Longman, and Roberts. 164 p.

Carl, Ph. 1864. Repertorium der cometen—astronomie. Munich: M. Rieger. 377 p.

Chambers, George F. 1909. The Story of Comets Simply Told for General Readers. Oxford: Clarendon Press. 256 p.

Guillemin, Amédée. 1877. The World of Comets, translated and edited by James Glaisher. London: Sampson Low, Marston, Searle, & Rivington. 548 p.

Grosser, Morton. 1962. The Discovery of Neptune. Cambridge, Mass.: Harvard University Press. 172 p.

Hind, J. Russell. 1852. The Comets: A Descriptive Treatise upon These Bodies. London: John W. Parker and Son. 184 p.

Marsden, Brian G. 1974. Cometary motions. Celestial Mechanics 9:303–314.

Chapter 9

The Introduction of Photography

Barnard, Edward E. 1895. On the photographic discovery of comet V 1892. Popular Astronomy 3:13–14.

Bond, George P. 1859. Letter to R.C. Carrington dated June 11, 1859. *In* Memorials of W.C. Bond and G.P. Bond by E.S. Holden. New York: 1897, p. 167. See also the account given in Royal Astronomical Society of London, Monthly Notices 19:138–139.

De Vaucouleurs, Gérard. 1961. Astronomical Photography. London: Faber and Faber. 94 p.

Draper, Henry. 1881. Note on the photographs of the spectrum of the comet of June, 1881. American Journal of Science, third series. 22:134–135.

Huggins, William. 1882. Preliminary note on the photographic spectrum of comet b 1881. Royal Society of London, Proceedings 33:1–3.

Norman, Daniel. 1938. The development of astronomical astronomy. Osiris 5:560–594.

Pickering, William H. 1895. Swift's comet 1892 I. Annals of the Astronomical Observatory of Harvard College 32(part 1):267–295.

Pluvinel, A. de la Baume. 1903. Le spectre de la comète 1902 b. Bulletin de la société astronomique de France 17:117–121.

Swings, Polydore. 1965. Cometary spectra. Royal Astronomical Society, Quarterly Journal 6:28–69.

Composition of the Cometary Atmosphere and Tail

Altenhoff, W.J., Batrla, W., Huchtmeier, W.K., Schmidt, J., Stumpff, P., and Walmsley, M. 1983. Radio observations of comet 1983d. Astronomy and Astrophysics 125:L19–L22.

Arago, Dominique F.J. 1820. Quelques nouveaux détails sur le passage de la comète découverte dans le mois de Juillet. 1819, devant le disque du soleil. Annales de chimie et de physique 13(séries 2):104–110.

Baldet, Fernand. 1926. Recherches sur la constitution des comètes et sur les spectres du carbone. Annales de l'observatoire d'astronomie physique de Paris sis parc de Meudon 7:1–109.

Becklin, Eric E. and Westphal, James A. 1966. Infrared observations of comet 1965 f. Astrophysical Journal 145:445–453.

Bertaux, J.L. and Blamont, Jacques. 1970. Observation de l'émission d'hydrogène atomique de la comète Bennett. Académie des sciences, Paris. Comptes rendus. 270 (séries B):1581–1584.

Bertaux, J.L., Blamont, Jacques, and Festou, Michel C. 1973. Interpretation of hydrogen Lyman-alpha observations of comets Bennett and Encke. Astronomy and Astrophysics 25:415–430.

Biermann, Ludwig and Trefftz, E. 1964. Über die mechanismen der ionisation und der anregung in kometenatmosphären. Zeitschrift für astrophysik 59:1–28.

Biraud, François, Bourgois, G., Crovisier, Jacques, Fillit, R., Gérard, Eric, and Kazès, Ilya. 1974. OH observations of comet Kohoutek (1973 f) at 18 cm wavelength. Astronomy and Astrophysics 34:163–166.

Black, John H., Chaisson, Eric J., Ball, John A., Penfield, H., and Lilley, A. Edward. 1971. Radio frequency emission from CH in comet Kohoutek (1973 f). Astrophysical Journal 191:L45–L47.

Blamont, Jacques and Festou, Michel C. 1974. Observations of the comet Kohoutek (1973f) in the resonance light of the OH radical. Icarus 23:538–544.

Bond, George P. 1862. Account of the great comet of 1858. Annals of the Astronomical Observatory of Harvard College 3:372 p.

Code, Arthur D., Houck, Theodore E., and Lillie, Charles F. 1970. International Astronomical Union Circular 2201, dated Jan. 21.

Code, Arthur D., Houck, Theodore E., and Lillie, Charles F. 1972. Ultraviolet observations of comets. *In* The Scientific Results from the Orbiting Astronomical Observatory, edited by Arthur D. Code. NASA SP-310. Washington, D.C.: National Aeronautics and Space Administration. pp. 109–114.

Copeland, Ralph and Lohse, J.C. 1882. Spectroscopic observations of comets III and IV, 1881, comet I 1882, and the great comet of 1882. Copernicus 2:225–244.

Coutrez, R., Hunaerts, J., and Koeckelenbergh, A. 1958. Radio emission from comet 1956h on 600 mc. Proceedings of the IRE (Institute of Radio Engineers) 46:274–279.

Curtis, G.W. 1966. Daylight observations of the 1965 f comet at Sacramento peak observatory. Astronomical Journal 71:194–196.

Deslandres, Henri and Bernard, A. 1907. Étude spectrale de la comète Daniel d 1907. Académie des sciences, Paris. Comptes rendus. 145:445–448.

Deslandres, Henri and Bernard, A. 1908. Recherches spectrales sur la comète Morehouse c 1908. Académie des sciences, Paris. Comptes rendus. 147:774–777.

Despois, D., Gérard, Eric, Crovisier, Jacques, and Kazès, Ilya. 1981. The OH radical in comets: observations and analysis of the hyperfine microwave transitions at 1667 Mhz and 1665 Mhz. Astronomy and Astrophysics 99:320–340.

Donati, Giovanni B. 1864. Schreiben des herrn professor Donati. Astronomische nachrichten 62:375–378.

Douglas, Alexander E. 1951. Laboratory studies of the λ4050 group of cometary spectra. Astrophysical Journal 114:466–468.

Dufay, Jean, Swings, Pol, and Fehrenbach, Charles H. 1965. Spectrographic observations of comet Ikeya-Seki (1965 f). Astrophysical Journal 142:1698–1699.

Evershed, John. 1907. The spectrum of comet 1907 d (Daniel). Royal Astronomical Society, Monthly Notices 68:16–18.

Fehrenbach, C. and Andrillat, Y. 1974. Le spectre de la comète Kohoutek (1973f). Académie des sciences, Paris. Comptes rendus. 278(series B):607–610.

Feldman, Paul D., Takacs, Peter Z., Fastie, William G., and Donn, Bertram. 1974. Rocket ultraviolet spectrophotometry of comet Kohoutek (1973f). Science 185:705–707.

Fowler, Alfred. 1910. Investigations relating to the spectra of comets. Royal Astronomical Society, Monthly Notices 70:484–496.

447

Greenstein, Jesse L. 1958. High resolution spectra of comet Mrkos (1957 d). Astrophysical Journal 128:106–113.

Herzberg, Gerhard. 1942. Laboratory production of the λ4050 group occurring in cometary spectra; further evidence for the presence of CH_2 molecules in comets. Astrophysical Journal 96:314–315.

Herzberg, Gerhard and Lew, H. 1974. Tentative identification of the H_2O^+ ion in comet Kohoutek. Astronomy and Astrophysics 31:123–124.

Huebner, Walter F., Snyder, Lewis E., and Buhl, David. 1974. HCN radio emission from comet Kohoutek (1973 f). Icarus 23:580–584.

Huggins, William. 1866. On the spectrum of comet I 1866. Royal Society of London, Proceedings 15:5–7.

Huggins, William. 1867. Note on the spectrum of comet II 1867. Royal Astronomical Society of London, Monthly Notices 27:288.

Huggins, William. 1868. Further observations on the spectra of some stars and nebulae. . . . Royal Society of London Philosophical Transactions 158:529–564.

Huggins, William. 1868. On the spectrum of Brorsen's comet 1868. Royal Society of London Proceedings 16:386–389.

Jackson, William M., Clark, Thomas A., and Donn, Bertram. 1976. Radio detection of H_2O in comet Bradfield (1974 b). The Study of Comets, edited by B. Donn, M. Mumma, W. Jackson, M. A'Hearn, and R. Harrington. NASA SP-393. Washington, D.C.: National Aeronautics and Space Administration. pp. 272–280.

Jenkins, Edward B. and Wingert, David W. 1972. The Lyman-alpha image of comet Tago-Sato-Kosaka (1969 g). Astrophysical Journal 174:697–704.

Keller, Horst Uwe. 1976. The interpretation of the ultraviolet observations of comets. Space Science Reviews 18:641–684.

Kraus, John D. 1958. Observations at a wavelength of 11 meters during the close approach of comet Arend-Roland. Astronomical Journal 63:55–58.

Liais, E. 1859. Observations faites dans l'hemisphere austral sur la comète de Donati. Académie des sciences, Paris. Comptes rendus. 48:624–627.

Liller, William C. 1960. The nature of grains in the tails of comets 1956 h and 1957 d. Astrophysical Journal 132:867–882.

Maas, R.W., Ney, Edward P., and Woolf, Neville J. 1970. The 10-micron emission peak of comet Bennett 1969 i. Astrophysical Journal 160:L101–L104.

Meisel, David D. and Berg, Richard A. 1974. High resolution spectrophotometry of selected features in the 1.1 μm spectrum of comet Kohoutek (1973f). Icarus 23:454–458.

Millman, Peter M. 1937. An analysis of meteor spectra. Annals of the Harvard College Observatory 82:113–146.

Millman, Peter M. 1937. An analysis of meteor spectra: second paper. Annals of the Harvard College Observatory 82:149–177.

Müller, H.G., Priester, W., and Fischer, G. 1957. Radioemission des kometen 1956h. Die naturwissenschaften 44:392–393.

Ney, Edward P. 1974. Infrared observations of comet Kohoutek near perihelion. Astrophysical Journal 189:L141–L143.

Ney, Edward P. 1974. Multiband photometry of comets Kohoutek, Bennett, Bradfield and Encke. Icarus 23:551–560.

Nicolet, M. 1938. Les bandes de CH et la présence de l'hydrogène dans les comètes. Zeitschrift für astrophysik 15:154–159.

O'Dell, Charles R. 1971. Nature of particulate matter in comets as determined from infrared observations. Astrophysical Journal 166:675–681.

Opal, Chet B., Carruthers, George R., Prinz, D.K., and Meier, R.R. 1974. Comet Kohoutek; ultraviolet images and spectrograms. Science 185:702–705.

Orlov, S.V. 1927. The spectrum of the comet 1882 II. Soviet Astronomical Journal 4:1–8.

Pickering, William H. 1895. Swift's comet 1892 I. Annals of the Astronomical Observatory of Harvard College 32(part 1):267–295.

Poey, A. 1859. Quelque observations physique faites à la Havane sur la comète Donati. Académie des sciences, Paris. Comptes rendus. 48:726–729.

Preston, George W. 1967. The spectrum of comet Ikeya-Seki (1965 f). Astrophysical Journal 147:718–742.

Righini, G. 1965. Early observations of cometary spectra. In Mémoires de la Société Royale des Sciences de Liège, 1965, fifth series, book XII, Nature et origine des comètes. pp. 129–131.

Secchi, Angelo. 1861. Observations faites à Rome de la comète du 29 Juin. Académie des sciences, Paris. Comptes rendus. 53:85–87.

Secchi, Angelo. 1862. Constitution physique de la seconde comète de 1862. Académie des sciences, Paris. Comptes rendus. 55:751–754.

Secchi, Angelo. 1866. Spectre de la comète de Tempel. Académie des sciences, Paris. Comptes rendus. 62:210.

Secchi, Angelo. 1868. Sur le spectre de la comète Brorsen. Académie des sciences, Paris. Comptes rendus. 66:881–884.

Schwarzschild, Karl and Kron, E. 1911. On the distribution of brightness in the tail of Halley's comet. Astrophysical Journal 34:342–352.

Sekanina, Zdenek. 1974. On the nature of the anti-tail of comet Kohoutek (1973 f) I. a working model. Icarus 23:502–518.

Slaughter, C.D. 1969. The emission spectrum of comet Ikeya-Seki 1965-f at perihelion passage. Astronomical Journal 74:929–943.

Swings, Polydore. 1941. Complex structures of cometary bands tentatively ascribed to the contour of the solar spectrum. Lick Observatory Bulletin 19:131–136.

Swings, Polydore. 1956. The spectra of comets. Vistas in Astronomy 2:958–981.

Swings, Polydore. 1965. Cometary spectra. Royal Astronomical Society Quarterly Journal 6:28–69.

Swings, Polydore, Elvey, C.T., and Babcock, Horace W. 1941. The spectra of comet Cunningham, 1940 c. Astrophysical Journal 94:320–343.

Swings, Polydore, McKellar, A., and Minkowski, R. 1943. Cometary emission spectra in the visible region. Astrophysical Journal 98:142–152.

Swings, Polydore and Page, Thornton. 1950. The spectrum of comet Bester (1947 k). Astrophysical Journal 111:530–554.

Swings, Polydore and Greenstein, Jesse L. Présence des raies interdites de l'oxygène dans les spectres comètaires. Académie des sciences, Paris. Comptes rendus. 246:511–513.

Sykes, M.V., Lebofsky, L.A., Hunten, D.M., and Low, F. 1986. The discovery of dust trails in the orbits of periodic comets. Science 232:1115–1117.

Thackeray, A.D., Feast, M.W., and Warner, B. 1965. Daytime spectra of comet Ikeya-Seki near perihelion. Astrophysical Journal 143:276–279.

Thollon, L. and Gouy, A. Sur une comète observée à Nice. Académie des sciences, Paris. Comptes rendus. 95:555–557.

Thollon, L. and Gouy, A. 1883. Sur le displacement des raies du sodium, observée dans le spectre de la grande comète de 1882. *In* Académie des sciences, Paris. Comptes rendus. 96:371–372.

Turner, Barry E. 1974. Detection of OH at 18-centimeter wavelength in comet Kohoutek (1973f). Astrophysical Journal 189:L137–L139.

Ulich, Bobby L. and Conklin, Edward K. 1974. Detection of methyl cyanide in comet Kohoutek. Nature 248:121–122.

Vogel, Hermann C. 1884. Einige beobachtungen über den cometen Pons-Brooks, insbesondere über das spectrum desselben. Astronomische nachrichten 108:21–26.

Walker, R.G., Aumann, H.H., Davies, J., Green, S., De Jong, T., Houck, J.R., and Soifer, B.T. 1984. Observations of comet IRAS-Araki-Alcock 1983d. Astrophysical Journal 278:L11–L14.

Wehinger, Peter A. and Wyckoff, Susan. 1974. H_2O^+ in spectra of comet Bradfield (1974 b). Astrophysical Journal 192:L41–L42.

Wehinger, Peter A. Wyckoff, Susan, Herbig, George H., Herzberg, Gerhard, and Lew, H. 1974. Identification of H_2O^+ in the tail of comet Kohoutek (1973 f). Astrophysical Journal 190:L43–L47.

Young, Charles A. 1881. Spectroscopic observations upon the comet b, 1881. American Journal of Science, third series. 22:135–137.

Zanstra, Herman. 1929. The excitation of line and band spectra in comets by sunlight. Royal Astronomical Society of London, Monthly Notices 89:178–197.

The Formation of Cometary Tails

Alfvén, Hannes. 1957. On the theory of comet tails. Tellus 9:92–96.

Arrhenius, Svante A. 1900. Über die ursache der nordlichter. Physikalische zeitschrift 2:81–87.

Bessel, Friedrich W. 1836. Bemerkungen über mögliche unzulänglichkeit der die anziehungen allein berücksichtigenden theorie der kometen. Astronomische nachrichten 13:345–350.

Bessel, Friedrich W. 1836. Beobachtungen über die physische beschaffenheit des Halley'schen kometen und dadurch veranlasste bemerkungen. Astronomische nachrichten 13:185–232.

Biermann, Ludwig. 1951. Kometenschweife und solar korpuskularstrahlung. Zeitschrift für astrophysik 29:274–286.

450

Biermann, Ludwig and Lüst, Rhea. 1963. Comets: structure and dynamics of tails. *In* The Moon, Meteorites and Comets, edited by B.M. Middlehurst and G.P. Kuiper. Chicago: University of Chicago Press. pp. 618–638.

Brandt, John C. 1967. Interplanetary gas. XIII. Gross plasma velocities from the orientations of ionic comet tails. Astrophysical Journal 147:201–219.

Brandt, John C., Roosen, Robert G., and Harrington, Robert S. 1972. Interplanetary gas. XVII. An astronomical determination of solar wind velocities from orientations of ionic comet tails. Astrophysical Journal 177:277–284.

Bredikhin, Fedor A. 1903. Prof. Th. Bredikhin's mechanische untersuchungen über cometenformen in systematischer darstellung. Compiled by R. Jaegermann. St. Petersburg.

Ershkovich, Alexander I. 1978. The comet tail magnetic field: large or small? Royal Astronomical Society of London, Monthly Notices 184:755–758.

Finson, Michael L. and Probstein, Ronald F. 1968. A theory of dust comets. I. Model and equations. Astrophysical Journal 154:327–352.

Finson, Michael L. and Probstein, Ronald F. 1968. A theory of dust comets. II. Results for comet Arend-Roland. Astrophysical Journal 154:353–380.

Fitzgerald, George F. 1883. On comet tails. Royal Dublin Society Scientific Proceedings 3:344–346.

Hyder, Charles L., Brandt, John C., and Roosen, Robert G. 1974. Tail structures far from the head of comet Kohoutek. I. Icarus 23:601–610.

Ip, W-H. and Mendis, D.A. 1976. The generation of magnetic fields and electric currents in cometary plasma tails. Icarus 29:147–151.

Jockers, Klaus, Lüst, Rhea, and Nowak, Th. 1972. The kinematical behavior of the plasma tail of comet Tago-Sato-Kosaka 1969 IX. Astronomy and Astrophysics 21:199–207.

Jockers, Klaus and Lüst, Rhea. 1973. Tail peculiarities in comet Bennett caused by solar wind disturbances. Astronomy and Astrophysics 26:113–121.

Lebedev, Pëtr N. 1901. Untersuchungen über die druckkrafte des lichtes. Annalen der physik 6:433–458.

Nichols, Ernest F. and Hull, Gordon F. 1901. A preliminary communication on the pressure of heat and light radiation. Physical Review 13:307–320.

Olbers, Heinrich W.M. 1812. Über den schweif des grossen kometen von 1811. *In* Monatliche correspondenz zur beförderung der erd und himmels kunde, herausgegeben Fr. von Zach. Volume 25. Gotha, Germany. pp. 3–22.

Peirce, Benjamin. 1859. American Academy of Arts and Sciences Proceedings 4:202–206.

Sekanina, Zdenek and Miller, Freeman D. 1973. Comet Bennett 1970 II. Science 179:565–567.

Schaeberle, John M. 1893. Preliminary note on a mechanical theory of comets. Astronomical Journal 13:151–153.

Schagrin, Morton L. 1974. Early observations and calculations on light pressure. American Journal of Science 42:927–940.

Schwarzschild, Karl. 1901. Der druck des lichts auf kleine kugeln und die

Arrheniussche theorie der cometenschweife. Sitzungsberichte der Bayerischen akademie der wissenschaften zu München, math-phys. Kl. 31:293–338.

Wurm, Karl. 1943. Die natur der kometen. Vierteljahrsschrift der astronomischen gesellschaft 78:18–87.

The Cometary Nucleus

A'Hearn, Michael F. 1988. Observations of cometary nuclei. Annual Review of Earth and Planetary Science 16:273–293.

A'Hearn, Michael F. and Cowan, John J. 1980. Vaporization in comets: The icy grain halo of comet West. Moon and Planets 23:41–52.

A'Hearn, Michael F., Campins, Humberto, Schleicher, David G., and Millis, Robert L. 1989. The nucleus of comet P/Tempel 2. Astrophysical Journal 347:1155–1166.

A'Hearn, Michael F., Feldman, Paul D., and Schleicher, David G. 1983. The discovery of S_2 in comet IRAS-Araki-Alcock 1983 d. Astrophysical Journal 274:L99–L103.

Baldet, Fernand. 1927. Sur le noyau de la comète Pons-Winnecke (1927 i). Académie des sciences, Paris. Comptes rendus. 185:39–41.

Bessel, Friedrich W. 1836. Bemerkungen über mögliche unzulänglichkeit der die anziehungen allein berücksichtigenden theorie der kometen. Astronomische nachrichten 13:345–350.

Bessel, Friedrich W. 1836. Schreiben des herrn geheimenraths und ritters Bessel an den herausgeber. Astronomische nachrichten 13:3–6.

Bobrovnikoff, Nicholas T. 1931. Halley's comet in its apparition of 1909–1911. Publications of the Lick Observatory 17(part II):309–482.

Bobrovnikoff, Nicholas T. 1954. Physical properties of comets. Astronomical Journal 59:357–358.

Bond, George P. 1862. Account of the great comet of 1858. Annals of the Astronomical Observatory of Harvard College 3:372 p.

Brin, G. David and Mendis, Devamitta A. 1979. Dust release and mantle development in comets. Astrophysical Journal 229:402–408.

Brownlee, Donald E., Horz, F., Tomandl, D.A., and Hodge, Paul W. 1976. Physical properties of interplanetary grains. In The Study of Comets, edited by B. Donn, M. Mumma, W. Jackson, M. A'Hearn, and R. Harrington. NASA SP-393. Washington, D.C.: National Aeronautics and Space Administration. pp. 962–982.

Campbell, Donald B., Harmon, John K., and Shapiro, Irwin I. 1989. Radar observations of comet Halley. Astrophysical Journal 338:1094–1105.

Campins, Humberto, A'Hearn, Michael F., and McFadden, Lucy-Ann A. 1987. The bare nucleus of comet Neujmin 1. Astrophysical Journal 316:847–857.

Delsemme, Armand H. and Swings, Polydore. 1952. Hydrates de gaz dans les noyaux comètaires et les grains interstellaires. Annales d'astrophysique 15:1–6.

Delsemme, Armand and Miller, David C. 1971. Physico-chemical phenomena in comets—III. The continuum of comet Burnham (1960 II). Planetary Space Science 19:1229–1257.

Delsemme, Armand and Rud, David A. 1973. Albedos and cross-sections for the nuclei of comets 1969 IX, 1970 II, and 1971 I. Astronomy and Astrophysics 28:1–6.

Dubiago, Alexander D. 1948. On the secular acceleration of motions of the periodic comets. Astronomical Journal of the U.S.S.R. 25:361–368.

Fay, Theodore D. and Wisniewski, Wieslaw. 1978. The light curve of the nucleus of comet d'Arrest. Icarus 34:1–9.

Feldman, Paul D. 1978. A model of carbon production in comet West (1975n). Astronomy and Astrophysics 70:547–553.

Feldman, Paul D. 1986. Carbon monoxide in cometary ice. In Asteroids, Comets, Meteors II, edited by C.-I. Lagerkvist, B.A. Lindblad, H. Lundstedt, and H. Rickman. Uppsala, Sweden: Uppsala University Press. pp. 263–267.

Fraundorf, P., Brownlee, Donald E., and Walker, Robert M. 1982. Laboratory studies of interplanetary dust. In Comets, edited by L.L. Wilkening. Tucson: University of Arizona Press. pp. 383–409.

Goldstein, Richard M., Jurgens, Raymond F., and Sekanina, Zdenek. 1984. A radar study of comet Iras-Araki-Alcock 1983 d. Astronomical Journal 89:1745–1754.

Hanner, M.S., Aitken, D.K., Knacke, R., McCorkle, S., Roche, P.F. and Tokunaga, A.T. 1985. Infrared spectrophotometry of comet Iras-Araki-Alcock (1983d): A bare nucleus revealed? Icarus 62:97–109.

Harmon, John K., Campbell, Donald B., Hine, A.A., Shapiro, Irwin I., and Marsden, Brian G. 1989. Radar observations of comet Iras-Araki-Alcock 1983d. Astrophysical Journal 338:1071–1093.

Hirn, G.A. 1889. Constitution de l'espace celeste. Paris: Gauthier-Villars. p. 254.

Hoag, Arthur A. 1984. First infrared observations of a comet. American Astronomical Society, Bulletin 16:942.

Jewitt, David and Luu, Jane. 1988. Periodic comet Tempel 2 (1987g). In International Astronomical Union Circular No. 4582, April 19.

Kamoun, Paul D., Campbell, Donald B., Ostro, Steven J., Petengill, Gordon H., and Shapiro, Irwin I. 1982. Comet Encke: Radar detection of nucleus. Science 216:293–295.

Kamoun, Paul D., Petengill, Gordon H., Shapiro, Irwin I., and Campbell, Donald B. 1982. Comet Grigg-Skjellerup: Radar detection of nucleus. American Astronomical Society, Bulletin 14:753.

Lampland, Carl O. 1928. Radiometric observations of Skjellerup's comet. Popular Astronomy 36:240.

Laplace, Pierre S. 1808. Exposition du système du monde. 3rd edition. Paris. 405 p.

Larson, Stephen M. and Minton, R.B. 1972. Photographic observations of comet Bennett, 1970 II. In Comets: Scientific Data and Missions, edited by G.P. Kuiper and E. Roemer. Tucson: University of Arizona. pp. 183–208.

Levin, Boris J. 1985. Laplace, Bessel, and the icy model of the cometary nuclei. The Astronomy Quarterly 5(18):113–118.

Luu, Jane and Jewitt, David. 1990. The nucleus of comet P/Encke. Icarus 86:69–81.

Lyttleton, Raymond A. 1953. The Comets and Their Origin. New York: Cambridge University Press. 173 p.

453

Lyttleton, Raymond A. 1972. Does a continuous solid nucleus exist in comets? Astrophysics and Space Science 15:175–184.

Marsden, Brian G. 1968. Comets and nongravitational forces. Astronomical Journal 73:367–379.

Marsden, Brian G. 1969. Comets and nongravitational forces. II. Astronomical Journal 74:720–734.

Marsden, Brian G., Sekanina, Z., and Yeomans, D.K. 1973. Comets and nongravitational forces. V. Astronomical Journal 78:211–225.

Millis, Robert L., A'Hearn, Michael F., and Campins, Humberto. 1988. An investigation of the nucleus and coma of comet P/Arend-Rigaux. Astrophysical Journal 324:1194–1209.

Newton, Isaac. 1687. Philosophiae naturalis principia mathematica. London.

Ostro, Steven J. 1985. Radar observations of asteroids and comets. Astronomical Society of the Pacific, Publications 97:877–884.

Peirce, Benjamin. 1859. American Academy of Arts and Sciences. Proceedings. 4:202–206.

Proctor, Mary and Crommelin, A.C.D. 1937. Comets, their Nature, Origin, and Place in the Science of Astronomy. London: The Technical Press. 204 p.

Roemer, Elizabeth. 1965. The dimensions of cometary nuclei. In Mémoires de la Société Royale des Sciences de Liège, 1965, fifth series, book XII, Nature et origine des comètes. pp. 23–28.

Schmidt, Johann F.J. 1863. Ueber Donati's cometen. Astronomische nachrichten 59:97–108.

Sekanina, Zdenek. 1979. Fan-shaped coma, orientation of rotation axis and surface structure of a cometary nucleus. I. Test of a model on four comets. Icarus 37:420–442.

Sekanina, Zdenek. 1988. Outgassing asymmetry of periodic comet Encke. I. Apparitions 1924–1984. Astronomical Journal 95:911–924, 970–971.

Sekanina, Zdenek. 1988. Outgassing asymmetry of periodic comet Encke. II. Apparitions 1868–1918 and a study of the nucleus evolution. Astronomical Journal 96:1455–1475.

Slipher, Vesto M. 1927. The spectrum of the Pons-Winnecke comet and the size of the cometary nucleus. Lowell Observatory Bulletin 3:135–137.

Vorontsov-Velyaminov, B. 1930. Imaginary contradiction of the phenomena observed in the comet 1893 IV (Brooks) with the mechanical theory of comets. Astronomical Journal of the U.S.S.R. 7:90–99.

Walker, R.G., Aumann, H.H., Davies, J., Green, S., DeJong, T., Houck, J.R., and Soifer, B.T. 1984. Observations of comet IRAS-Araki-Alcock (1983d). Astrophysical Journal 278:L11–L14.

Whipple, Fred L. 1950. A comet model. I. The acceleration of comet Encke. Astrophysical Journal 111:375–394.

Whipple, Fred L. 1951. A comet model. II. Physical relations for comets and meteors. Astrophysical Journal 113:464–474.

Whipple, Fred L. 1976. Background of modern cometary theory. Nature 263(5572):15–19.

Whipple, Fred L. 1982. Rotation of comet nuclei. *In* Comets, edited by L.L. Wilkening. Tucson: University of Arizona Press. pp. 227–250.

Whipple, Fred L. and Sekanina, Zdenek. 1979. Comet Encke: Precession of the spin axis, nongravitational motion, and sublimation. Astronomical Journal 84: 1894–1909.

Wisniewski, Wieslaw. 1988. Periodic comet Tempel 2 (1987g). *In* International Astronomical Union Circular No. 4603, May 25.

Woods, T.N., Feldman, P.D., Dymond, K.F., and Sahnow, D.J. 1986. Rocket ultraviolet spectroscopy of comet Halley and abundance of carbon monoxide and carbon. Nature 324:436–438.

Wurm, Karl. 1934. Beitrag zur deutung der vorgänge in kometen. I. Zeitschrift für astrophysik 8:281–291.

Wurm, Karl. 1935. Beitrag zur deutung der vorgänge in kometen. II. Zeitschrift für astrophysik 9:62–78.

General Works

Arpigny, Claude. 1965. Spectra of comets and their interpretation. Annual Review of Astronomy and Astrophysics 3:351–376.

Bobrovnikoff, Nicholas T. 1951. Comets. *In* Astrophysics: A Topical Symposium, edited by J.A. Hynek. New York: McGraw-Hill. pp. 302–356.

Marsden, Brian G. 1974. Comets. Annual Review of Astronomy and Astrophysics 12:1–21.

Richter, N.B. 1963. The Nature of Comets. Translated and revised by Arthur Beer. London: Methuen and Company. 221 p.

Sekanina, Zdenek. 1981. Rotation and precession of cometary nuclei. Annual Review of Earth and Planetary Science 9:113–145.

Spinrad, Hyron. 1987. Comets and their composition. Annual Review of Astronomy and Astrophysics 25:231–269.

Swings, Polydore. 1948. Le spectre de la comète d'Encke, 1947i. Annales d'astrophysique 11:124–136.

Whipple, Fred L. and Huebner, Walter E. 1976. Physical processes in comets. Annual Review of Astronomy and Astrophysics 14:143–172.

Wyckoff, Susan. 1982. Overview of comet observations. *In* Comets, edited by L.L. Wilkening. Tucson: University of Arizona Press. pp. 3–55.

Chapter 10

The Returns of Comet Halley

Biot, Édouard C. 1843. Recherches faites dans la grande collection des historiens de la Chine, sur les anciennes apparitions de la comète de Halley. *In* Connaissance des tems. . . . pour l'an 1846. Additions. Paris. pp. 69–84.

Brady, Joseph L. 1972. The effect of a trans-Plutonian planet on Halley's comet. Astronomical Society of the Pacific. Publications. 84:314–322.

Brady, Joseph L. 1982. Halley's comet: A.D. 1986 to 2647 B.C. British Astronomical Association Journal 92:209–215.

Brady, Joseph L. and Carpenter, Edna. 1967. The orbit of Halley's comet. Astronomical Journal 72:365–369.

Brady, Joseph L. and Carpenter, Edna. 1971. The orbit of Halley's comet and the apparition in 1986. Astronomical Journal 76:728–739.

Broughton, R. Peter. 1979. The visibility of Halley's comet. Royal Astronomical Society of Canada Journal 73:24–36.

Burckhardt, Johann K. 1804. Fortsetzung der untersuchungen über ältere cometen. Monatliche correspondenz zur beförderung der erd- und himmelskunde 10: 162–167.

Chang, Yü-che. 1979. Halley's comet: Tendencies in its orbital evolution and its ancient history. Chinese Astronomy 3:120–131.

Chirikov, B.V. and Vecheslavov, V.V. 1989. Chaotic dynamics of comet Halley. Astronomy and Astrophysics 221:146–154.

Cowell, Philip H. and Crommelin, Andrew C.D. 1908. The perturbations of Halley's comet. Royal Astronomical Society of London, Monthly Notices 67:174, 386–411, 511–521.

Cowell, Philip H. and Crommelin, Andrew C.D. 1907. The perturbations of Halley's comet in the past. First paper. The period 1301 to 1531. Royal Astronomical Society of London, Monthly Notices 68:111–125.

Cowell, Philip H. and Crommelin, Andrew C.D. 1907. The perturbations of Halley's comet in the past. Second paper. The apparition of 1222. Royal Astronomical Society of London, Monthly Notices 68:173–179.

Cowell, Philip H. and Crommelin, Andrew C.D. 1908. The perturbations of Halley's comet in the past. Third paper. The period from 1066 to 1301. Royal Astronomical Society of London, Monthly Notices 68:375–378.

Cowell, Philip H. and Crommelin, Andrew C.D. 1908. The perturbations of Halley's comet in the past. Fourth paper. The period 760 to 1066. Royal Astronomical Society of London, Monthly Notices 68:510–514.

Cowell, Philip H. and Crommelin, Andrew C.D. 1908. The perturbations of Halley's comet in the past. Fifth paper. The period B.C. 240 to A.D. 760. Royal Astronomical Society of London, Monthly Notices 68(supplementary no.): 665–670.

Cowell, Philip H. and Crommelin, Andrew C.D. 1908. The perturbations of Halley's comet, 1759–1910. Royal Astronomical Society of London, Monthly Notices 68:379–395.

Cowell, Philip H. and Crommelin, Andrew C.D. 1910. Investigation of the motion of Halley's comet from 1759 to 1910. *In* Publikation der Astronomischen Gesellschaft, no. 23.

Damoiseau, Théodore, baron de. 1820. Mémoire sur l'époque du retour au périhélie de la comète de année 1759. Accademia delle scienze di Torino, memorie 24:1–76.

Damoiseau, Théodore, baron de. 1829. Sur les perturbations des comètes. *In* Connaissance des tems. . . . pour l'an 1832. Paris. Additions. pp. 25–34.

Dumouchel, Étienne. 1835. Auszug aus einem schreiben des Herrn Dumouchel, Directors der Sternwarte des Collegio Romano in Rom. Astronomische nachrichten 12:415–416.

Hind, John Russell. 1850. On the past history of the comet of Halley. Royal Astronomical Society of London, Monthly Notices 10:51–58.

Kiang, Tao. 1972. The past orbit of Halley's comet. *In* Royal Astronomical Society, Memoirs 76(part 2):27–66.

Know-Shaw, Harold. 1911. Positions of Halley's comet and of comet 1910 a from photographs taken at the Khedivial observatory, Helwan. Royal Astronomical Society of London, Monthly Notices 71:573–577.

Landgraf, Werner. 1986. On the motion of comet Halley. Astronomy and Astrophysics 163:246–260.

Laugier, Paul A.E. 1842. Note sur la première comète de 1301. Académie des sciences, Paris. Comptes rendus. 15:949–951.

Laugier, Paul A.E. 1843. Note sur la apparition de la comète de Halley en 1378. Académie des sciences, Paris. Comptes rendus. 16:1003–1006.

Laugier, Paul A.E. 1846. Mémoire sur quelques anciennes apparitions de la comète de Halley, inconnues jusqu'ici. Académie des sciences, Paris. Comptes rendus. 23:183–189.

Lehmann, Jacob W.H. 1835. Versuch, die berechnungen zur bestimmung der wiederkehr des Halleyschen kometen aufs reine zu bringen. Astronomisches nachrichten 12:369–400.

Marsden, Brian G. 1968. Comets and nongravitational forces. Astronomical Journal 73:367–379.

Marsden, Brian G. 1969. Comets and nongravitational forces. II. Astronomical Journal 74:720–734.

Marsden, Brian G., Sekanina, Zdenek, and Yeomans, Donald K. 1973. Comets and nongravitational forces. V. Astronomical Journal 78:211–225.

Michielsen, Herman F. 1968. A rendezvous with comet Halley in 1985–1986. Journal of Spacecraft and Rockets 5:328–334.

Pingré, Alexandre G. 1783–1784. Cométographie; ou, traité historique et théorique des comètes. 2 volumes. Paris.

Pontécoulant, Philippe Gustave Le Doulcet, comte de. 1830. Détermination du prochain retour au périhélie de la comète de 1759. *In* Connaissance des tems. . . . pour l'an 1833. Paris. Additions, pp. 104–113.

Pontécoulant, Philippe Gustave Le Doulcet, comte de. 1834. Note sur la détermination du prochain retour de la comète de 1759. *In* Connaissance des tems. . . . pour l'an 1837. Paris. Additions, pp. 102–104.

Pontécoulant, Philippe Gustave Le Doulcet, comte de. 1835. Mémoire sur le calcul des perturbations et le prochain retour á son périhélie de la comète de Halley. *In* Académie des sciences, Paris. Mémoires présentés par divers savans. Science mathématiques et physiques. Ser. 2. 6:875–947.

Pontécoulant, Philippe Gustave Le Doulcet, comte de. 1864. Notice sur la comète de Halley et ses apparitions successives de 1531 à 1910. *In* Académie des sciences. Comptes rendus. 58:825–828, 915.

Rasmusen, Hans Q. 1967. Table 11. Perturbations from 0.001 of Jupiter on comet Halley. *In* The Definitive Orbit of Comet Olbers for the Periods 1815–1887–1956 with a Correction to the Mass of Jupiter. Publikationer og mindre meddelelser, no. 194. Copenhagen: Universitet, Astronomisk observatorium. pp. 96–99.

Rasmusen, Hans Q. 1979. Three revolutions of the comets Halley and Olbers 1759–2024. Copenhagen Det Kongelige Danske videnskabernes selskab. Matematisk-fysiske meddelelser 40(4):54 p.

Rosenberger, Otto A. 1830. Elemente des Halleyschen kometen bei seiner letzten sichbarkeit. Astronomisches nachrichten 8:221–250.

Rosenberger, Otto A. 1831. Elemente des Halleyschen cometen bei seiner vorletzten erscheinung im jahre 1682. Astronomisches nachrichten, 9:53–68.

Rosenberger, Otto A. 1833. Ueber die störungen des Halleyschen Kometen während seines umlaufs um die sonne vom 15ten September 1682 bis zum 13ten März 1759. Astronomisches nachrichten 11:157–180.

Rosenberger, Otto A. 1835. Berichtigung und notiz, die störungen des Halleyschen Kometen betreffend. Astronomische nachrichten 12:187–194.

Sitarski, G. 1988. On the nongravitational motion of comet Halley. Acta astronomica 38:253–268.

Stephenson, F.R., Yau, K.K.C., and Hunger, H. 1985. Records of Halley's comet on Babylonian tablets. Nature 314:587–592.

Wolf, Max. 1909. Photographische beobachtungen des Halleyschen kometen 1909 c. Astronomische nachrichten 182:211–212.

Yeomans, D.K. 1977. Comet Halley—the orbital motion. Astronomical Journal 82:435–440.

Yeomans, Donald K. and Kiang, Tao. 1981. The long-term motion of comet Halley. Royal Astronomical Society of London, Monthly Notices 197:633–646.

The Halley Spacecraft Armada, the International Halley Watch, and the International Cometary Explorer

Belton, M.J.S. (chairman). 1977. Scientific Rationale and Strategies for a First Comet Mission: Report of the Comet Halley Science Working Group. NASA Technical Memorandum 78420. Washington, D.C.: National Aeronautics and Space Administration.

Brandt, J.C. (chairman). 1980. The International Halley Watch: Report of the Science Working Group. NASA Technical Memorandum 82181. Edited by J.C. Brandt, L.D. Friedman, R.L. Newburn, and D.K. Yeomans. Washington, D.C.: National Aeronautics and Space Administration.

Brandt, J.C., Farquhar, R.W., Maran, S.P., Niedner, M.B. Jr., and von Rosenvinge, T.T. 1988. The International Cometary Explorer. Exploration of Halley's Comet, edited by M. Grewing, F. Praderie, and R. Reinhard. New York: Springer-Verlag. pp. 969–980.

Friedman, Louis D. 1981. The International Halley watch. *In* International Astronautical Congress, Applications of Space Developments. Selected Papers from the XXXI International Astronautical Congress, Tokyo, 21–28 September 1980. Edited by L.G. Napolitano. New York: Pergamon Press. pp. 273–290.

Hirao, K. and Itoh, T. 1988. The Sakigake/Suisei missions to Halley's Comet. *In* Exploration of Halley's Comet, edited by M. Grewing, F. Praderie, and R. Reinhard. New York: Springer-Verlag. pp. 965–968.

Logsdon, John M. 1989. Missing Halley's comet: The politics of big science. Isis 80(302):254–280.

McLaughlin, William I. 1987. International Halley Watch. Spaceflight 29:434–435.

Ogilvie, K.W., Coplan, M.A., Bochsler, P., and Geiss, J. 1986. Ion composition results during the International Cometary Explorer encounter with Giacobini-Zinner. Science 232:374–377.

Reinhard, Rudiger. 1988. The Giotto mission to Halley's comet. *In* Exploration of Halley's Comet, edited by M. Grewing, F. Praderie, and R. Reinhard. New York: Springer-Verlag. pp. 949–958.

Sagdeev, R.Z. 1988. The Vega mission to Halley's comet. *In* Exploration of Halley's Comet, edited by M. Grewing, F. Praderie, and R. Reinhard. New York: Springer-Verlag. pp. 959–964.

Sanderson, T.R., Wenzel, K.-P., Daly, P.W., Cowley, S.W.H., Hynds, R.J., Richardson, I.G., Smith, E.J., Bame, S.J., and Zwickl, R.D. 1986. Observations of heavy energetic ions far upstream from comet Halley. *In* Proceedings, 20th ESLAB Symposium on the Exploration of Halley's Comet, edited by B. Battrick, E.J. Rolfe, and R. Reinhard. ESA SP-250. Noordwijk, Netherlands: European Space Agency Publications. Volume 1. pp. 105–108.

Scarf, F.L., Coroniti, F.V., Kennel, C.F., Sanderson, T.R., Wenzel, K.-P., Hynds, R.J., Smith, E.J., Bame, S.J., and Zwickl, R.D. 1986. ICE plasma wave measurements in the ion pick-up region of comet Halley. Geophysical Research Letters 13: 857–860.

Smith, E.J., Tsurutani, B.T., Slavin, J.A., Jones, D.E., Siscoe, G.L., and Mendis, D.A. 1986. International Cometary Explorer encounter with Giacobini-Zinner: Magnetic field observations. Science 232:382–385.

Tsurutani, B.T., Brinca, A.L., Smith, E.J., Thorne, R.M., Scarf, F.L., Gosling, J.T., and Ipavich, F.M. 1986. MHD waves detected by ICE at distances $> 28 \times 10^6$ km from comet Halley: Cometary or solar wind origin? *In* Proceedings, 20th ESLAB Symposium on the Exploration of Halley's comet edited by B. Battrick, E.J. Rolfe, and R. Reinhard. ESA SP-250. Noordwijk, Netherlands: European Space Agency Publications. Volume 3. pp. 451–456.

Veverka, Joseph (chairman). 1979. Report of the Comet Science Working Group. NASA Technical Memorandum 80543. Washington, D.C.: National Aeronautics and Space Administration.

Halley's Dust and Gas Atmosphere

A'Hearn, M.F., Hoban, S., Birch, P.V., Bowers, C., Martin, R., and Klinglesmith, D.A. 1986. Gaseous jets in comet P/Halley. *In* Proceedings, 20th ESLAB Symposium on the Exploration of Halley's comet, edited by B. Battrick, E.J. Rolfe, and R. Reinhard. ESA SP-250. Noordwijk, Netherlands: European Space Agency Publications. Volume 1. pp. 483–486.

Allen, D.A. and Wickramasinghe, D.T. 1987. Discovery of organic grains in comet Wilson. Nature 329:615–616.

Allen, M., Delitsky, M., Huntress, W., Yung, Y., Ip, W.-H., Schwenn, R., Rosenbauer, H., Shelley, E., Balsiger, H., and Geiss, J. 1987. Evidence for methane and ammonia in the coma of comet P/Halley. Astronomy and Astrophysics 187:502–512.

Axford, W.I. 1964. The interaction of the solar wind with comets. Planetary and Space Science 12:719–720.

Belton, M.J.S. 1985. P/Halley: The quintessential comet. Science 230:1229–1236.

Boyarchuk, A.A., Grinin, V.P., Sheikhet, A.I., and Zvereva, A.M. 1987. Pre- and post-perihelion Astron ultraviolet spectrophotometry of comet Halley: A comparative analysis. Soviet Astronomical Letters 13:92–96.

Brandt, J.C. and Niedner, M.B. 1986. Plasma structures in comet Halley. In Proceedings, 20th ESLAB Symposium on the Exploration of Halley's comet, edited by B. Battrick, E.J. Rolfe, and R. Reinhard. ESA SP-250. Volume 1. Noordwijk, Netherlands: European Space Agency Publications. pp. 47–52.

Bregman, J.D., Campins, H., Witteborn, F.C., Wooden, D.H., Rank, D.M., Allamandola, L.J., Cohen, M., and Tielens, A.G.G.M. 1987. Airborne and groundbased spectrophotometry of comet P/Halley from 5–13 micrometers. Astronomy and Astrophysics 187:616–620.

Combes, M., Moroz, V.I., Crifo, J.F., Lamarre, J.M., Charra, J., Sanko, N.F., Soufflot, A., Bibring, J.P., Cazes, S., Coron, N., Crovisier, J., Emerich, C., Encrenaz, T., Gispert, R., Grigoryev, A.V., Guyot, G., Krasnopolsky, V.A., Nikolsky, Yu V., and Rocard, F. 1986. Infrared sounding of comet Halley from Vega 1. Nature 321:266–268.

Combes, M., Moroz, V.I., Crovisier, J., Encrenaz, T., Bibring, J.-P., Grigoriev, A.V., Sanko, N.F., Coron, N., Crifo, J.F., Gispert, R., Brockelée-Morvan, D., Nikolsky, Yu. V., Krasnopolsky, V.A., Owen, T., Emerich, C., Lamarre, J.M., and Rocard, F. 1988. The 2.5–12 μm spectrum of comet Halley from the IKS-VEGA experiment. Icarus 76:404–436.

Craven, J.D. and Frank, L.A. 1987. Atomic hydrogen production rates for comet P/Halley from observations with Dynamics Explorer 1. Astronomy and Astrophysics 187:351–356.

Despois, D., Crovisier, J., Brockelée-Morvan, D., Schraml, J., Forveille, T., and Gérard, E. 1986. Observations of hydrogen cyanide in comet Halley. Astronomy and Astrophysics 160:L11–L12.

Eberhardt, P., Krankowsky, D., Schulte, W., Dolder, U., Lämmerzahl, P., Berthelier, J.J., Woweries, J., Stubbemann, U., Hodges, R.R., Hoffman, J.H., and Illiano, J.M. 1986. On the CO and N_2 abundance in comet Halley. In Proceedings, 20th ESLAB Symposium on the Exploration of Halley's comet, edited by B. Battrick, E.J. Rolfe, and R. Reinhard. ESA SP-250. Noordwijk, Netherlands: European Space Agency Publications. Volume 1, pp. 383–386.

Festou, M.C., Feldman, P.D., A'Hearn, M.F., Arpigny, C., Cosmovici, C.B., Danks, A.C., McFadden, L.A., Gilmozzi, R., Patriarchi, P., Tossi, G.P., Wallis, M.K., and Weaver, H.A. 1986. IUE observations of comet Halley during the Vega and Giotto encounters. Nature 321:361–363.

Gérard, E., Brockelée-Morvan, D., Bourgois, G., Colom, P., and Crovisier, J. 1987.

18-cm wavelength radio monitoring of the OH radical in comet P/Halley 1982i. Astronomy and Astrophysics 187:455–461.

Greenberg, J.M. 1982. What are comets made of? A model based on interstellar dust. *In* Comets, edited by Laurel L. Wilkening. Tucson: University of Arizona Press. pp. 131–163.

Huebner, W.F. 1987. First polymer in space identified in comet Halley. Science 237:628–630.

Kissel, J., Brownlee, D.E., Büchler, K., Clark, B.C., Fechtig, H., Grün, E., Hormung, K., Igenbergs, E.B., Jessberger, E.K., Krueger, F.R., Kuczera, H., McDonnell, J.A.M., Morfill, G.M., Rahe, J., Schwehm, G.H., Sekanina, Z., Utterback, N.G., Völk, H.J., and Zook, H.A. 1986. Composition of comet Halley dust particles from Giotto observations. Nature 321:336–337.

Kissel, J. and Krueger, F.R. 1987. The organic component in dust from comet Halley as measured by the PUMA mass spectrometer on board Vega 1. Nature 326:755–760.

Kissel, J., Sagdeev, R.Z., Bertaux, J.L., Angarov, V.N., Audouze, J., Blamont, J.E., Buckler, K., Evlanov, E.N., Fechtig, H., Fomenkova, M.N., von Hoerner, H., Inogamov, N.A., Khromov, V.N., Knabe, W., Frueger, F.R., Langevin, Y., Leonas, V.B., Levasseur-Regourd, A.C., Managadz, G.G., Podkolzin, S.N., Shapiro, V.D., Tabaldyev, S.R., and Zubkov, B.V. 1986. Composition of Halley dust particles from Vega observations. Nature 321:280–282.

Krasnopolsky, V.A., Gogoshev, M., Moreels, G., Moroz, V.I., Krysko, A.A., Gogosheva, Ts., Palazov, K., Sargoichev, S., Clairemidi, J., Vincent, M., Bertaux, J.L., Blamont, J.E., Troshin, V.S., and Valnícek, B. 1986. Spectroscopic study of comet Halley by the Vega 2 three-channel spectrometer. Nature 321:269–271.

McCoy, R.P., Opal, C.B., and Carruthers, G.R. 1986. Far-ultraviolet spectral images of comet Halley from sounding rockets. Nature 324:439–441.

McDonnell, J.A.M., Kissel, J., Grun, E., Grard, R.J.L., Langevin, Y., Olearczyk, R.E., Perry, C.H., and Zarnecki, J.C. 1986. Giotto dust impact detection system DIDSY and particulate impact analyzer PIA: Interim assessment of the dust distribution and properties within the coma. *In* Proceedings, 20th ESLAB Symposium on the Exploration of Halley's comet, edited by B. Battrick, E.J. Rolfe, and R. Reinhard. ESA SP-250. Noordwijk, Netherlands: European Space Agency Publications. Volume 2. pp. 25–38.

McDonnell, J.A.M., Lamy, P.L., and Pankiewicz, G.S. 1991. Physical properties of cometary dust. *In* Comets in the Post-Halley Era, edited by Ray L. Newburn, Jr., Marcia M. Neugebauer, and Jürgen Rahe. Dordrecht, Netherlands: Kluwer Academic Publishers.

Mitchell, D.L., Lin, R.P., Anderson, K.A., Carlson, C.W., Curtis, D.W., Korth, A., Rème, H., Sauvaud, J.A., d'Uston, C., and Mendis, D.A. 1987. Evidence for chain molecules enriched in carbon, hydrogen, and oxygen in comet Halley. Science 237:626–628.

Mumma, M.J., Weaver, H.A., Larson, H.P., and Davis, D.S. 1986. Detection of water vapor in Halley's comet. Science 232:1523–1528.

Mumma, M.J. and Reuter, D.C. 1989. On the identification of formaldehyde in Halley's comet. Astrophysical Journal 344:940–948.

Niedner, M.B. and Brandt, J.C. 1978. Interplanetary gas. XXIII. Plasma tail disconnection events in comets: Evidence for magnetic field line reconnection at interplanetary sector boundaries? Astrophysical Journal 223:655–670.

Niedner, M.B. and Schwingenschuh, K. 1987. Plasma-tail activity at the time of the Vega encounters. Astronomy and Astrophysics 187:103–108.

Neubauer, F.M., Glassmeier, K.H., Pohl, M., Raeder, J., Acuna, M.H., Burlaga, L.F., Ness, N.F., Musmann, G., Mariani, F., Wallis, M.K., Ungstrup, E., and Schmidt, H.U. 1986. First results from the Giotto magnetometer experiment at comet Halley. Nature 321:352–355.

Reinhard, Rudiger. 1988. The Giotto mission to Halley's Comet. In Exploration of Halley's Comet, edited by M. Grewing, F. Praderie, and R. Reinhard. New York: Springer-Verlag. pp. 949–958.

Schloerb, F.P., Kinzel, W.M., Swade, D.A., and Irvine, W.M. 1986. HCN production from comet Halley. Astrophysical Journal 310:L55–L60.

Stewart, A.I.F. 1987. Pioneer Venus measurements of H, O, and C production in comet P/Halley near perihelion. Astronomy and Astrophysics 187:369–374.

Snyder, L.E., Palmer, P., and de Pater, I. 1989. Radio detection of formaldehyde emission from comet Halley. Astronomical Journal 97:246–253.

Snyder, L.E., Palmer, P., and de Pater, I. 1990. Observations of formaldehyde in comet Machholz (1988j). Icarus 86:289–298.

Winnberg, A., Ekelund, L., and Ekelund, A. 1987. Detection of HCN in comet P/Halley. Astronomy and Astrophysics 172:335–341.

Woods, T.N., Feldman, P.D., Dymond, K.F., and Sahnow, D.J. 1986. Rocket ultraviolet spectroscopy of comet Halley and abundance of carbon monoxide and carbon. Nature 324:436–438.

Wyckhoff, S., Wagner, R.M., Wehinger, P.A., Schleicher, D.G., and Festou, M.C. 1985. Onset of sublimation in comet Halley (1982i). Nature 316:241–242.

Comet Halley's Nucleus

Belton, Michael J.S. 1991. Characterization of the rotation of cometary nuclei. In Comets in the Post-Halley Era, edited by Ray L. Newburn, Jr., Marcia M. Neugebauer, and Jürgen Rahe. Dordrecht, Netherlands: Kluwer Academic Publishers.

Belton, M.J.S. and Julian, William H. 1990. A model for the spin rate of P/Halley. American Astronomical Society, Bulletin 22(3):1089–1090.

Brockelée-Morvan, D., Crovisier, J., Despois, D., Forveille, T., Gérard, E., Schraml, J., and Thum, C. 1987. Molecular observations of comets P/Giacobini-Zinner 1984e and P/Halley 1982i at millimeter wavelengths. Astronomy and Astrophysics 180:253–262.

Combes, M., Moroz, V.I., Crifo, J.F., Lamarre, J.M., Charra, J., Sanko, N.F., Soufflot, A., Bibring, J.P., Cazes, S., Coron, N., Crovisier, J., Emerich, C., Encrenaz, T., Gispert, R., Grigoryev, A.V., Guyot, G., Krasnopolsky, V.A., Nikolsky, Yu. V., and Rocard, F. 1986. Infrared sounding of comet Halley from Vega 1. Nature 321:266–268.

Cruikshank, D.P., Hartmann, W.K., and Tholen, D.J. 1985. Colour, albedo and nucleus size of Halley's comet. Nature 315:122–124.

Delsemme, A.H. 1991. Nature and history of the organic compounds in comets: An

462

astrophysical view. *In* Comets in the Post-Halley Era, edited by Ray L. Newburn, Jr., Marcia M. Neugebauer, and Jürgen Rahe. Dordrecht, Netherlands: Kluwer Academic Publishers.

Divine, N., Fechtig, H., Gombosi, T.I., Hanner, M.S., Keller, H.U., Larson, S.M., Mendis, D.A., Newburn, R.L., Jr., Reinhard, R., Sekanina, Z., and Yeomans, D.K. 1986. The comet Halley dust and gas environment. Space Science Reviews 43:1–104.

Donn, Bertram. 1977. A comparison of the composition of new and evolved comets. *In* Comets, Asteroids, Meteorites: Interrelations, Evolution and Origin, edited by Armand H. Delsemme. Toledo, Ohio: University of Toledo Press. pp. 15–23.

Feldman, P.D. and A'Hearn, M.F. 1985. Ultraviolet albedo of cometary grains. *In* Ices in the Solar System, edited by J. Klinger et al. Dordrecht, Netherlands: D. Reidel Publishers. pp. 453–461.

Jewitt, David and Danielson, G. Edward. 1984. Charged-coupled device photometry of comet Halley. Icarus 60:435–444.

Kaneda, E., Ashihara, O., Shimizu, M., Takagi, M., and Hirao, K. 1986. Observation of comet Halley by the ultraviolet imager of Suisei. *In* Nature 321:297–299.

Keller, H.U., Arpigny, C., Barbieri, C., Bonnet, R.M., Cazes, S., Coradini, M., Cosmovici, C.B., Delamere, W.A., Huebner, W.F., Hughes, D.W., Jamar, C., Malaise, D., Reitsema, H.J., Schmidt, H.U., Schmidt, W.K.H., Seige, P., Whipple, F.L., and Wilhelm, K. 1986. First Halley multicolour camera imaging results from Giotto. Nature 321:320–326.

Millis, R.L. and Schleicher, D.G. 1986. Rotation period of comet Halley. Nature 324:646–649.

Peale, S.J. 1989. On the density of Halley's comet. Icarus 82:36–49.

Reitsema, H.J., Delamere, W.A., Huebner, W.F., Keller, H.U., Schmidt, W.K.H., Wilhelm, K., Schmidt, H.U., and Whipple, F.L. 1986. Nucleus morphology of comet Halley. *In* Proceedings, 20th ESLAB Symposium on the Exploration of Halley's comet, edited by B. Battrick, E.J. Rolfe, and R. Reinhard. ESA SP-250. Noordwijk, Netherlands: European Space Agency Publications. Volume 2. pp. 351–354.

Rickman, H. 1986. Masses and densities of comets Halley and Kopff. *In* The Comet Nucleus Sample Return Mission, proceedings of a workshop held in Canterbury, U.K., July 15–17, 1986. ESA SP-249. Noordwijk, Netherlands: European Space Agency Publications. pp. 195–205.

Rickman, H. 1989. The nucleus of comet Halley: Surface structure, mean density, gas and dust production. Advances in Space Research 9(3):59–71.

Sagdeev, R.Z., Szabó, F., Avanesov, G.A., Cruvellier, P., Szabó, L., Szegó, K., Abergel, A., Balazs, A., Barinov, I.V., Bertaux, J.-L., Blamont, J., Detaille, M., Demarelis, E., Dul'nev, G.N., Endröczy, G., Gardos, M., Kanyo, M., Kostenko, V.I., Krasikov, V.A., Nguyen-Trong, T., Nyitrai, Z., Reny, I., Rusznyak, P., Shamis, V.A., Smith, B., Sukhanov, K.G., Szabó, F., Szalai, S., Tarnopolsky, V.I., Toth, I., Tsukanova, G., Valnicek, B.I., Varhalmi, L., Zaiko, Yu. K., Zatsepin, S.I., Ziman, Ya. L., Zsenei, M., and Zhukov, B.S. 1986. Television observations of comet Halley from Vega spacecraft. Nature 321:262–266.

Sagdeev, R.Z., Elyasberg, P.E., and Moroz, V.I. 1988. Is the nucleus of comet Halley a

463

low density body? Nature 331:240–242.

Schleicher, D.G. and Bus, Schelte J. 1990. Comet P/Halley's periodic brightness variations in 1910. American Astronomical Society, Bulletin 22(3):1089.

Schlosser, Wolfhard, Schulz, Rita, and Koczet, Paul. 1986. The cyan shells of comet P/Halley. *In* Proceedings, 20th ESLAB Symposium on the Exploration of Halley's comet, edited by B. Battrick, E.J. Rolfe, and R. Reinhard. ESA SP-250. Noordwijk, Netherlands: European Space Agency Publications. Volume 3. pp. 495–498.

Sekanina, Z. and Larson, S.M. 1986. Coma morphology and dust-emission pattern of periodic comet Halley: IV. Spin vector refinement and map of discrete dust sources for 1910. Astronomical Journal 92:462–482.

Sekanina, Z. and Larson, S.M. 1986. Dust jets in comet Halley observed by Giotto and from the ground. Nature 321:357–361.

Stewart, A.I.F. 1987. Pioneer Venus measurements of H, O, and C production in comet P/Halley near perihelion. Astronomy and Astrophysics 187:369–374.

Wilhelm, K., Cosmovici, C.B., Delamere, W.A., Huebner, W.F., Keller, H.U., Reitsema, H., Schmidt, H.U., and Whipple, F.L. 1986. A three-dimensional model of the nucleus of comet Halley. *In* Proceedings, 20th ESLAB Symposium on the Exploration of Halley's Comet, edited by B. Battrick, E.J. Rolfe, and R. Reinhard. ESA SP-250. Noordwijk, Netherlands: European Space Agency Publications. Volume 2. pp. 367–369.

Wyckoff, S., Lindholm, E., Wehinger, P.A., Petersson, B.A., Zucconi, J.-M., and Festou, M.C. 1989. The $^{12}C/^{13}C$ abundance ratio in comet Halley. Astrophysical Journal 339:488–500.

General Works

Balsiger, Hans, Fechtig, Hugo, and Geiss, Johannes. 1988. A close look at Halley's comet. Scientific American 259(3):62–69.

Delsemme, Armand H. 1991. Nature and history of the organic compounds in comets: an astrophysical view. *In* Comets in the Post-Halley Era, edited by Ray L. Newburn, Jr., Marcia M. Neugebauer, and Jürgen Rahe. Dordrecht, Netherlands: Kluwer Academic Publishers.

Freitag, Ruth S. 1984. Halley's Comet: A Bibliography. Washington, D.C.: U.S. Government Printing Office. 555 p.

Mendis, D.A. 1988. A postencounter view of comets. Annual Review of Astronomy and Astrophysics 26:11–49.

Spinrad, Hyron. 1987. Comets and their composition. Annual Review of Astronomy and Astrophysics 25:231–269.

Chapter 11

Interstellar and Solar System Sources of Comets

Bobrovnikoff, Nicholas T. 1929. On the disintegration of comets. Lick Observatory Bulletin 14:28–37.

Halley, Edmond. 1705. A Synopsis of the Astronomy of Comets. London.

Herschel, William. 1783. On the proper motion of the Sun and solar system; with an account of several changes that have happened among the fixed stars since the time of Mr. Flamsteed. *In* Philosophical Transactions of the Royal Society of London 73:247–283.

Herschel, William. 1808. Observations of a comet made with a view to investigate its magnitude and the nature of its illumination. Philosophical Transactions of the Royal Society of London 98:145–163.

Herschel, William. 1812. Observations of a comet. Philosophical Transactions of the Royal Society of London 102:115–143.

Herschel, William. 1812. Observations of a second comet. Philosophical Transactions of the Royal Society of London 102:229–237.

Jaki, Stanley L. 1978. Planets and Planetarians. New York: John Wiley and Sons. 266 p.

Kant, Immanuel. 1755. Allgemeine naturgeschichte und theorie des himmels. Konigsberg and Leipzig.

Lagrange, Joseph Louis. 1812. Mémoire sur l'origine des comètes. Journal de physique, de chimie, d'histoire naturelle et des arts. 74:228–235.

Laplace, Pierre-Simon, Marquis de. 1796. Exposition du système du monde. Paris.

Laplace, Pierre-Simon, Marquis de. 1813. Sur les comètes. *In* Connaissance des temps pour l'an 1816, additions. 216:213–220.

Lyttleton, Raymond A. 1948. On the origin of comets. Royal Astronomical Society of London, Monthly Notices 108:465–475.

Lyttleton, Raymond A. 1951. On the structure of comets and the formation of tails. Royal Astronomical Society of London, Monthly Notices 111:268–277.

Newton, Hubert Anson. 1878. On the origin of comets. American Journal of Science and Arts 16:165–179.

Newton, Hubert Anson. 1893. On the capture of comets by planets, especially their capture by Jupiter. National Academy of Sciences, Memoirs 6:7–23.

Proctor, Richard A. 1884. The capture theory of comets. Knowledge: A Monthly Record of Science 6:111–112, 126–128.

Russell, Henry Norris. 1920. On the origin of periodic comets. Astronomical Journal 33:49–61.

Schaffer, Simon. 1980. The great laboratory of the universe: William Herschel on matter theory and planetary life. Journal for the History of Astronomy 11: 81–111.

Schiaparelli, Giovanni V. 1866–1867. Intorno al corso ed all'origine probabile delle stelle meteoriche. Five letters to Angelo Secchi published in Dal bullettino meteorologico dell'osservatorio del Collegio Romano, 5(8, 10, 11, 12) and 6(2).

Strömgren, Svante Elis. 1914. Ueber den ursprung der kometen. Publikationer og mindre meddelelser fra Kobenhavns observatorium 19:193–250.

Thraen, Anton Karl. 1894. Untersuchung über die vormalige bahn des cometen 1886 II. Astronomische nachrichten 136:133–138.

Tisserand, François Félix. 1890. Hypothèses de Lagrange sur l'origine des comètes et des aérolithes. Bulletin astronomique 7:453–461.

Vsekhsvyatskij, Sergey K. 1930. Zur frage des ursprunges der kurzperiodischen kometen. Astronomische nachrichten 240:273–280.

465

Vsekhsvyatskij, Sergey K. 1931. Uber die bildung der kometen. Astronomische nachrichten 243:281–300.

Vsekhsvyatskij, Sergey K. 1970. The Nature and Origin of Comets and Meteors. A 1970 NASA translation of the original 1967 Russian work, NASA TT F-608. Washington, D.C.: National Aeronautics and Space Administration.

The Oort Cloud of Comets

Oort, Jan H. 1950. The structure of the cloud of comets surrounding the solar system, and a hypothesis concerning its origin. Bulletin of the Astronomical Institutes of the Netherlands 11:91–110.

Oort, Jan H., and Schmidt, Maarten. 1951. Differences between old and new comets. Bulletin of the Astronomical Institutes of the Netherlands 11:259–270.

Öpik, Ernst J. 1932. Note on stellar perturbations of nearly parabolic orbits. Proceedings of the American Academy of Arts and Sciences 67:169–183.

Van Woerkom, Adrianus Jan Jasper. 1948. On the origin of comets. Bulietin of the Astronomical Institutes of the Netherlands 10:445–472.

Further Refinements and Alternatives to the Oort Cloud

Antonov, V.A. and Latyshev, I.N. 1972. Determination of the form of the Oort cometary cloud as the Hill surface in the galactic field. In The Motion, Evolution of Orbits, and Origin of Comets, edited by G.A. Chebotarev, E.I. Kazimirchak-Polonskaya, and B.G. Marsden. New York: Springer-Verlag. pp. 341–345.

Biermann, Ludwig, Huebner, Walter F., and Lüst, Rhea. 1983. Aphelion clustering of "new" comets: Star tracks through Oort's cloud. National Academy of Sciences of the United States of America, Proceedings 80:5151–5155.

Biermann, Ludwig and Lüst, Reimar. 1978. Über durchgänge de "Oortschen wolke" von kometenkernen durch dichte interstellare wolken. In Bayerische akademie der wissenschaften mathematisch-naturwissenschaftliche klasse, sitzungsberichte jahrgang 1978. Munich. pp. 5–9.

Bogart, Richard S. and Noerdlinger, Peter D. 1982. On the distribution of orbits among long period comets. Astronomical Journal 87:911–917.

Byl, John. 1983. Galactic perturbations on nearly-parabolic cometary orbits. The Moon and the Planets 29:121–137.

Carrington, Richard C. 1861. On the distribution of the perihelia of the parabolic comets in relation to the motion of the solar system in space. Royal Astronomical Society of London, Monthly Notices 21:42–43.

Chebotarev, G.A. 1965. On the dynamical limits of the solar system. Soviet Astronomy 8:787–792.

Chebotarev, G.A. 1966. Cometary motion in the outer solar system. Soviet Astronomy 10:341–344.

Clube, S.V.M., and Napier, W.M. 1982. Spiral arms, comets and terrestrial catastrophism. Royal Astronomical Society of London, Quarterly Journal 23:45–66.

Clube, S.V.M., and Napier, W.M. 1984. Comet capture from molecular clouds: A dynamical constraint on star and planet formation. Royal Astronomical Society of London, Monthly Notices 208:575–588.

466

Delsemme, Armand H. 1987. Galactic tides affect the Oort cloud: An observational confirmation. Astronomy and Astrophysics 187:913–918.

Delsemme, Armand H. 1989. Whence come comets? Sky and Telescope 77: 260–264.

Eddington, Arthur Stanley. 1913. The distribution of cometary orbits. Observatory 36:142–146.

Everhart, Edgar. 1969. Close encounters of comets and planets. Astronomical Journal 74:735–750.

Everhart, Edgar. 1972. The origin of short-period comets. Astrophysical Journal Letters 10:131–135.

Everhart, Edgar. 1973. Examination of several ideas of comet origins. Astronomical Journal 78:329–337.

Everhart, Edgar. 1977. The evolution of comet orbits as perturbed by Uranus and Neptune. *In* Comets, Asteroids, Meteorites: Their Interrelations, Evolution and Origins, edited by Armand H. Delsemme. Toledo, Ohio: University of Toledo Press. pp. 99–105.

Harrington, Robert S. 1985. Implications of the observed distributions of very long period comet orbits. Icarus 61:60–62.

Hasegawa, Ichiro. 1976. Distribution of the aphelia of long period comets. Publications of the Astronomical Society of Japan 28:259–276.

Heisler, Julia and Tremaine, Scott. 1986. The influence of the galactic tidal field on the Oort comet cloud. Icarus 65:13–26.

Hut, Piet and Tremaine, Scott. 1985. Have interstellar clouds disrupted the Oort comet cloud? Astronomical Journal 90:1548–1557.

Kazimirchak-Polonskaya, E.I. 1972. The major planets as powerful transformers of cometary orbits. *In* The Motion, Evolution of Orbits, and Origin of Comets, edited by G.A. Chebotarev, E.I. Kazimirchak-Polonskaya, and B.G. Marsden. New York: Springer-Verlag. pp. 373–397.

Marsden, Brian G., Sekanina, Zdenek, and Everhart, Edgar. 1978. New osculation orbits for 110 comets and analysis of original orbits for 200 comets. Astronomical Journal 83:64–71.

Weissman, Paul R. 1979. Physical and dynamical evolution of long-period comets. *In* Dynamics of the Solar System, edited by R.L. Duncombe. Dordrecht, Netherlands: D. Reidel. pp. 277–282.

Weissman, Paul R. 1985. Dynamical evolution of the Oort cloud. *In* Dynamics of Comets: Their Origin and Evolution, edited by Andrea Carusi and Giovanni B. Valsecchi. Dordrecht, Netherlands: D. Reidel. pp. 87–96.

A Possible Inner Oort Cloud and a Comet Belt Beyond Neptune

Aumann, H.H., Gillett, F.C., Beichman, C.A., DeJong, T., Houck, J.R., Low, F.J., Neugebauer, G., Walker, R.G., and Wesselius, R.R. 1984. Discovery of a shell around Alpha Lyrae. Astrophysical Journal 278:L23–L27.

Bailey, Mark E. 1983. Theories of cometary origin and the brightness of the infrared sky. Royal Astronomical Society of London, Monthly Notices 205:47P–52P.

Bailey, Mark E. 1986. The near-parabolic flux and the origin of short-period comets. Nature 324:350–352.

467

Duncan, Martin, Quinn, Thomas, and Tremaine, Scott. 1987. The formation and extent of the solar system comet cloud. Astronomical Journal 94:1330–1338.

Duncan, Martin, Quinn, Thomas, and Tremaine, Scott. 1988. The origin of short-period comets. Astrophysical Journal 328:L69–L73.

Duncan, Martin, Quinn, Thomas, and Tremaine, Scott. 1989. The long-term evolution of orbits in the solar system: A mapping approach. Icarus 82:402–418.

Fernández, Julio A. 1980. On the existence of a comet belt beyond Neptune. Royal Astronomical Society of London, Monthly Notices 192:481–491.

Hamid, S.E., Marsden, Brian G., and Whipple, Fred L. 1968. Influence of a comet belt beyond Neptune on the motions of periodic comets. Astronomical Journal 73:727–729.

Harper, D.A., Lownstein, R.F., and Davidson, J.A. 1984. On the nature of the material surrounding Vega. Astrophysical Journal 285:808–812.

Mills, Jack G. 1981. Comet showers and the steady-state infall of comets from the Oort cloud. Astronomical Journal 86:1730–1740.

Low, F.J., Beintema, D.A., Gautier, T.N., Gillett, F.C., Beichman, C.A., Neugebauer, G., Young, E., Aumann, H.H., Boggess, N., Emerson, J.P., Habing, H.J., Hauser, M.G., Houck, J.R., Rowan-Robinson, M., Soifer, B.T., Walker, R.G., and Wesselius, P.R. 1984. Infrared cirrus: New components of the extended infrared emission. Astrophysical Journal 278:L19–L22.

Stagg, C.R. and Bailey, M.E. 1989. Stochastic capture of short-period comets. Royal Astronomical Society of London, Monthly Notices 241:507–541.

Weissman, Paul R. 1984. The Vega particulate shell: Comets or asteroids? Science 224:987–989.

Whipple, Fred L. 1964. Evidence for a comet belt beyond Neptune. National Academy of Sciences, Proceedings 51:711–718.

Cometary End States

Alvarez, Luis W., Alvarez, Walter, Asaro, Frank, and Michel, Helen V. 1980. Extraterrestrial cause for the Cretaecous-Tertiary extinction. Science 208:1095–1108.

Bardwell, Conrad. 1988. International Astronomical Union Circular 4600, dated May 19. Cambridge, Massachusetts.

Cameron, A.G.W. 1962. The formation of the Sun and planets. Icarus 1:13–69.

Cameron, A.G.W. 1973. Accumulation processes in the primitive solar nebula. Icarus 18:407–450.

Cameron, A.G.W. 1978. The primitive solar accretion disk and the formation of the planets. In The Origin of the Solar System, edited by S.F. Dermott. New York: John Wiley and Sons. pp. 49–74.

Carusi, Andrea, Kresák, Lubor, Perozzi, E., and Valsecchi, Giovanni B. 1985. First results of the integration of motion of short-period comets over 800 years. In Dynamics of Comets: Their Origin and Evolution, edited by Andrea Carusi and Giovanni B. Valsecchi. Dordrecht, Netherlands: D. Reidel. pp. 319–340.

Clube, S.V.M., and Napier, W.M. 1982. Spiral arms, comets and terrestrial catastrophism. Royal Astronomical Society of London. Quarterly Journal 23:45–66.

Clube, S.V.M., and Napier, W.M. 1984. Comet capture from molecular clouds: A dynamical constraint on star and planet formation. Royal Astronomical Society of London, Monthly Notices 208:575–588.

Drummond, Jack D. 1982. Theoretical meteor radiants of Apollo, Amor, and Aten asteroids. Icarus 49:143–153.

Ganapathy, Ramachandran. 1983. The Tunguska explosion of 1908: Discovery of meteoritic debris near the explosion site and at the south pole. Science 220:1158–1161.

Hajduk, Anton. 1985. The past orbit of comet Halley and its meteor stream. *In* Dynamics of Comets: Their Origin and Evolution, edited by Andrea Carusi and Giovanni B. Valsecchi. Dordrecht, Netherlands: D. Reidel. pp. 399–403.

Halley, Edmond. 1688. An account of some observations lately made at Nurenburg by Mr. P. Wurtzelbaur. Philosophical Transactions of the Royal Society, London 16:402–406.

Halley, Edmond. 1726. Some considerations about the cause of the universal deluge, laid before the Royal Society, on the 12th of December 1694. Philosophical Transactions of the Royal Society, London 33:118–119.

Hartmann, William K., Tholen, David J., and Cruikshank, Dale P. 1987. The relationship of active comets, "extinct" comets, and dark asteroids. Icarus 69:33–50.

Hartmann, William K., Tholen, David J., Meech, Karen J., and Cruikshank, Dale P. 1990. 2060 Chiron: Colorimetry and cometary behavior. Icarus 83:1–15.

Hills, Jack G. 1982. The formation of comets by radiation pressure in the outer protosun. Astronomical Journal 87:906–910.

Hoffman, Antoni. 1985. Patterns of family extinction depend on definition and geological timescale. Nature 315:659–662.

Hoyle, Fred and Wickramasinghe, Nalin Chandra. 1979. Diseases from Space. New York: Harper & Row. 196 p.

Kirkwood, Daniel. 1880. On the great southern comet of 1880. Observatory 3:590–592.

Kresák, Lubor. 1978. The Tunguska object: A fragment of comet Encke? Bulletin of the Astronomical Institute of Czechoslovakia 29:129–134.

Kresák, Lubor. 1981. Evolutionary aspects of the splits of cometary nuclei. Bulletin of the Astronomical Institute of Czechoslovakia 32:19–40.

Kresák, Lubor. 1985. The aging and lifetimes of comets. Dynamics of comets: Their origin and evolution, edited by Andrea Carusi and Giovanni B. Valsecchi. Dordrecht, Netherlands: D. Reidel. pp. 279–302.

Kresák, Lubor. 1986. On the aging process of periodic comets. *In* 20th ESLAB Symposium on the Exploration of Halley's Comet, edited by B. Battrick, E.J. Rolfe, and R. Reinhard. European Space Agency SP-250. Volume 2. Noordwijk, Netherlands: European Space Agency Publications. pp. 433–438.

Kresák, Lubor and Kresáková, M. 1987. The absolute total magnitude of periodic comets and their variations. *In* Symposium on Diversity and Similarity of Comets. Brussels: European Space Agency Publications. ESA SP-278. pp. 37–42.

Kreutz, Heinrich. 1888. Untersuchungen über das system der cometen 1843 I, 1880 I and 1882 II: der grosse Septembercomet 1882 II. *In* Publication der Sternwarte in Kiel 3:1–110.

Kreutz, Heinrich. 1891. Untersuchungen über das system der cometen 1843 I, 1880 I and 1882 II: der grosse Septembercomet 1882 II. *In* Publication der Sternwarte in Kiel 6:1–66.

Kreutz, Heinrich. 1901. Untersuchungen über das system der cometen 1843 I, 1880 I and 1882 II. Astronomische Abhandlungen, Kiel 1:1–90.

Kuiper, Gerard P. 1951. On the origin of the solar system. *In* Astrophysics: A Topical Symposium, edited by J. A. Hynek. New York: McGraw-Hill. pp. 357–424.

McCrea, W.H. 1975. Solar system as space probe. Observatory 75:239–255.

Marsden, Brian G. 1967. The sungrazing comet group. Astronomical Journal 72:1170–1183.

Marsden, Brian G. 1989. The sungrazing comet group. II. Astronomical Journal 98:2306–2321.

Maupertuis, Pierre Louis Moreau de. 1750. Essai de cosmologie. Berlin.

Meech, Karen J. and Belton, Michael J.S. 1989. (2060) Chiron. *In* International Astronomical Union Circular No. 4770, dated April 11. Cambridge, Massachusetts.

Olsson-Steel, Duncan I. 1988. Identification of meteoroid streams from Apollo asteroids in the Adelaide radar orbit surveys. Icarus 75:64–96.

Öpik, Ernst J. 1961. The survival of comets and cometary material. Astronomical Journal 66:381–382.

Oró, John. 1961. Comets and the formation of biochemical compounds on the primitive Earth. Nature 190:389–390.

Raup, David M. and Sepkoski, J. John Jr. 1984. Periodicity of extinctions in the geologic past. National Academy of Sciences, Proceedings 81:801–805.

Safronov, V.S. 1969. Ehvolyutsiya doplanetnogo oblaka i obrazovanie zemli i planet. Moscow. Also published as NASA technical translation NASA TT F-677. Washington, D.C.: National Aeronautics and Space Administration.

Safronov, V.S. 1972. Ejection of bodies from the solar system in the course of the accumulation of the giant planets and the formation of the cometary cloud. *In* The Motion, Evolution of Orbits, and Origin of Comets, edited by G.A. Chebotarev, E.I. Kazimirchak-Polonskaya, and B.G. Marsden. New York: Springer-Verlag. pp. 329–334.

Sekanina, Zdenek. 1976. A probability of encounter with interstellar comets and the likelihood of their existence. Icarus 27:123–133.

Sekanina, Zdenek. 1982. The problem of split comets in review. *In* Comets, edited by Laurel L. Wilkening. Tucson: University of Arizona Press. pp. 251–287.

Sekanina, Zdenek. 1983. The Tunguska event: No cometary signature in evidence. Astronomical Journal 88:1382–1414.

Sekanina, Zdenek and Yeomans, Donald K. 1984. Close encounters and collisions of comets with the Earth. Astronomical Journal 89:154–161.

Sekanina, Zdenek and Yeomans, Donald K. 1985. Orbital motion, nucleus precession, and splitting of periodic comet Brooks 2. Astronomical Journal 90: 2335–2352.

Smith, Bradford A. and Terrile, Richard J. 1984. A circumstellar disk around β Pictoris. Science 226:1421–1424.

Valtonen, Mauri J. and Innanen, Kimmo A. 1982. The capture of interstellar comets. Astrophysical Journal 255:307–315.

470

Van Flandern, Thomas C. 1978. A former asteroidal planet as the origin of comets. Icarus 36:51–74.

Weissman, Paul R. 1979. Physical and dynamical evolution of long-period comets. *In* Dynamics of the Solar System, edited by R.L. Duncombe. Dordrecht, Netherlands: D. Reidel. pp. 277–282.

Weissman, Paul R. 1990. Are periodic comets bombardments real? Sky and Telescope 79(3):266–270.

Wetherill, George W. 1988. Where do the Apollo objects come from? Icarus 76:1–18.

Whipple, Fred L. 1950. A comet model. I. The acceleration of comet Encke. Astrophysical Journal 111:375–394.

Whipple, Fred L. 1983. 1983 TB and the Geminid meteors. *In* International Astronomical Union Circular 3881. Cambridge, Massachusetts.

Wisdom, Jack. 1985. Meteorites may follow a chaotic route to Earth. Nature 315:731–733.

General Works

Hut, Piet, Alvarez, Walter, Elder, William P., Hansen, Thor, Kauffman, Erle G., Keller, Gerta, Shoemaker, Eugene M., and Weissman, Paul R. 1987. Comet showers as a cause of mass extinctions. Nature 329:118–126.

Shoemaker, Eugene M. and Wolfe, Ruth F. 1986. Mass extinctions, crater ages and comet showers. *In* The Galaxy and the Solar System, edited by Roman Smoluchowski, John N. Bahcall, and Mildred S. Matthews. Tucson: University of Arizona Press. pp. 338–386.

Weissman, Paul R. 1990. The Oort cloud. Nature 344(6269):825–830.

Wilkerson, M. Susan and Worden, Simon P. 1978. On egregious theories—the Tunguska event. Royal Astronomical Society of London, Quarterly Journal 19:282–289.

Appendix: *Naked-Eye Comets Reported Through* A.D. *1700*

Barrett, A.A. 1978. Observations of comets in Greek and Roman sources before A.D. 410. Royal Astronomical Society of Canada Journal 72:81–106.

Berberich, A. 1888. Der comet des jahres 1672. Astronomische nachrichten 118:49–71.

Bessel, Friedrich Wilhelm. 1804. Berechnung der Harriot'schen und Torporley'schen beobachtungen des cometen von 1607. Monatliche correspondenz 10:425–440.

Broughton, Peter. 1988. Arthur Storer of Maryland: His astronomical work and his family ties with Newton. Journal for the History of Astronomy 19:77–96.

Carl, Ph. 1864. Repertorium der cometen-astronomie. Munich. 377 p.

Chambers, George F. 1889. A handbook of descriptive and practical astronomy. 4th ed. Oxford. 1:511–588.

Crommelin, A.C.D. 1917. The observations by Fabricius and Heller of the comet of

1556. Royal Astronomical Society of London, Monthly Notices 77:633–643.

Halley, Edmond. 1705. A Synopsis of the Astronomy of Comets. London.

Halley, Edmond. 1752. Astronomical Tables. London.

Hasegawa, Ichiro. 1979. Orbits of ancient and medieval comets. Publications of the Astronomical Society of Japan 31:257–270.

Hasegawa, Ichiro. 1980. Catalogue of ancient and naked eye comets. Vistas in Astronomy 24:59–102.

Hellman, Clarice Doris. 1971. The Comet of 1577: Its Place in the History of Astronomy. Reprint of the 1944 edition with addenda, errata, and a supplement to the appendix. New York: AMS Press. 488 [i.e., 502] p.

Henderson, Thomas. 1843. On the comet of 1668. Astronomische nachrichten, 20:333–336.

Hind, J.R. 1876. Our astronomical column (Comet of 1698). Nature 14:152.

Hind, J.R. 1877. Our astronomical column (Comet of 1402). Nature 16:49–50.

Hind, J.R. 1879. Our astronomical column (Comet of 1699). Nature 20:481–482.

Hind, J.R. 1879. Our astronomical column (Comet of 1652). Nature 21:164–165.

Ho Peng Yoke. 1962. Ancient and medieval observations of comets and novae in Chinese sources. Vistas of Astronomy 5:127–225.

Ho Peng-Yoke and Ang Tian-Se. 1970. Chinese astronomical records on comets and "guest stars." In Oriens extremus. Book 1/2. Wiesbaden. December. pp. 63–99.

Hunger, Herman. 1989. Personal communication to F. Richard Stephenson dated November 13. Fragments of Babylonian diaries relating to comets were translated from clay tablets in the British Museum collection in London.

Jervis, Jane L. 1980. Vögelin on the comet of 1532: Error analysis in the 16th century. Centaurus 23(3):216–229.

Jervis, Jane L. 1985. Cometary theory in fifteenth-century Europe. Dordrecht, Netherlands: D. Reidel Publishing Co. 209 p.

Kiang, Tao. 1972. The past orbit of Halley's comet. Royal Astronomical Society, London, Memoirs 76:27–66.

Knobel, E.B. 1897. On some original unpublished observations of the comet of 1652. Royal Astronomical Society of London, Monthly Notices 57:434–438.

Kronk, Gary W. 1984. Comets, A Descriptive Catalog. Hillside, New Jersey: Enslow Publishers. 331 p.

Landgraf, Werner. 1977. Ober die bahn des ersten kometen von 1577. Mitteilungen astronomischer vereinigungen südwestdeutschlands, 16:68–69.

Landgraf, Werner. 1985. Über die bahn des zweiten kometen von 1618. Die Sterne 61:351–353.

Marsden, Brian G. 1967. The sungrazing comet group. Astronomical Journal 72:1170–1183.

Marsden, Brian G. 1989. A Catalogue of Cometary Orbits. Cambridge, Massachusetts: Smithsonian Astrophysical Observatory.

Marsden, Brian G. 1989. The sungrazing comet group. II. Astronomical Journal 98:2306–2321.

McIntyre, Donald. 1949. Comets in Old Cape Records. Cape Town, South Africa: Cape Town Limited. 15 p.

Marcus, Joseph N. and Seargent, David A.J. 1986. Dust forward scatter brightness enhancement in previous apparitions of Halley's comet. Proceedings, 20th

472

ESLAB Symposium on the Exploration of Halley's Comet Volume 3, edited by B. Battrick, E.J. Rolfe, and R. Reinhard. ESA SP-250. Noordwijk, Netherlands: European Space Agency Publications: pp. 359–362.

Maunder, A.S.D. 1934. Old cometary records. Observatory 57:278–281.

Needham, Joseph, Beer, Arthur, and Ho Ping Yü. 1957. Spiked Comets in Ancient China. Observatory 77:137–138.

Newton, Isaac. 1687. Philosophiae naturalis principia mathematica. London.

Newton, Robert R. 1972. Medieval Chronicles and the Rotation of the Earth. Baltimore, Maryland: Johns Hopkins University Press. Appendix II.

Olbers, W. 1824. Ueber einen im jahre 1625 erschienenen cometen. Astronomische nachrichten 2:101–104.

Olbers, W. 1831. Ueber einen im jahre 1639 erschienenen cometen. Astronomische nachrichten 8:58–59.

Pankenier, David W. 1983. Private communication. January.

Pedersen, O. and Pihl, M. 1974. Early Physics and Astronomy. New York: American Elsevier.

Petit, Pierre. 1665. Dissertation sur la nature des comètes. Paris.

Pingré, Alexandre Guy. 1783–1784. Cométographie ou traité historique et théorique des comètes. Paris. 2 volumes.

Roche, John J. 1985. Thomas Harriot's observations of Halley's comet in 1607. *In* The Light of Nature: Essays in the History and Philosophy of Science, presented to A.C. Crombie. Dordrecht, Netherlands: M. Nijhoff. pp. 175–191.

Rodgers, R.F. 1952. Newly discovered Byzantine records of comets. Royal Astronomical Society of Canada Journal 46:177–180.

Ruffner, James A. 1966. The Background and Early Development of Newton's Theory of Comets. Ph.D. Thesis, Indiana University, Bloomington. 363 p.

Selga, M. 1930. Astronomical observations made in the Philippines prior to 1927. Publications of the Manila Observatory, 1. The Philippines.

Schubart, J. 1965. International Astronomical Union Circular No. 1907. Cambridge, Massachusetts.

Stein, J. 1910. A newly discovered document on Halley's comet in 1066. Observatory 33:234–238.

Stephenson, F. Richard and Yau, Kevin K.C. 1985. Far eastern observations of Halley's comet: 240 B.C. to A.D. 1368. British Interplanetary Society Journal 38:195–216.

Thorndike, Lynn. 1945. Peter of Limoges on the Comet of 1299, Isis 36:3–6.

Thorndike, Lynn. 1950. Latin Treatises on Comets between 1238 and 1368 A.D. Chicago: University of Chicago Press. 274 p.

White, Andrew D. 1910. A History of the Warfare of Science with Theology in Christendom. 2 volumes. New York: D. Appleton and Co.

Yeomans, D.K. 1977. Comet Halley—the orbital motion. Astronomical Journal 82:435–440.

Yeomans, D.K. 1981. Comet Tempel-Tuttle and the Leonid meteors. Icarus 47:492–499.

Yeomans, D.K. and Kiang T. 1981. The long term motion of comet Halley. Royal Astronomical Society of London, Monthly Notices 197:633–646.

Index